AF262870

Fight Like a Machine

FIGHT LIKE A MACHINE

The Combat Future of the Human Body

THOMAS M. HUNT

UNIVERSITY OF OKLAHOMA PRESS : NORMAN

Publication of this book is made possible through the generosity of Edith Kinney Gaylord.

Library of Congress Cataloging-in-Publication Data
ISBN: 9780806197081
Library of Congress Control Number: 2025054611

The views and opinions expressed herein are solely those of the individual author(s) and do not reflect the policy, opinions, or positions of the University of Oklahoma, its regents, officers, or employees.

Portions of this work originally appeared in *Stratfor* and are reprinted here with permission of RANE Network Inc.

The paper in this book meets the guidelines for permanence and durability of the Committee on Production Guidelines for Book Longevity of the Council on Library Resources, Inc. ∞

The manufacturer's authorized representative in the EU for product safety is Mare Nostrum Group B.V., Mauritskade 21D, 1091 GC Amsterdam, The Netherlands, email: gpsr@mare-nostrum .co.uk.

1 2 3 4 5 6 7 8 9 10

To my grandfathers

Contents

ACKNOWLEDGMENTS

A number of individuals supported me as I conducted the research and writing for this book. My former department chair and "big brother" Dr. John Bartholomew was a *constant* source of enthusiasm. Literally thousands (and perhaps tens of thousands) of texts, phone calls, video conferences, and in-person conversations were held—talk about someone who cares. He treated me like a family member, and so did his lovely wife, Pam. Also closer to brothers than just friends, Dr. Matthew T. Bowers and Dr. Tolga Ozyurtcu have kept my spirits up virtually every single day of the many years that I have known them. Dr. Jan Todd has been a cherished mentor and the matriarch of our Longhorn family. Moreover, many passages of what I think of as my best writing were crafted with the memory of her husband, Dr. Terry "Doc" Todd, in mind. Patty Coffman is among the most admirable individuals I've ever encountered. She is a model of generosity and capability. Dr. Josephine "Tres" Hinds, a former Army officer and assault helicopter pilot, was the inspiration for my chapter on women in the US military. She is the bravest person I know and the closest I will ever have to a little sister. Her wife, Dr. Audrey Stone, is also one of my favorite people. So too is Dr. Emily Sparvero.

A relative newcomer to the department, Dr. Charles Stocking proved himself a loyal friend and a source of sage advice in short order. Dr. Brian Mills shared my enthusiasm for craft beer and offered words of encouragement in numerous instances. Kyle R. Martin read several chapters and gave insightful advice on both content and potential visuals to use.

Other past and former members of my department who deserve mention include Dr. Lawrence D. Abraham, Dr. Darla Castelli, Dr. Keryn Pasch, Dr. Eddie Coyle, Dr. Kimberly A. Beckwith, Dr. Roger P. Farrar, Dr. Lisa Griffin, Dr. Carole Holahan, Dr. Jody Jensen, Dr. John L. Ivy, Dr. Sophie Laland, Dr. Dorothy Lambdin, Dr. Alexandra Loukas, Dr. Emalee Nelson, Dr. Rachel M. Ozerkevich, Dr. Keryn Pasch, Dr. Deborah Salvo, Dr. Dixie Stanforth and her husband Phil, Dr. Mary A. Steinhardt, and Dr. Hiro-fumi Tanaka. The office of Dr. Dean Charles Martinez provided key support during the initial stages of this project. I was much enabled as well by several student research assistants. These include my good friend Brett Cook, a former naval aviator who today works at the University of Texas at Austin's Clements Center for National Security. Kaleigh Haynie provided fresh insights into the interests and perspectives of today's young Americans. And in her penetrating reads and rereads of my manuscript, Maddie Hebert demonstrated qualities normally only associated with experienced professional editors. I am forever grateful to them all.

Outside of UT-Austin, my brother Dr. Jonathan R. Hunt and his wife, Dr. Vivien Chang, gave me numerous pieces of advice along the way. Jonathan also rendered his support by making available to me the professional network of colleagues gained during his time as a faculty member at the Air War College in Montgomery, Alabama, and then later at the US Naval War College in Newport, Rhode Island. Among those at the latter, I already knew Dr. Nicholas Evan Sarantakes, a historian with a long and impressive list of publications. Nick provided thoughts on potential publishers and always responded when I needed a bit of advice. Jacqulyn Teoh provided an early suggestion on the order of chapters that proved quite useful. Nicholas Solis offered feedback on my initial proposal for the book project. My dear friends Anthony Daywood, Eric Perlmutter, and Arthur Jacob provided many instances of much-needed joy and laughter as I worked on the manuscript. In addition, my good friend Dr. Jörg Krieger chewed over troublesome issues with me whenever needed. Dr. Jaime Schultz and I have been close for many years. She has always been one of the first to whom I could turn when navigating the inevitable challenges of balancing academia and family life. Other individuals who deserve thanks include Lisa Benton, Skyler Borque, Drs. Mark and JoDee Dyreson and their son McClane, Laura Feldman, Hailey Horton, Dr. Sarah K. Fields, Sarah Jackson, Will Kunesh, Reid Minot, Miranda Moore, Dr. Dan Nathan, Tyler Paige, Matt Pittman, Dr. Toby C. Rider, Neftali Santiago, Dr. Mouna Sfeir, Bing VanderKam, Brea Wakefield, and Dr. Kevin Witherspoon.

In thinking through the various alternatives on the choice of publishers, I increasingly saw the appeal of the University of Oklahoma Press. My brilliant law school roommate, Chase Griffith, was a proud OU grad, and his recent death after a long battle with cancer bestowed some personal meaning to the idea. From the moment of our very first interaction, Oklahoma Press Editorial Director Andrew Berzanskis demonstrated the qualities of an ideal partner. This initial impression was time and again proven accurate, as Andrew maintained a steady hand on the publication process as well as proving a trusted source of fresh ideas and encouragement. Wherever the future takes us, he will always be my first choice as an editor. OU Press Acquisitions Coordinator Riley Hines scheduled meetings, managed reviews from outside experts, and kept track of all the many small details that go along with getting a book to market. Others who were involved at the press include Ben Folger, editorial fellow; Katie Baker, director of marketing and sales; Amy Hernandez, promotions manager and inventory coordinator; Tony Roberts, director; and Joe Schiller, acquisitions editor. Estelle Lloyd at Westchester Publishing Services saw the project through the last stages of the editing process. Karin Kipp performed the copyediting for the manuscript, and Brad Allard created the index.

Hilary Bentsen will forever remain a deeply treasured member of my family. Our boys, Pierce and Bentsen, are the lights of our lives. We cannot wait to experience the next chapter with them. My mother and father, Dr. Thomas L. and Mrs. Laurie P. Hunt, have supported me in so many ways that words fall short of anything that could come close to acknowledgment. I hope they know just how much I love them. Finally, I would like to thank my two grandfathers, Tom Hunt and Paul Pierce. They were my heroes growing up, and this book is dedicated to their memory.

Introduction

In February 2022, Russia invaded neighboring Ukraine, starting the first major European war in a generation. In the days following, tremendous acts of bravery by members of the Ukrainian population fended off the country's ostensibly stronger adversary. A particularly dramatic example took place when fifteen-year-old Andriy Pokrasa launched his small personal drone into the skies near the embattled city of Kyiv by means of a video-game-style handheld controller. In doing so, the teenager was able to identify and track the movements of nearby Russian troops and vehicles, which he and his father Stanislav then passed on to Ukrainian artillery units. Within minutes of one such transmission, a devastating artillery strike rained down on the Russian formation. The event was perhaps unprecedented in history. Without military training and at his own behest using the modest technologies of a video game, a teenager had contributed to the destruction of over twenty vehicles of one of the world's great military powers.[1]

After its report in the American press, I spent a considerable amount of time reflecting on the episode and what it might say about the nature of modern warfare. Were the United States and its allies truly prepared for the type of combat being waged in Ukraine, I wondered? And if not, what should be done about the situation? This book is the result of my thinking on those questions. During the initial stages of the project, I pondered a great deal over what new ideas I might bring to the table on national security issues. A robust network of defense intellectuals exists in the United States and allied countries. These include individuals who have spent decades thinking about the evolving nature of war and how it will be fought in the future. "Why should they take me seriously?" I fretted. I'd never served in the military or worked at a defense-oriented think tank. And I therefore worried that I might not be taken seriously by the sorts of readers whom I hoped to reach.

Given my legal training and scholarly background in the field of kinesiology, however, I in time came to see myself as perhaps uniquely positioned to make an important contribution on the evolving nature of war. The first phase of my academic career was devoted to examining the legal and political history of sport in global political affairs, particularly on the issue of performance enhancement in the international regulation of athletics. A monograph of mine titled *Drug Games: The International Olympic Committee and the Politics of Doping, 1960–2008* (University of Texas Press, 2011) offers a representative publication of this line of work.[2] I am pleased that my efforts on the subject have made at least a modest impact on how it is discussed both within and beyond academia. Moreover, a number of connections exist between the study of performance enhancement in sport and the study of the human body in war. I was reassured on this point when, as I worked on this book, my ideas received thoughtful attention from defense experts in academia and the military as well as other organizations in the national security community. I am especially grateful on this point to my brother Dr. Jonathan R. Hunt for giving me opportunities to gain insights from the professional network that he had developed during his time as a faculty member at the US Naval War College in Newport, Rhode Island.

As for the specifics of the book, *Fight Like a Machine: The Combat Future of the Human Body* argues that a fundamental shift has taken place in the relationship between the human body and warfare. Supported by an assortment of digital tools and robotic equipment, today's warriors confront one another across multiple dimensions of physical and non-physical activity. As I contemplated the various implications of this reality, a 2015 speech on the future of warfare by then US Deputy Secretary of Defense Robert Work struck me as particularly insightful. The key to winning on the battlefield in the decades ahead, he said, "[will be] what we call human-machine collaboration and combat teaming. . . . And our work suggests that artificial intelligence and autonomy will allow entirely new levels of what we refer to as man-machine symbiosis on the battlefield."[3] Work also understood that considerable thought should be devoted to the ethical dimensions of the framework—especially so given what America's rivals were doing on the matter. "Our adversaries," he warned, "quite frankly are pursuing enhanced human operations. And it scares the crap out of us, really. We're going to have to have a big, big decision on whether or not we are comfortable going that way."[4]

The waging of war has of course always been influenced by technology.[5] This was as true for the warriors of antiquity as it will be for the troops who fight the battles of the coming centuries. And yet every once in a great while the technological changes are so great as to constitute a true "revolution in military affairs."[6] Exactly such an occurrence is now unfolding—one described by defense analyst Andrew Krepinevich as "enabled by advances in a range of commercial technologies—among them artificial intelligence, additive ('3D') manufacturing, synthetic biology, and quantum computing—as well as military-driven technologies, including directed energy and hypersonic weapons."[7] Given the scale of these progressions, this book asserts, American and allied military leaders would do well to modify the intellectual paradigm through which they conceptualize how to best optimize the human body within the man-machine team on the battlefield.

This is a large topic. And it is one that should be addressed within the context of the relative strengths and weaknesses of American society. Every state in the country now has an adult population in which at least one-fifth are obese; twenty-three of them feature rates of 35 percent or higher.[8] Some 14.5 million American children and adolescents are today afflicted with the disease.[9] With weight problems and other medical/physical conditions constituting the single largest reason for disqualification, less than a quarter of the country's young population is at this point even eligible for military service.[10] The recruiting challenges posed by this reality are made more difficult by the fact that over half of Americans under thirty years of age possess a negative perception of US military institutions.[11] In light of these interrelated matters, it should be of no surprise that the nation's armed forces are having an increasingly difficult time fulfilling their annual recruiting quotas. "The all-volunteer force faces one of its greatest challenges since inception," asserts former Under Secretary of Defense for Personnel and Readiness Ashish S. Vazirani.[12]

How then within the ongoing revolution in military affairs to minimize the negative impact of these problems while simultaneously making best use of the positive attributes of those Americans who can and do serve? In an ideal world, of course, a way would be found to dramatically improve the health standing of the American population. After so many years of failure, however, such a result at this point seems decidedly unlikely. This said, in some areas of military activity, one must be careful not to overstate the relationship between the physical capacities of the human body and *overall* battlefield effectiveness. Yes, frontline troops engaging in direct contact with the enemy will always need substantial physical capabilities. However, the corporeal traits of speed, strength, and endurance play relatively minor roles in many crucial aspects of modern warfare. Most cyber and space operations, for example, take place far from any actual shooting. Even for "traditional" combatants such as infantry on the ground, success will depend just as much on cognitive agility and technological adeptness as it will on physiological conditioning.[13] In a future war with a peer or near-peer adversary, the United States will need to very quickly train large numbers of troops to utilize sophisticated weapons technologies.[14] In light of this reality, considerable thought should be given to identifying appropriate physical standards for each military occupational specialty within the military services.

Moreover, when consideration is given to enhancing performance in the ongoing revolution in military affairs, it should be kept in mind that optimizing the man-machine *system* should be the priority—not the physical capacity of the individual human soldier.[15] Such a mode of thinking would act as an extension of the way in which weapons today are valued more for their contributions to the larger networked force structure as opposed to what they bring to the table as discrete platforms.[16] It should moreover be understood that American doctrine will increasingly separate the human form from the most dangerous parts of the battle-field. Uncrewed machinery will in the future, for example, provide the primary means of direct assault. As described in a February 2024 over-view of the emerging operating concept of the US Marine Corps, "combining increased ratios of precision indirect fires with uncrewed systems, enabled by a wide diversity of sensors, allows for the development of a new warfighting system that renders the advantages of the defense substantially less relevant."[17]

The skills needed to operate such a system deserve much attention. Generation Z, defined as encompassing those individuals born between the years 1997 and 2012, today helps fill the ranks of young officers and enlisted in the US military; it moreover serves as the primary population source of new recruits. Contrary to what one might expect of these first "digital natives" in society, the generation features sizable gaps in its technical abilities. Indeed, skills pertaining to such things as hardware, machinery, and even traditional office computing lag considerably.[18] Instead, this is a group in possession of a fairly narrow sort of technological adeptness—one that is centered on software-based systems designed from the beginning to be intuitive and user-friendly. Smartphone apps, video games, and social media platforms constitute perhaps the most prominent examples of the types of tools with which Generation Z is most comfortable.[19] No doubt the members of the younger age group that follows will feature their own unique traits in these areas.[20]

Given this context, it seems logical that military leaders should try to incorporate the traits featured in those systems as much as possible going forward. As has been demonstrated in the ongoing conflict in Ukraine, success in the wars of the future will depend at least in part on who can most efficiently train and equip new troops. In addition, the armed services must find ways to ensure that their ranks can remain filled with effective personnel in the aftermath of significant losses in a future major conflict. The war in Ukraine has shown that troop numbers matter and that combat forces will at times require large-scale regeneration. Leveraging the technological strengths of US military personnel possesses considerable potential from this perspective. More specifically, solutions borrowed from the digital gaming world offer cost-efficient means of leveraging the skills of today's technology-enabled younger generation.

In addition, a review of underutilized US population groups indicates that women likely hold the most potential for expansion in the nation's armed forces. Women comprise just over half of the American population, and they as a group currently outperform men in a range of important areas. And yet at this point women constitute just 17.5 percent of active-duty military personnel.[21] Several reasons underpin this reality. Cases of sexual assault and harassment have for far too long remained endemic problems within the nation's armed services.[22] Less well known is the fact that military equipment and uniform choices have historically been made with the male body solely in mind. Even when embedded in frontline units during the War on Terror, for example, women were issued ill-fitting body armor that limited their movements and gave inadequate protection from ballistic threats.[23] Even today, engineering specifications are only slowly being adapted to mixed-gender usage.[24] The broader point, however, is that we are in an era of profound changes in the way that war is conducted. In many areas of combat, the physical power of individual human combatants has lessened in importance. As American military leaders grapple with this transition period, it will be important for them to enact reforms where possible to allow for the largest and most talented pool of potential recruits. Women should be one of the main targets of their efforts.

In tracing these connections, this book has little to say about such "high-end" weaponry as hypersonic missiles or new thermonuclear warheads. The focus of the book is instead centered on those individuals and technologies found at the "point of contact" in warfighting.[25] Chapter 1 begins by outlining the long-standing concerns about physical fitness and obesity within the American military. Declining fitness measures on the part of military draftees played an important role in prompting President Dwight D. Eisenhower in 1956 to establish the President's Council on Youth Fitness.[26] Military leaders share many of the same concerns today. Indicators show overall fitness to be substantially below the levels measured during the Cold War era. The interconnected issues of health and inadequate physical fitness constitute the two largest reasons for this military ineligibility.[27] The Pentagon has struggled to find effective solutions to the problem, including how to deal with active-duty troops who fall below fitness and obesity standards or are categorized as non-deployable for related health reasons.[28] Indeed, current and former high-level military officers describe the issue as constituting no less than a national security crisis.[29]

Chapter 2 builds on the preceding section by showing that military organizations have for many years engaged in efforts to strengthen their forces through an assortment of pharmaceutical and non-pharmaceutical mechanisms. Military deployment of performance-enhancing substances, including stimulants, was widespread during the Second World War—most famously in Germany's military forces.[30] While the United States never officially sanctioned the use of pharmaceutical stimulants during the conflict, American servicemen also used them in large numbers.[31] In the postwar era, the US military came to formally embrace the distribution of stimulants, first during the Korean War. Several years after the Korean conflict ended, the US Air Force's Strategic Air Command approved the use of amphetamines to address the increasingly long mission times its aircrews faced; tactical units followed suit in 1962. But it was in Vietnam that American military personnel began to use performance-enhancing (as well as recreational) drugs at notably increased rates.

Over time, the use of performance-enhancing technologies by the military continued to be refined. In a survey of US tactical aviators conducted in the aftermath of the 1991 Gulf War, 60 percent of respondents who used stimulants during the conflict considered the drugs "essential" to mission success.[32] More recently, performance-enhancing drugs have been considered for use in the US special operations community.[33] The chapter concludes with a warning that alongside the benefits of alertness and combat stamina comes a set of potential dangers in the employment of such substances. In making this point, the text focuses on a 2002 friendly-fire incident in which four Canadian soldiers died in Afghanistan after US pilots on a ten-hour mission mistakenly attacked their position while under the effect of stimulants.[34]

Chapter 3 examines how technological changes in digital form have fueled a fundamental shift in how we should think about the "military body." During the late 1940s and early 1950s, military and civilian strategists began using computers to play out various war scenarios.[35] But it was not until the 1990s that the true battlefield potential of digital technologies was realized. Early in the decade, during Operation Desert Storm, the United States employed a number of information-age technologies in its command-and-control and targeting networks. These capabilities enabled American and allied forces to quickly overwhelm an Iraqi army that was, at the time, one of the largest in the world.[36] Observers of the conflict argued that the US military's performance demonstrated no less than a revolution in the way war could be conducted.[37] Given these lessons, US officials became ever more creative in their efforts to take advantage of digital tools. It was in this atmosphere that the US Marine Corps modified a popular off-the-shelf video game called *Doom II* into a virtual training tool for small-unit tactics.[38] In 1999, US Army Colonel Casey Wardynski began working on a project with the goal of creating a digital platform to provide participants a virtual military experience. The release three years later of the first-person shooter video game *America's Army* gave the service a powerful new recruiting tool aimed at the country's technology-minded youth.[39] The dramatic rise of digital gaming technologies that followed reinforced this paradigm shift.

The military employment of digital gaming technologies has accelerated ever since. A 2014 documentary titled *Drone* revealed that specially designed video games were being used at US military recruiting fairs to attract prospective pilots for unmanned aerial vehicles.[40] According to the Air Force, the initiative was motivated by the fact that studies had shown that gamers possessed skill sets that made them better candidates for the position than traditional pilots.[41] The service even announced that it was developing a virtual flight simulator for youth to try at home; the IP addresses of especially skilled participants would be tracked so that recruiting material could be sent to their homes.[42] Not to be left behind, defense contractors emulated the gaming culture in their product designs to create interfaces familiar to today's youth. The

Boeing Company employed an Xbox control pad as the command mechanism for its High Energy Laser Mobile Demonstrator, a weapons system designed to shoot down mortar rounds and unmanned aerial vehicles.[43] Similar control devices have been used to operate robotic ground vehicles, unmanned airframes, and even US Navy periscopes.[44] Digital gaming technologies, the chapter asserts, will only grow in importance as cost-effective solutions that leverage the skills of today's technology-enabled younger generation.

Chapter 4 considers how the aforementioned issues reflect on the place of women in the American military. Physiological differences intrinsically exist between men and women. That said, the chapter explains that any analysis of the human body in combat should begin with an appreciation for the fact that much of the historically dominant mode of thinking on the subject has been greatly influenced by the way in which military physicality is customarily framed as being based largely on strength, speed, and endurance.

Most combat arms specialties are now open to women.[45] Within these forces, however, women continue to face numerous challenges, including ill-fitting equipment and vehicles designed for men. The fact remains that women constitute the largest available group through which to address the shortcomings in US military recruitment and retention. The chapter posits that the ongoing revolution in robotics and autonomous technologies may well constitute a true game changer for the place of women in combat.[46] Motorized exoskeletons (first proposed in the 1960s), for example, have been cited by some observers as potentially leveling the physical differences in strength of men and women on the battlefield.[47]

Chapter 5 addresses the ongoing war in Ukraine with an eye toward fleshing out the lessons that it may offer for America's military forces. In the months after the Russian invasion began, the Ukrainian military and associated forces engaged in around-the-clock, desperate efforts to hold off an ostensibly superior Russian army.[48] In the short term, stemming the tide required significant cunning alongside a willingness to rapidly train with unfamiliar equipment supplied in intermittent waves by external supporters. These requisites were eased by the fact that Western military advisors had been engaged with the country ever since the 2014 Russian intervention into the Crimean region of southern Ukraine.[49] Over the ensuing years, resourcefulness became enmeshed as a defining value in Ukrainian military culture at virtually every level (including its volunteer militias). Indeed, once it became clear that the country would withstand the initial assault, national military leaders began to engage with ideas on a longer time horizon—among them ones that might never occur to countries in less difficult situations.[50]

It seems clear at this point that the conflict in Ukraine reflects a fundamental shift in the relationship between technology, the human body, and war. Indeed, new forms of combat interface have emerged by means of linking the human form to a network of digital tools and robotic equipment. In this way, according to former Google CEO Eric Schmidt, "Ukraine offers a preview of future conflicts: wars that will be waged and won by humans and machines working together."[51] Indeed, columnist David Ignatius echoed in a December 2022 issue of the *Washington Post*, "That's the central fact of the extraordinary drama the world has been watching since Russia invaded so recklessly in February. This is a triumph of man and machine, together."[52]

While the war in Ukraine has helped weaken one of America's most long-standing rivals, chapter 6 focuses on the fact that the conflict has also served to highlight an important set of ongoing challenges facing US and other Western military leaders. In the 2021–22 time frame, for example, the US armed services entered a dire recruiting position—the worst such situation in half a century.[53] By early summer 2022 worries were mounting that each of the military's six branches might fail to reach its recruiting goals for the year.[54] The chapter goes on to assert that any solution to the matter will by necessity be multipronged and thoughtful.[55] One idea (which has a long history) is to reopen the door for a number of foreign nationals to receive citizenship in return for military service.[56] But perhaps the most difficult challenge that the country will face is the fact that the physical activity and obesity trends that are the focus of chapter 1 in this book show no signs of abating.[57] In fact, the interconnected problems became even more acute over the course of the Covid-19 pandemic that spread across the globe.[58]

The chapter concludes by asserting that the issue should be seen in the context of a rapidly changing technological environment—one that Klaus Schwab of the World Economic Forum called "the Fourth Industrial Revolution.[59] In a perfect world, of course, a way would be found to reverse the deteriorating state of US fitness and concurrent rise in obesity levels in society at large. Given the unlikelihood of successfully reversing this trend anytime soon, however, the chapter offers an argument that standards for military personnel in certain key technology areas should be reconsidered and that further exploration of the man-machine interface be undertaken with sensitivities toward identifying its appropriate moral boundaries.

Place in the Literature

Fictional accounts of the future of warfare have helped inform these perspectives.[60] As it did for many my age, the classic science fiction writing of Orson Scott Card, Robert A. Heinlein, Joe Haldeman, and John Scalzi helped shape my thinking on the subject.[61] This was also the case for Craig Thomas's 1977 novel *Firefox*, as well as the film adaptation of its mind-controlled Soviet fighter aircraft.[62] Much as they influenced Western leaders in office at the time, such Cold War novels as *The Third World War: A Future History* by General Sir John Hackett and *Red Storm Rising* by Tom Clancy played their own roles.[63] It is perhaps relevant that the Modern War Institute at the US Military Academy at West Point now regularly features posts on what fiction can tell us about the evolving nature of war.[64] So too does the US Naval Institute's *Proceedings*, the flagship professional journal of the nation's maritime services.[65] An Art of Future Warfare project was initiated in November 2014 by the Atlantic Council think tank.[66] The effort produced an anthology the following year composed of fictional accounts relating to the subject.[67] This was followed in 2017 with the holding of the first in a series of science fiction writing contests by the US Army Training and Doctrine Command.[68]

The winner of the 2019 competition was particularly impactful on my thinking. Written by Army Colonel Jasper Jeffers and titled "AN41," the short story featured a future of AI-enhanced, digitally linked technological "super-soldiers" fighting alongside human troops.[69]

Peter Singer and August Cole's 2015 bestseller *Ghost Fleet: A Novel of the Next World War* is perhaps the most prominent recent title in this vein.[70] Equally impactful on me, however, has been a six-book series written by former naval officer David Poyer on a potential future conflict between China and the United States (and their allies) in the Pacific.[71] "My overall impression . . . ," Poyer writes, "is that if the U.S. and China should go to war, raw staying power may count far more toward final victory than whatever force is in being at the commencement of hostilities. Weapons will swiftly become outmoded, if history is any guide. Technology will make astonishing leaps. Tactics and strategy will advance under forced draft."[72] A partnership between a pair of former military officers, ex-Marine special operator Elliot Ackerman and retired Admiral James Stavridis (who notably was NATO supreme allied commander Europe from 2009 to 2013), produced the prescient *2034: A Novel of the Next World War.*[73] Mick Ryan's *White Sun War* provides a nearer-term title in this tradition.[74] In each of these titles, the relationship between man and machine has been explored as a critical component of the future technological nature of warfare. And their fictional accounts possess real-world lessons. As US Secretary of the Army Dan Driscoll has recently remarked, "You'll start to see this mixture of human and machine in a way that used to be kind of science fiction, but we're right on the edge of it."[75]

In terms of academic scholarship, relatively few sophisticated monographs have been published about US military physical training policies. Garrett Gatzemeyer's recent *Bodies for Battle: US Army Physical Culture and Systematic Training, 1885–1957* constitutes a notable exception.[76] The text ties the history of Army physical training to the evolution of military technology and strategy from the late nineteenth to the mid-twentieth century. In Gatzemeyer's view, the development of physical training for soldiers during this period also reflected broader societal concerns that the American populace was growing weaker over time. The text, how-ever, addresses only one of the nation's military services. And it centers on a time period that significantly preceded the events that constitute the analytical focus of this book. As Gatzemeyer himself writes in a bibliographic essay included at the end of his monograph, "There is more to be told beyond 1957. . . . For instance, future researchers might usefully apply the concept of physical cultures to learn more about the US Army's development [or that of the broader American military] and its relationship to American society in the latter half of the twentieth century."[77]

"The Army Isn't All Work": Physical Culture and the Evolution of the British Army, 1860–1920 by James D. Campbell offers an insightful assessment of how, in the time between the Crimean War and the conclusion of the First World War, its titular subject developed into one of the most proficient military organizations in the world.[78] In doing so, the book also makes a compelling argument that the service's sporting programs made a profound cultural impact on many of the native peoples within Britain's overseas empire. Moreover, and although it does so less explicitly than Gatzemeyer's text, the book touches on how the technological implements of war related to the physical training ideas of the British Army during the period. *Fit to Serve: A History of US Army Physical Readiness* by Whitfield B. East offers a straightforward (and quite useful) chronology of its subject going back to the period of the American Revolution.[79] It again, however, attends to only one of the US armed services, and it pays scant attention to the types of technological and cultural matters that are a real strength of *Bodies for Battle* and, albeit to a lesser degree, *"The Army Isn't All Work."*

In contrast to this modest historiography, the scholarship that grapples with the history of war from the perspective of gender is considerably more expansive. Kristin L. Hoganson's *Fighting for American Manhood: How Gender Politics Provoked the Spanish-American and Philippine-American Wars* constitutes a seminal work of this type of scholarship. Alongside it, *The Male Body at War: American Masculinity During World War II* by Christina S. Jarvis offers a nuanced, well-researched analysis of the ways by which military training in the Second World War helped define a generation of American manhood.[80] Rachel Louise Moran's impressive 2018 *Governing Bodies: American Politics and the Shaping of the Modern Physique* offers a related work that examines the respective physical cultures of the First World War, Second World War, and Cold War eras.[81] With its concerns fixed squarely on the broader American society, however, the work has little to say about physical training as a policy issue within the US military. Moreover, and like Hoganson's *Fighting for American Manhood* and Jarvis's *The Male Body at War*, here again is a text that says little about contemporary and future military affairs.

This is not to say that thoughtful literature connecting gender dynamics to current defense issues is entirely absent. In *Male Armor: The Soldier-Hero in Contemporary American Culture*, Jon Robert Adams evaluates the cultural impact (in terms of literature, theater, and cinema) of masculine norms within the US military.[82] Moreover, the place of women in modern military affairs has received important attention in recent years.[83] To take just one example, in *Fight Like a Girl: The Truth Behind How Female Marines Are Trained* (Prometheus Books, 2018), former Lieutenant Colonel Kate Germano recounts with co-author Kelly Kennedy her time as the commander of the Marine Corps's Fourth Recruit Training Battalion.[84] In that capacity, she was able to greatly narrow a long-standing, considerable basic training performance gap between the sexes. And yet, Germano relates, she was fired because those results conflicted with the position taken by a multimillion-dollar gender-integration study that all-male Marine combat units performed better than those that included members of both sexes. Much can be gleaned from Germano's words about the relationship between gender and military physical training policies. The same is true of the continuing fight against bias and discrimination in American military culture. Even so, *Fight Like a Girl* has little to say on such technological issues as how the digital tools of today's society might best be leveraged in military recruitment and training. The dearth of analysis on such matters is especially noteworthy considering the profound changes taking place in the technologies of war.

Indeed, this is a subject that has been written about extensively, almost to a degree that escapes comprehensive description of even the monograph-length literature.[85] Compared to this body of scholarship, the number of works that focus on the military employment of performance-enhancing drugs (and related mechanisms) has been relatively limited. Lukasz Kamienski's *Shooting Up: A Short History of Drugs and War* (Oxford University Press, 2016) contains a wealth of useful material.[86] Nevertheless, its global focus going back to the ancient world limits its insightfulness about contemporary policy issues within the military. Moreover, it does little to connect drug usage to the wider technological and operational imperatives of the military forces that it covers. Among those titles that have focused explicitly on today's debates regarding combatant performance enhancement, a common theme revolves around the coming potential of combat "Supermen" enabled by a variety of performance-enhancing mechanisms. These include Jai Galliott and Mianna Lotz's *Super Soldiers: The Ethical, Legal and Social Implications* (Routledge, 2016) and Jean-François Caron's *A Theory of the Super Soldier: The Morality of Capacity-Increasing Technologies in the Military* (Manchester University Press, 2018).[87]

Within this body of work, Andrew Bickford's recently published *Chemical Heroes: Pharmaceutical Supersoldiers in the U.S. Military* constitutes an especially rich text.[88] Like the bulk of the literature outlined above, those who have written from this viewpoint (including the three books mentioned here) have, however, given little consideration to how their ideas connect to the physical activity and technological behaviors of American youth (who, after all, constitute the next generation of US military personnel). While the military impact of these behaviors goes unaddressed in the work, Jean Twenge's influential 2017 book *iGen* sketches out a number of their potential implications.[89] Although it has little to do with military employment of such technologies, University of Kent historian John Wills offers a fascinating cultural overview of digital gaming in *Gamer Nation: Video Games and American Culture*.[90] Matthew B. Caffrey Jr.'s *On Wargaming: How Wargames Have Shaped History and How They May Shape the Future* superbly fills this space in the literature.[91] Even so, the relationship of war-gaming to American physical culture is absent within the text.

Additional works of course can be found in each of these areas. In terms of the connections between national health and physical fitness, performance enhancement, digital gaming, gender, and the realities of the modern battlefield, a significant gap nevertheless exists in the current body of literature—one that *Fight Like a Machine: The Combat Future of the Human Body* aspires to fill.

PHYSICAL FITNESS, OBESITY, AND THE AMERICAN NATIONAL SECURITY CRISIS

The impact of physical fitness levels on US military operations has long been a point of concern among the nation's leaders.[1] Reflecting on American forces of the early 1900s, President Theodore Roosevelt wrote that "many of the older officers were so unfit physically that their condition would have excited laughter, had it not been so serious to think that they belonged to the military arm of the Government."[2] For the preponderance of the country's involvement in the First World War, US military physical fitness efforts remained suboptimal—as was the case for its broader military training efforts.[3] Recruits often engaged in only an hour of calisthenics per day—and even then, the sessions were usually optional rather than mandatory.[4] In combination with shortcomings in equipment, doctrine, and logistical support, the result was an army underprepared for the battlegrounds in France.[5] An analysis prepared just after the war's conclusion by the intelligence section of the American Expeditionary Forces General Staff painted a very different picture of combatant engagement, however. Enemy interrogations show considerable respect for US ground and air units as adversaries. A summary of the statements made by a sergeant of the German 28th Infantry Regiment offers a representative example in this regard. "Before going into action," the soldier was noted as saying, "all the men who had never had any experience with the American troops were told that the Americans were green, untrained and cowardly, and that they would not fight. It was not long before his opinion in this regard underwent a considerable change."[6] Whatever the weaknesses in their training and equipment, American combat personnel were generally viewed as resolute and capable on the battlefield.

The generation of young officers who witnessed the challenges that ensued in mobilizing for the Great War vowed to do better going forward.[7] In a 1935 speech before Illinois National Guardsmen, then Colonel (and later US Army Chief of Staff General) George C. Marshall asserted that "you cannot train without planning. You cannot impart instruction without preparation. . . . Instruction [in the past] has been a failure and a waste of time due to lack of forethought and preparation."[8] And yet at the start of the ensuing Second World War, military officers again expressed dismay at the seemingly poor levels of physical preparedness among new inductees. Writing in a January 1942 issue of the Navy's most prestigious professional journal, Admiral Jonas H. Ingram announced that "since the [First] World War there has been a steady retrogression in American physical fitness affecting all ages. This in spite of the fact that medical science has been successful in greatly improving the national health. In 1890 life expectancy was only 43 years. Since then it has been raised to 61. Today we are on the road that leads beyond our long accepted 'three score and ten' limitation. Yet as a nation we are soft."[9]

In describing his assessment after assuming command of the newly reconstituted 82nd Division (90 percent of which was manned by fresh enlistees), US Army General Omar N. Bradley asserted similarly that

the rudest shock we experienced with the draftees was the discovery that they, the prime youth of America, were generally in appallingly poor physical condition. After the war, I read many disturbing analyses about this—that a third or more of all inductees were found to be in physically poor condition. If our own experience was typical, then the analyses were valid. Some of our draftees could not walk a mile with a pack without keeling over. Most were overweight and soft as marshmallows. Only a very few were capable of the hard sustained physical exertion that we knew they would experience in combat.[10]

These trans-service observations perhaps should not have been surprising. Medical examinations of potential Selective Service inductees that took place in the immediate prewar period resulted in a rejection rate of over 40 percent.[11]

The services engaged in a variety of fitness programming during the war to prepare servicemen for the battlefield.[12] In 1941, the army adopted a new manual for physical training. Among its guidelines was a stipulation of one and a half hours of physical training per day for new recruits.[13] By the time D-Day took place on the shores of Normandy in June 1944, American military personnel were in a significantly enhanced state of combat readiness.[14] All was not ideal, however. The scale of mobilization was so vast as to preclude anything near absolute optimization. Even the US Marine Corps, which prided itself on the toughness of its standards, exhibited suboptimal training processes. A 2021 historical analysis described the reality that while Marine boot camp remained effective throughout the war, training within the service at the reserve troop and combat unit levels left much to be desired.[15] And yet the nation prevailed despite whatever shortcomings existed in the national health, physical fitness, and training, as these were outweighed by other factors, including the ability to manufacture war materials on a truly massive scale, logistical ability, and technological ingenuity.[16]

Cold War Years

Concern over the physical preparedness of the American population continued in the aftermath of the Second World War.[17] Ironically, beliefs dating from earlier in the century that Americans were at risk of undernourishment contributed to the country's continuing problems with health and fitness. The 1946 National School Lunch Act thus focused on the provision of high-calorie meals rather than meals optimally nutritious for the nation's children.[18]

The army revised its physical training manual in 1946, which resulted in a document that would remain in place for decades. It featured twelve conditioning exercises designed to increase strength, flexibility, and cardiovascular fitness.[19] Even so, the physical fitness test that was adopted at the time was administered only in basic training. It took until 1957 before combat units would have an analogous assessment in place.[20] National leaders realized in time that waiting until someone either was drafted or volunteered for service risked being too late to make a meaningful difference regarding their physical conditioning. Indeed, declining fitness measures on the part of military draftees played an important role in prompting President Dwight Eisenhower in 1956 to establish the President's Council on Youth Fitness.[21] The former World War II supreme allied commander in Europe was particularly motivated by the 1954 publication of an international study (the so-called Kraus-Weber test), which found that European children were significantly outpacing their American counterparts in a series of strength and flexibility evaluations.[22]

Shortly after Eisenhower's decision and citing the trend toward lower fitness levels in both new recruits and active-duty officers and enlistees, acting Marine Corps commandant Lieutenant General Vernon Megee ordered the establishment of the first service-wide annual fitness test in the force's history. The components and associated minimum standards were (1) three chin-ups, (2) twenty-one push-ups, (3) twenty-five sit-ups within two minutes, (4) fifteen squat thrusts within sixty seconds, (5) a six-foot broad jump, (6) an uninterrupted duck walk of fifty feet, and (7) a half-mile run or jog without stopping. Even then, exceptions were put in place for certain individuals. Female Marines were exempt from the testing requirements, for example, and slightly different criteria were put in place for males who were thirty years of age or older.[23]

In 1960, John F. Kennedy, then the president-elect, felt so strongly about the subject of the state of the citizenry's health and fitness that he wrote a multipage article, "The Soft American," for *Sports Illustrated* magazine. In it, he stated:

Throughout our history we have been challenged to armed conflict by nations which sought to destroy our independence or threatened our freedom. The young men of America have risen to those occasions, giving themselves freely to the rigors and hardships of warfare. But the stamina and strength which the defense of liberty requires are not the product of a few weeks' basic training or a month's conditioning. These only come from bodies which have been conditioned by a lifetime of participation in sports and interest in physical activity. Our struggles against aggressors throughout our history have been won on the playgrounds and corner lots and fields of America.[24]

The problem, Kennedy asserted, constituted nothing less than a national crisis. "Thus, in a very real and immediate sense," he said, "our growing softness, our increasing lack of physical fitness, is a menace to our security."[25]

Recognizing the evolving nature of modern armed conflict, the US Army created different fitness and health standards for combatant and noncombatant personnel during the Vietnam War.[26] While the application of different standards to noncombatant personnel during the war was noteworthy, military physical fitness policies changed even more dramatically in the years after the conflict. In 1976, the US Navy adopted from the civilian world a program of training pioneered by physician Dr. Kenneth Cooper, popularly known as "the Father of Aerobics."[27] Four years later, the US Army replaced its combat-focused fitness examinations and training with ones based on a more holistic framework of individual fitness and health.[28] Within this structure, a new army physical readiness test went into effect that had only three components: (1) two minutes of sit-ups, (2) two minutes of push-ups, and (3) a two-mile run.[29] Notably, the activities could be done in gym clothes rather than in combat equipment.[30] Critics called into question the utility of the results given the clear disconnection with battlefield realities. Nevertheless, the Physical Readiness Test remained in place for the next four decades.

A New Century and the Current Environment

The decade of the 2000s witnessed a realization that physical standards needed to be tied to what was happening in actual combat.[31] The focus up until this point was to train individual soldiers to simply meet the minimum requirements of a set of tasks—none of which featured rapid changes of direction or sudden bursts of movement. Lieutenant Colonel William Rieger, the commandant of the Army's Physical Fitness School, emphasized this point in contrasting the required events against his past experiences as a paratrooper. "When I jumped out of an airplane, I landed, I bundled up my chute and moved to cover," he said. "It was all anaerobic exercise. It was not running two miles."[32] In an experiment designed to better link the testing of contemporary soldiers to battlefield realities, fitness specialist Edward Thomas put enlisted soldiers at the school through the battery of fitness tests used during the Second World War. The results were less than impressive. "I can't say we were surprised," Thomas said. "But we were intrigued by how poorly our soldiers did compared to the soldiers of World War II." Thomas ascribed the drift away from functional training as a product of the fact that it had simply been a long time since American forces had been engaged in a major war. Operation Desert Storm had been the most recent—and the land campaign in that conflict only lasted around a hundred hours. "Peacetime has a tendency to make the doctrine less focused on what is needed on the battlefield," he said.[33]

Military leaders have shared the same concerns in recent years. Indicators show overall fitness measures to be substantially below the levels that existed during the Eisenhower and Kennedy eras. The Pentagon found in 2017 that only 29 percent of the 34 million Americans between the ages of seventeen to twenty-four were eligible for military service. The interconnected issues of impaired health as expressed primarily by obesity and inadequate physical fitness constituted the two largest reasons for these numbers.[34] In October 2018, Major General Frank Muth, then commander of US Army Recruiting Command, explained that "out of all the reasons that we have future soldiers disqualify, the largest—31 percent—is obesity."[35] By 2022, the number of eligible Americans had narrowed further. Indeed, by this point some 77 percent of young Americans would fail to meet the threshold requirements for entrance into the nation's armed forces.[36] The increasing popularity of online entertainment activities and the de-prioritization of school-based physical education programs have been identified as playing important roles in the current fitness crisis—facts that stand to have long-term effects on US military preparedness.[37]

By 2015, Frank Palkoska, who was then the head of the US Army Physical Fitness School, was also observing that the problem was leading to higher musculoskeletal injury rates among those who even managed to qualify for enlistment. At least one report cited these rates as being as high as 44 percent in the service's initial entry training.[38] "You acquire most of your basic movement patterns by first grade," Palkoska explained, "and our youth today just aren't getting the physical education time they need. Lack of fitness is a societal problem. The injury rate (among recruits in basic training) is developing into a taxpayer concern in terms of medical care and lost training expenses. And the lack of qualified recruits is becoming a national security issue."[39] Those already in the ranks were affected as well, with 21.6 percent having been identified as obese in 2021.[40] By 2024, the problem had become so acute as to cause an article published in a leading military professional journal to declare, "The Army's ability to regenerate and maintain combat power is heavily dependent on its ability to manage MSKI [musculoskeletal injuries]."[41] From a broad perspective, scholars at the Modern War Institute at West Point outlined the gravity of the situation. "The stakes are clear: The greatest national security emergency tied to our declining health is not having a population capable of defending our nation in a crisis."[42]

Making matters worse was the fact that key US military recruiting areas consistently remained among the worst performers in physical fitness and obesity assessments.[43] A 2019 study led by Military College of South Carolina (popularly known as The Citadel) faculty member Daniel B. Bornstein found, for example, that male and female US Army recruits from ten southern and southeastern states (which have high rates of enlistment) were significantly less fit than their counterparts from other regions. Not coincidentally, those recruits were also between 22 percent and 28 percent (depending on sex) more likely to sustain a training-related injury (TRI) during basic combat training compared with those who grew up in states designated as among the "most fit" in the country.[44] The study thus concluded that "given the economic and tactical impact of TRIs on military readiness, results from this study demonstrate the disproportionate burden that certain states are having on national security."[45]

The military has struggled to find a solution to the problem, including deciding what to do with those among active-duty ranks who fall below fitness standards. In February 2018, the Pentagon announced a policy whereby personnel who remained nondeployable for twelve months, whatever the reason, would be forced to leave the military.[46] Substandard physical fitness was among the reasons that some 286,000 troops were so categorized. Given force requirements, budgetary considerations, and the continuing rapid pace of operations overseas, the decision on the surface made sense. Even so, the reality is that obesity standards have made it difficult to field enough troops. The Covid-19 pandemic made matters still worse, pushing some 10,000 personnel into obesity in the Army alone.[47] According to a 2022 document issued by the Centers for Disease Control and Prevention, some 658,000 workdays were lost every year from overweight or obese active-duty troops. In the following year, it was reported that nearly seven out of ten servicemen exceeded the threshold for being overweight or obese.[48] The amount spent by the defense department on obesity-related health care problems has been substantial. Indeed, with the costs of replacing personnel with such conditions included, the annual amount comes to roughly $1.5 billion.[49]

Even for those who remain deployable, the time frame through which the military conducts operational training is ill-suited to maintaining a physically fit force. Writing for the Modern War Institute at West Point, Major Matt Clark has asserted that

> the Army routinely deploys its soldiers when they are at their lowest level of physical fitness....
>
> Most soldiers spend three to four months of the year conducting field training and other events that interfere with organized physical training...and most units do not conduct deliberate, consistent physical training in the field. This inconsistent training leads to a "seesaw" effect where soldiers are continually losing and rebuilding fitness. The overall impact on readiness is compounded by the fact that units spend increasingly more time in the field as they progress through their collective

training cycle. . . . The result: a unit's lowest dip in physical readiness coincides with its peak in tactical readiness. Maximum combat readiness—a combination of both—is never achieved.[50]

An October 2023 report issued by the Washington, DC–based American Security Project declared that "rapid and sustained recurrence of obesity across all services, ranks and positions now poses a dire threat, especially for at-risk populations and those in critical combat roles."[51]

Fitness Testing and Battlefield Realities

Significant thought has been put into reforming the military's physical fitness programs so that they better align with battlefield actualities. In October 2012, the Marine Corps implemented a high-intensity tactical training program as a complement to its existing fitness efforts.[52] Later that same month, the Army issued a new field manual for physical training that offered a sharp departure from the service's previous approach, which focused overwhelmingly on cardiovascular conditioning over other fitness traits. Combat preparedness was now central to the service's efforts in the area. "The objective," asserted the new manual, "is to develop Soldiers' physical capabilities to perform their duty assignments and combat roles. Army PRT [Physical Readiness Training] incorporates those types of training activities that directly support war-fighting tasks within full spectrum operations."[53] In 2017, the service began pilot tests on a new combat fitness test that was meant to place battlefield realities at the heart of the service's physical activity goals while also minimizing musculoskeletal injuries; particularly significant was the fact that the same standards were applied to both men and women.[54]

After it was reported that women failed at a rate 54 percent higher than men, the test underwent a several-year period of refinement.[55] The test was finally delineated in 2022.[56] The new examination required a ninety-pound sled pull, deadlifting, and a backward medicine ball toss. The new test was perhaps easier to pass than what had existed previously (which emphasized running to a much greater extent), although it was more difficult to maximize one's score.[57] "This fitness test is hard—no one should be under any illusions about it," said Army chief of staff General Mark Milley. "We really don't want to lose soldiers on the battlefield. We don't want young men and women to get killed in action because they weren't fit."[58] And yet the test has been criticized as being overly complicated and prone to producing an increased number of musculoskeletal injuries.[59]

It is difficult to criticize the goal of creating physical training programs that prepare US combatants for the real-world situations they will face. Nevertheless, a key part of the test's history is that its linkage to battlefield realities was removed after a 2022 independent assessment by the RAND Corporation called the benefit of that association into question.[60] Infantry officers have as a result found the test inadequate for foot troops. A February 2025 essay penned by two of them for the Modern War Institute on infantry training in the Indo-Pacific declared that "the Army Combat Fitness Test and road marches . . . do not achieve realistic preparation for this operational environment." Instead, the authors asserted that "what we needed was a time under load for longer durations on tough terrain. There is no replacing it. The only way to be ready to walk all night is to do it."[61] In 2025, Secretary of Defense Pete Hegseth ordered that sex-neutral fitness tests be instituted for direct combat troops.[62] Such efforts to more closely link military fitness programming to combat environments make good sense for troops in the ground combat arms branches.

On the other hand, extending what is most helpful to such units to all areas of the nation's military is an overly narrow approach. To a large degree lost in media discussions of Hegseth's order was a nuanced passage in this regard. "I am directing the Secretaries of the Military Departments to develop comprehensive plans to distinguish combat arms occupations from non-combat arms occupations," Hegseth wrote. "This effort will ensure that our standards are clear, mission-focused, and reflective of the unique physical demands placed on our Service members in various roles."[63] Here the secretary of defense demonstrated an understanding that the *full* breadth of modern battlefield needs should be a component of any decision regarding physical fitness policies and standards. Policymakers have increasingly, at least since the Vietnam War, conceptualized the future of combat through the lens of high technology. Fighting the wars of the future will require crucial investments in cyberwarfare, electronic warfare, and other Information Age technologies. How to recruit personnel with critical skills in these areas and, equally important, retain them in those roles has long posed a challenge for the Pentagon. With robust growth in the private-sector technology industry projected to continue and plentiful high-paying jobs available in the field, the military must find a way to compete. But the armed forces have lagged in the development of a clearly defined career path for cyber professionals. To take but one example, all 126 of the Air Force specialists who recently served a tour with the Pentagon's Cyber Mission Force ended up filling roles in the service outside the cyberwarfare realm after their tour with the command concluded.[64]

Outlining a clear career path for technological specialists will be an important step in retaining their services. But retaining talent in high-technology military missions will take much more than such an action alone. Policies designed to deal with physical fitness and obesity trends will play a role in determining the success of those efforts.[65] In a 2020 statement that reflected the situation across the various military services, former Chairman of the Joint Chiefs of Staff General Mark Milley said, "The capability of the Total Army on today's battlefield is threatened by poor health and lack of physical readiness."[66] In analyzing any potential changes on the issue, steps should be taken to identify faulty historical assumptions. A representative instance can be found in the Marine Corps's original 1956 physical fitness test. In contrast to contemporary beliefs on the comparative toughness of earlier standards, the fact is that the Marine physical fitness test in place today is considerably more difficult—particularly in its respective twenty-eight- and thirty-one-minute or under-three-mile timed run requirements for men and women.[67]

Officials have asserted interest in more comprehensive approaches to the matter than those just focused on physical training.[68] In 2020, for example, the Army launched its Holistic Health and Fitness (H2F) system.[69] Within this paradigm, nutritionists have worked to overhaul the food provided to service personnel. More needs to be done on the matter.

To give one example, vitamin D deficiency constitutes a key risk factor for stress fractures, and it is widespread across the United States, including in its armed forces. A program to address the problem would go a long way toward reducing the high musculoskeletal injury rates among military personnel, which were outlined earlier in this chapter.[70] The dining facilities afforded to university athletes offer a model for the services to follow. According to University of Oregon sports dietician Nikki Jupe, "Incorporating the basic nutrition principles will build a foundation for mission readiness, cognitive performance as well as endurance performance. Using different nutritional strategies [also helps warfighters] prepare for deployment."[71] As of November 2023, the Army's food standards remained suboptimal, for instance. "The most up-to-date nutritional research suggests that the Army's standard, carbohydrate-heavy diet is undermining soldiers' health and combat readiness," wrote Brigadier General Stephen L. Rhoades. "Updating our chow-hall menus to feature fewer starches along with more healthy fats and proteins would quite literally improve our national security. . . . Military leaders have an opportunity to improve troops' health—now and for the rest of their lives—by making healthy eating part of their training. Changing the chow hall could help spark a revolution in military readiness."[72]

In addition, the Pentagon began a sweeping review of its military fitness requirements at around the same time the decision on nondeployable forces was made.[73] A precedent already exists for reducing physical training requirements for personnel in high-technology areas. As early as 2010, concern over the need to attract sufficient numbers of technically proficient recruits induced the US Navy to implement a direct commissioning program for professionals with expertise sufficient to be cyber warfare engineer officers.[74] The US Army has recently opened a similar initiative, which allows those experienced in the field to become officers without going through its basic combat training course.[75] Two enlisted Air Force personnel notably received officer appointments in 2018 because of their proficiency in cyber operations.[76] Even the tradition-bound Marine Corps contemplated in 2017 whether an exception for tech-savvy recruits should be made to its core belief in the credo "Every Marine a rifleman."[77] The initial proposal, which was not adopted, would have allowed enlisted recruits and officer candidates possessing critical computer skills to respectively skip boot camp and the Basic School. In a 2021 guidance document, Marine Commandant David H. Berger revisited the idea. "The rapid rise in importance of the cyber domain, for instance," he wrote, "has challenged us to find creative ways to quickly build critical skills at mid-career and senior levels. Unless we find a means to quickly infuse expertise into the force—at the right ranks—I am concerned that advances in artificial intelligence and robotics, among other fields where the speed of technological change is exponential, will force us into a reactive posture."[78]

The Army has also struggled with whether the standards found in its new combat fitness test should apply uniformly across the service. And although they remain in the minority, a growing number of defense intellectuals have begun to question the utility of current medical and fitness standards in the context of the recruiting difficulties the armed forces face.[79] Taryn Sylvester of the Center for New American Security states that "people who are in more support MOSs [Military Occupational Specialties] like cyber don't necessarily focus on the same kind of fitness as people who are in combat MOSs. So, trying to strike a balance between those competing priorities is not easy."[80] And the fact is that the military needs far more individuals in non-combat-oriented occupational specialties. Toward the end of 2024, for instance, the Army announced that it was looking for several hundred combat arms officers to voluntarily move to support fields.[81] Defense intellectual Emma Moore offers from this perspective that "physical fitness is an outdated measure of readiness for the majority of Army jobs, yet the Army has doubled down on emphasizing just that," she writes. "The Army should be deemphasizing fitness as the cornerstone of every soldier's identity and instead focusing on testing relevant to specific job categories. The need to broaden the recruiting pool to meet the increasingly technical realities of war runs counter to the endless pursuit of fitness as a measure of readiness."[82]

Jacquelyn Schneider, a faculty member at the US Naval War College and a former Air Force officer, has made a case for rethinking military regulations based on job specialty. "Fitness requirements should reflect the needs of the mission and therefore the future warrior's health standards should better represent their combat and day-to-day duties," she argued. "In general, there is a baseline of good health that should be required whether you're sitting behind a computer terminal or on a foot patrol. But beyond good health, many of the more technical mission sets probably don't require two miles, 50 push-ups, and 50 sit-ups."[83] Decorated retired naval officer John Cordle adds that "a drone operator or cyber warfare specialist plays a critical role in increasing the Navy's lethality through technological expertise and precise real-time decision-making. These roles demand sharp cognitive skills, situational awareness, and technical proficiency—not muscular strength. Turning away such talent . . . is a loss to the Navy's overall capability."[84] Joining them is Colonel John G. Sotos, the State Air Surgeon of the California Air National Guard, who in June 2019 wrote that "the exclusive focus on conditioning heart and skeletal muscle overlooks the easily deconditioned organ most critical to modern military success: the brain. In essence, the physical fitness programs of today's armed forces are designed for the battles Homer's Achilles fought at the gates of Troy, not for the battles the United States will fight at the gates of air, sea, space, and cyberspace."[85]

In a February 2017 essay titled "Physical Fitness Is Not the Key to Winning America's Future Wars," Lieutenant Colonel Jahara Matisek, the director of the US Air Force Academy's Center for Airpower Studies, took the argument still further. In doing so, he asserted that "while . . . [critics] are correct about poor physical fitness undermining productivity and increasing health care costs, neither issue affects the warfighting capacity of the US military, nor has either variable determined the outcome of any recent wars. Unfortunately, their argument misdiagnoses why America stopped winning its wars, and what it will take to win future conflicts." Indeed, Matisek declared, "It is not poor physical fitness that threatens our military readiness, but poor intellectual fitness."[86] Skepticism has been expressed within the military at the expansiveness of these contentions.[87] No less a technologist than the head of the service's lead technology incubator, General James Rainey of the US Army Futures

Command, has asserted that the accelerated pace of operations seen in Ukraine demonstrates the continuing relevance of individual physical endurance. "What are the effects on the humans operating at that kind of tempo?" he asks. Considering the absence of extended opportunities to rest, he says, a successful ground assault may well require a numerical advantage three times larger than what was needed in the past.[88] With such matters in mind, the National Defense Authorization Act for Fiscal Year 2024 directed the Army to increase the physical fitness standards for close combat units under the Army combat fitness test.[89]

*   *   *

A host of benefits come with physical activity.[90] Cognitive performance improves alongside physiological measures of health.[91] The United States has in response devoted considerable resources toward the issue. Even so, and despite many years of research on the matter, little improvement has occurred on the national level. If anything, matters are becoming worse.[92] Today approximately 40 percent of Americans are obese; soon that number will encompass nearly half of population.[93]

As will be discussed in the forthcoming sections of this book, however, a variety of methods exist by which to moderate the impact of the problem on current and future military operations. Each of these possesses strengths and weaknesses as a potential choice to pursue. Discerning the best path forward will in many ways dictate America's prospects on the battlefields of the future. With this in mind, the following chapter considers performance-enhancing drugs and other biomedical means to improve the physical performance levels of the human body. After a thorough review of their potential, it concludes that better options exist by which to increase war-fighting effectiveness.

Performance Enhancement and the American Military Body

It has long been believed that the warriors of antiquity imbibed certain materials that they hoped would improve their physical performance levels.[1] Given the breadth of time that has passed, ascertaining the veracity of such claims is virtually impossible. What we do know is that the inhabitants of the ancient world were aware that the body could be profoundly impacted by any number of substances. In the city-state of Sparta, senior leaders at each training site were tasked with ensuring that those under their command exercised at a level that would at a minimum balance out the amount of food that they had eaten.[2] According to scholar Michael Rinella, the Greek word *pharmakon* to a fair degree mirrors our contemporary definition for "drug" (broadly construed).[3] Understanding on the subject developed within the context of a long history of advancements in medical knowledge in the ancient world.[4] Indeed, any number of specialized medical practices took shape over time.[5] The writings of Roman physician Galen include descriptions of stimulants used by competitors in the ancient Olympics.[6] Some even speculate that the athletes and warriors of the era ate the severed testicles of sheep in order to enhance their performance levels through a rudimentary understanding of what we now know is the hormone testosterone.[7]

Anthropological evidence can also be found that the gladiators of ancient Rome employed a special diet and drank a liquid solution containing bone ash as a mineral supplement.[8] Similar studies of mummified remains found in northern Chile demonstrated that the peoples of the Andes were using coca as a stimulant some 3,000 years ago.[9] No doubt the warriors of the Middle Ages used analogous ingredients in their tournaments and jousts. Indeed, in his 2015 doctoral dissertation, Georgia Southern University professor Grant Gearhart details a possible instance of just such a practice. According to Gearhart, a medieval epic regarding the life and exploits of the legendary Castilian knight Rodrigo Díaz de Vivar includes an episode in which the figure "turns to a stimulant—a potion in this case—to provoke the unnatural fury within" before fighting an opponent.[10] It is also known that for thousands of years the ma huang plant (in the Western world called ephedra) was prized for its stimulating effects (as well as for other reasons) in China. Guardsmen of the Great Wall employed it to stay alert at night, for example.[11]

The present day may seem so far removed from the ancient world or the Middle Ages as to render these early examples meaningless. Even so, contemporary debates on the ethical dimensions of military performance enhancement remain very much rooted in the myths of the deep past. In describing the central theme of the 2019 conference of the American Kinesiology Academy, University of Maryland scholar Bradley Hatfield asserted, "How exciting to think of the preparation of the ancient athletes in the Olympic arena and of the warriors for the field of battle. These early practices persist to the present in some form."[12] The old influences the current and future, it seems.

Performance Enhancement and the
Modern Military World

During the 1820s, General William Miller (an English-born soldier who fought in multiple conflicts in Latin America) noted the incredible impact of chewing coca leaves on human performance levels: "Their [native combatants'] every day [*sic*] pedestrian feats are truly astonishing. Guides perform a long journey at the rate of twenty or twenty-five leagues a day. A battalion, eight hundred strong, has been known to march thirteen or fourteen leagues in one day, without leaving more than ten or a dozen stragglers on the road."[13] Even so, knowledge on the topic remained relatively scarce in the West. University of Stirling scholar Paul Dimeo traces the modern history of doping in the Western world to an article that appeared in an 1876 issue of the *British Medical Journal*—one that detailed the curiosity of a Scottish physician named Sir Robert Christison regarding the physiological effects of South American coca leaves.[14] Awareness of the potential impact of performance-enhancing substances arose, of course, in the context of a long-running, wider movement in society toward an interest in chemical innovation as a tool to improve all manner of physical and cognitive functioning.[15]

By this point, caffeine was already in use by military forces. Greater access to coffee has even been cited as constituting a noteworthy advantage that Union troops possessed over their Confederate counterparts in the American Civil War, for example.[16] "Nobody can 'soldier' without coffee," wrote one soldier in his diary.[17] During the late nineteenth century, other types of physiological and psychological stimulants became available to military forces. Among these, cocaine especially was seen as offering significant military potential, primarily through its ability to depress hunger symptoms among deployed personnel and thereby reduce the logistical challenge of supplying them with food. The First World War saw widespread adoption of the drug by armies on both sides of the conflict for another reason—to enhance the physical performance levels of troops in the field.[18] Instant coffee also became a staple among the American units that deployed to Europe during the conflict.[19]

In 1887 at a Japanese research facility, the active ingredient in the ma huang plant was for the first time extracted in the form of ephedra. It took another three decades for a synthetic version to be developed (which took place again in Japan). This was followed in 1927 by the synthesis of a commercially viable form of amphetamine—one that five years later would be introduced to the masses by the Smith, Kline, & French pharmaceutical company.[20] It would have significant military implications in the years ahead.

World War II

Military deployment of performance-enhancing techniques became widespread during the Second World War. According to Norman Ohler's pathbreaking book *Blitzed: Drugs in the Third Reich*, at the outset of the war German researchers began looking into the battlefield potential of Pervitin, an amphetamine-based drug sold by the Temmler pharmaceutical company. As the head of Germany's Research Institute of Defense Physiology, Dr. Otto Ranke asserted that the substance served as "an excellent substance for rousing a weary squad. . . . We may grasp what far-reaching military significance it would have if we managed to remove natural tiredness using medical methods. . . . [This is a] militarily valuable substance."[21] Informed by the stimulant's utility in Germany's invasion of Poland the previous year, an April 1940 memorandum signed by German army Commander-in-Chief Walther von Brauchitsch declared that issuing Pervitin was a military necessity given the demands of the modern battlefield. In preparation for the campaign against France that would soon begin, an order went out for some 35 million tablets.[22]

The astonishing success of the Wehrmacht's blitzkrieg tactics might thus be at least partially explained by the pharmaceutical advantage its soldiers enjoyed. After reviewing reports on the matter, historians Nicolas Rasmussen and James Pugh wrote that British authorities began in short order to study the potential benefits that might be afforded by Benzedrine (essentially Pervitin under a different brand name). The Royal Navy authorized its use in September 1941, and the British Army and its Army Air Corps did so the following year.[23] By the end of the conflict, British forces had distributed some 72 million doses of the stimulant. For its part, the United States never officially sanctioned the use of pharmaceutical stimulants during World War II. Rasmussen asserts, however, that through off-the-record means, US servicemen were able to obtain amphetamines, using them in perhaps even larger numbers than their British counterparts.[24] And as for caffeine, in the words of writer Mark Pendergrast, "the American soldier became so closely identified with his coffee that G.I. Joe gave his name to the brew."[25] Historians have confirmed amphetamine usage during the war by Japanese and Finnish forces as well. According to political scientist Lukasz Kamienski, the Soviet Union was the sole exception on the matter of wartime adoption of amphetamines.[26]

The Cold War Era and Beyond

During the postwar era, the US military came to formally embrace the distribution of stimulants, first during the Korean War—in this case mostly Dexedrine from the Smith, Kline & French pharmaceutical company. Several years after the conflict, the US Air Force's Strategic Air Command approved the use of amphetamines to address the increasingly long mission times its aircrews faced; tactical units followed suit in 1962. But it was in Vietnam, says Kamienski, that American military personnel began to use performance-enhancing drugs in truly astonishing numbers. A 1971 governmental report noted that 225 million doses of stimulants had been handed out during that conflict between 1966 and 1969 alone.[27] Another study released two years later based on interviews with 451 Vietnam veterans noted that 45 percent of the individuals asserted that amphetamines were available within their units; a smaller set of 15 percent described access to cocaine.[28]

As pharmaceutical research continued, the use of performance-enhancing technologies by the military became more refined. Comparing his experience in the Vietnam conflict with the situation that he faced later during the 1989 US intervention in Panama in what was dubbed "Operation Just Cause," US Air Force pilot Paco Geisler asserted, "I don't know if the difference is dose or drug formulation or what. But there were no noticeable side effects during Just Cause; we just felt wide awake. But there was none of the nervousness—no feeling 'wired' like I remember in Vietnam."[29] US aircrews who deployed to the Middle East in Operations Desert Shield and Desert Storm after the August 1990 Iraqi invasion of Kuwait expressed similar findings. In one survey, approximately 60 percent of respondents who used amphetamines declared the drug "essential" to mission success.[30] After reviewing this history, a 1997 article in *Airpower Journal* (the US Air Force's flagship professional journal) drew a sharp distinction between pharmacological preparedness on the battlefield and doping by athletes on the playing field: "Using drugs to enhance performance in sports may be 'immoral,' but war is not a sporting event. Success in combat is not a question of fairness but of power; our weapons and training are designed to maximize combat power. We do not seek to equalize each side's chance of success prior to initiating contact (as we do in sports), but we do seek to obtain every advantage for our forces."[31] Aircrew surveys conducted in the early stages of Operation Iraqi Freedom found analogous beliefs regarding the utility of stimulants during long-duration flights. In one of them that examined crew members of the B-2 Spirit stealth bomber who had used dextroamphetamine in combat sorties, 97 percent expressed that the stimulant had provided a benefit to their performance.[32]

While amphetamines could help boost a soldier's alertness, they could also extract a cost. Physiological side effects accompanied their use, and despite the positive reviews from personnel who used them, they were at least potentially addictive. They also threatened to interrupt sleep once aircrew members finished their missions. This, in turn, has led to the issuance to military pilots of such "no-go" sleep pills as Ambien and Restoril.[33] A 2002 friendly-fire incident in which four Canadian soldiers died in Afghanistan after US pilots mistakenly attacked their position brought the matter to public attention. During disciplinary hearings regarding the event, the pilots argued that they had been compelled to take Dexedrine tablets for the ten-hour mission.[34] The informed consent forms that the individuals had signed indicated that personnel could be removed from flying status in the case of refusal.[35] In an interview with *Chicago* magazine, Major Harry Schmidt bitterly recalled that the Air Force "gave them to us like they were nothing. . . . I don't know what the effect was supposed to be. All I know is something [was] happening to my body and brain."[36]

The same is true of other forms of performance enhancement. Noted social and cultural historian John Hoberman asserted in his seminal book *Dopers in Uniform* that, for some time, "the US Army has been unable to enforce its zero-tolerance policy regarding anabolic steroids and has instead tolerated the massive use of dietary supplements as an alternative to male hormone doping."[37] The long-term social and medical risks of steroids and other performance-enhancing drugs may negate their short-term impact of improved strength and endurance. In terms of the former—and although it remains difficult to assess the degree to which the drugs might have affected him in the field—US Army Staff Sergeant Robert Bales admitted that he had been using anabolic steroids prior to his 2012 murder of sixteen noncombatant civilians in Afghanistan.[38] As for the latter, a 2022 investigation of a student's death at the US Navy's infamously brutal Basic Underwater Demolition/SEAL course (in which the student's class had only a 10 percent passage rate) revealed that a significant percentage of attendees were using steroids and other performance-enhancing substances.[39] Public scrutiny on the subject led to the implementation of random drug testing for Naval Special Warfare personnel the following year.[40]

Performance Optimization and the US Military

Pentagon leaders expressed interest in new forms of performance enhancement during the first decade of the twenty-first century.[41] In the aftermath of an encouraging 2003–4 study sponsored by the US Air Force Research Laboratory, a sleep-fighting drug called Modafinil gained popularity as an amphetamine replacement in the US military.[42] The British followed suit the next year. The drug, it is worth noting, offers cognitive enhancement across several dimensions of brain activity.[43] This was just a small step, however. A January 2004 policy memorandum issued by Deputy Secretary of Defense Paul Wolfowitz directed the US military services to "develop the next generation of . . . programs designed to optimize human performance and maximize fighting strength."[44] In May of the following year, the Office of Net Assessment attempted to define such an undertaking in terms of "the relatively precise, controlled and combined application of certain substances and devices over the short and long-term to achieve optimization in a person or unit's performance overall."[45]

Rivals Abroad

In November 2008, Islamist militants used cocaine during an approximately sixty-hour battle in Mumbai with Indian governmental forces.[46] In the 2014–15 time frame, it was reported that militants associated with the Islamic State were using an amphetamine-based stimulant called Captagon. Western reports described a drug-crazed, relentless fighter seething with rage and unencumbered by the needs of rest. A November 2015 report by Peter Holley of the *Washington Post* offers a representative example in this vein. The article asserted that the stimulant being used "quickly produces a euphoric intensity in users, allowing Syria's fighters to stay up for days, killing with a numb, reckless abandon." In Holley's description, the drug constituted nothing less than "the tiny pill fueling Syria's war and turning fighters into superhuman soldiers.[47] More sober analyses offered a different picture of the situation. The stimulant being used, it turns out, possessed less potency than Adderall and several other ADHD drugs that are widely used in Western populations. As put by *Vox* journalist German Lopez (who now writes for the *New York Times*), "Captagon won't turn anyone into the Hulk or even give him superhuman abilities."[48] Pharmacological modes of performance enhancement remain limited in their wartime potential. "Fenthyline could not have delivered Baghdadi's caliphate when the more potent Pervitin failed to save Hitler's Third Reich," wrote medical researcher Joseph El Khoury.[49] Even so, a fascination exists within Western outlets on the subject. Intelligence officials claimed that Hamas militants were taking Captagon during their October 2023 attack on Israel.[50]

What Lies Ahead

The 2018 National Defense Strategy of the United States reflected a paradigm shift within the national leadership on the issue of how power is waged in international affairs. "For decades," the document asserted, "the United States has enjoyed uncontested or dominant superiority in every operating domain. We could generally deploy our forces when we wanted, assemble them where we wanted, and operate how we wanted.

Today, every domain is contested—air, land, sea, space, and cyberspace."[51] In light of this reality, said the document, multiple areas of new investment were needed. Although written in the narrow context of US special operations forces (SOF), the words of one set of defense researchers also captured what was needed in the broader force. US combatants, they said, "must simultaneously dominate the physical, virtual, and cognitive domains in tomorrow's future operating environment. The side that cognitively dominates the situation by making informed decisions faster will have the most influence in competition and will win in conflict."[52]

Performance-enhancing substances have been discussed as potential tools for making progress in these areas—especially so when it comes to the nation's special operations community.[53] Ben Chitty, a senior project manager in US Special Operations Command's Science and Technology office at USSOCOM, asserted in 2017 on this point that "if there are . . . different ways of training, different ways of acquiring performance that are non-material, that's preferred but in a lot of cases we've exhausted those areas."[54] Chitty confirmed a year later that "we [at USSOCM] are in the initial stages of researching safe and effective solutions to maintain peak performance of our operators throughout their careers." Much work needed to be done before anything would be implemented, however. "We will continue to work with Defense health research partners to expand our understanding of the risks and benefits of these technologies," the project manager said. "Once we have identified a safe and effective means of increasing our operator human cognitive and physical performance, we will develop a fielding strategy that optimizes the benefit to the operator while minimizing risk."[55]

Frustration was expressed around this time with the state of US military effort on performance enhancement. "The Army's current approach to human enhancement is ad hoc, allowing unregulated use of over-the-counter supplements but forgoing research on a number of potentially promising enhancements," said the Washington, D.C.-based Center for a New American Security think tank. "This is the least optimal strategy," it explained, "one that maximizes the risks and minimizes the potential benefit of treatments. The Army should institute a high-level policy review of potentially promising methods to enhance soldier physical and cognitive performance."[56] Within this paradigm, calls have been made to build physically and cognitively enhanced "Super Soldiers."[57]

The ethical dimensions of the matter are complicated.[58] Moreover, human performance can only be improved so far through pharmacology. Indeed, despite the ubiquity of performance-enhancing substances, an analysis of running times at over 28,000 sporting events shows that Americans are on average becoming slower.[59] Observers of elite sport have for some time also known that the pace by which athletic records are being broken has become far more gradual than was the case in the past.[60] As such, speculation has been made that a biological limit exists to further increases in human performance levels (at least on any meaningful scale) and that further progress on the matter will only be made through the application of new scientific and technological innovations.[61] At the individual level, powered exoskeletons serve as an example of the type of performance-enhancing technologies that are being developed under this framework.[62]

It is worth mentioning that far more transformative possibilities are now on the horizon consequent to advances that have been made in several key biomedical areas. More specifically, it is now possible for modifications to human genomes to be made.[63] The military applications of this development are potentially profound. A 2021 report from an independent commission tasked by the US Congress with evaluating the national security ramifications of artificial intelligence included a warning that "AI, when applied to biology, could optimize for the physiological enhancement of human beings, including intelligence and physical attributes."[64]

A point of connection between the military and sporting world in the context of performance-enhancing technologies now and in the future was provided by the Australian Army Research Center's Dr. Jason Mazanov, one that he says merits reconsideration:

> Thinking on human enhancing technology is dominated by international sport's anti-doping policy, the World Anti-Doping Code (the Code). Like

other market sectors, the Australian Defence Force (ADF) has adopted parts of the Code to manage human-enhancing technologies, and in particular, human enhancing drugs. . . . While the Prohibited List may have been a useful starting point for the ADFs engagement with human enhancing technology policy, it is clearly time for the ADF to begin maturing towards "second generation" policy.[65]

Mazanov is correct in saying that societal engagement on the issue of performance enhancement has largely been relegated to the world of sport. In light of the military technologies now in development, a broader view of the subject certainly seems warranted.

Areas for Investment

Performance-enhancing drugs may be useful in a handful of tactical situations. The ability to keep combatants awake does offer an advantage in long-duration operations, for example. As put in a spring 2001 issuance from US Air Force Chief of Staff General John P. Jumper, "We have learned that 24-hour, seven-day-a-week (24/7) persistence is required for large-scale sustained operations."[66] Indeed, caffeine-laden energy drinks have even been labeled "the unsung hero of the Global War on Terror" by one member of America's ground forces.[67] As for more aggressive pharmacological interventions, dextroamphetamine remains approved for use by Army, Navy, Marine, and Air Force aviators; the latter has also kept Modafinil as a potential alternative. Fighter pilots assert that the small aircraft they fly simply weren't designed for the four-plus-hour missions they often found themselves in during the Global War on Terror.[68] The stimulants they turn to, however, are unlikely to provide anything close to a truly war-winning edge. A 2008 study funded by the Pentagon's Office of Defense Research and Engineering made just this point in comparing military performance-enhancement to that which exists in the sporting world: "The consequences of gaining a small performance advantage, even if it is highly statistically significant, are likely quite different as regards force-on-force engagements than as regards Olympic competition. In brief, a small performance advantage in force-on-force should generally result in a small change in the outcome, while in Olympic competition it can result in a large change in the outcome."[69]

Moreover, the relatively limited benefits of pharmaceutical interventions should be weighed against their potential downsides. A set of 2018 assessment by the Center for a New American Security think tank found that pharmaceutical interventions and better training might be able to "modestly improve" the physical effectiveness of soldiers while emerging technologies such as powered exoskeletons possessed the potential to "radically transform" combat performance.[70] Heat-mitigation solutions such as self-cooling combat uniforms would also be of significant value to troops operating in high-heat/high-humidity environments.[71] These would be especially relevant in land operations in the Indo-Pacific; a 2024 US Army training event in the region found that a high heat index and daily thunderstorms together posed the single largest risk faced by participating troops.[72] In any event, comparative evaluations of risks and rewards appear especially relevant on the potential use of genetic manipulation to optimize human performance levels. According to former Defense Department official Dr. Carol Kuntz, such an ability will take multiple generations to achieve, and the results may well disappoint in the end. In a 2022 report from the Center for Strategic and International Studies, she wrote that in comparison to curing a genetic disease, "enhancing genetic potential, with its greater risk of unforeseen interaction effects [compared to curing a genetic disease], poses a much greater technical challenge. Ambitions of creating an adult human with a particular preselected set of physical or intellectual traits would likely be the work of many generations, if it could even be accomplished. And as so many fairy tales and myths have warned, one fears that those wishes, once fulfilled, may prove vastly less satisfying than anticipated."[73]

In addition, effective and much nearer-term alternative options exist. As Kuntz explains, "Many of the objectives that are posited for HHGE [heritable human genome editing] in the military context—extending endurance, improving eyesight, increasing strength—are probably better advanced in the near and medium term through other technical strategies. For enhanced strength, one should remember that this is an era where robotic exoskeletons seem not only plausible but inevitable. For enhanced vision, one can consider augmented reality headsets. These strategies seem very promising and are likely to be successful on a vastly shorter timeline than HHGE."[74] And the stakes for achieving these non-pharmacological breakthroughs are high. Speaking at an April 2021 event, US Space Force chief scientist Joel Mozer declared it essential to improve human performance in the future through a combination of nerve stimulation and virtual/enhanced reality systems: "In the last century, Western civilization transformed from an industrial-based society to an information-based society, but today we're on the brink of a new age: the age of human augmentation. . . . In our business of national defense, it's imperative that we embrace this new age, lest we fall behind our strategic competitors."[75]

Sleep constitutes perhaps the most obvious area through which to improve the performance levels of military personnel. Department of Defense[76] manuals as well as guidelines declared by civilian experts cite seven hours of sleep as the minimum amount necessary for optimal health and performance levels.[77] A review of the literature found ample support for the fact that poor sleep degrades the physical and cognitive effectiveness of military personnel.[78] To take one real-world example, a project comparing an artillery unit that received seven hours of sleep per night against one that received four found that the former was nearly perfectly proficient while the latter was categorized as "dangerous."[79] And yet only around one-third of American soldiers obtain an average of at least seven hours of sleep per night.[80] Moreover, some 88.2 percent of the 110 American military personnel participating in a 2013 study were categorized as suffering from one form of sleep disorder or another.[81] Even in 2024, troops on staff duty with the Army's 18th Airborne Corps had to literally be ordered to take four-hour sleep breaks.[82] "A lack of sleep is breaking the US military," writes Army officer Dave Nixon."[83] A 2017 study involving one hundred soldiers suffering from chronic insomnia at Fort Hood in Texas found that both in-person and (though less effective) internet-based cognitive-behavioral therapy offered useful solutions.[84]

Moreover, the US military services have failed to make best use of research on how to employ optimal work shifts for sleep.[85] To take an operational example, US naval ship crews have historically worked in watch schedules that virtually ensured their nearly constant exhaustion at sea.[86] This led to suboptimal performance levels that, in the worst scenarios, contributed to fatal accidents that nearly resulted in a total loss of ship.[87] The service's leaders should have realized the benefits of a more thoughtful sleep cycle years earlier; guidelines designed to ensure adequate sleep for naval aircrews had been in place for decades, after all. The mandatory implementation of new watch schedules in 2017 improved matters dramatically.[88] For their part, the US Army's guidelines on sleep remain inadequate and/or underappreciated by service leaders.[89] A 2021 report on sleep deprivation in the military by the US Department of Defense acknowledged that "sleep may be the most important biological factor that determines Service member health and combat readiness."[90] It nevertheless found that the issue was a problem in all of the country's service branches.[91] Studies show that poor sleep accelerates muscle loss and impairs cognitive performance significantly.[92] Others have found that the ability to distinguish between friendly and hostile troops decreases over time in the context of diminished sleep among military marksmen.[93] A 2021 essay penned by a retired naval ship commander and a former Navy SEAL proposed that naval ships should be reimagined through insights gleaned from the latest human-factors research. Better rest would be enabled, to give one example of their thinking, by noise-, light-, and temperature-controlled sleep quarters for crew.[94] At a minimum, others argue, the service should mandate significant uninterrupted sleep periods.[95]

The broader point, however, is that military leaders should reconsider the intellectual paradigm through which they make decisions about the human body and its optimization for battle. For many decades, the central framework through which the subject has been considered has been based largely on physical strength, speed, and endurance. Profound changes are taking place in terms of how humans physically and intellectually interact with the world, however. In contemplating how to optimize combatant performance most effectively, military leaders would do well to focus on the interfaces through which these already-dominant modes of activity increasingly occur. Failing to take advantage of the opportunities they afford would be like remaining committed to the wireless telegraph after satellite communications became possible. They may well be the keys to victory in the wars of the future. The nation's rivals abroad are certainly doing so. In a 2019 article published in the National Defense University's quarterly peer-reviewed *PRISM*, US Army War College's quarterly peer-reviewed *Parameters*, defense intellectual Elsa B. Kania outlined the result of an extensive review of Chinese sources:

> The Chinese People's Liberation Army (PLA) has been actively exploring a range of new theories, capabilities, and technologies that are believed to be critical to future operational advantage. . . . In particular, Chinese innovation is poised to pursue synergies among brain science, artificial intelligence (AI), and biotechnology that may have far-reaching implications for its future military power and aggregate national competitiveness. Chinese military leaders appear to believe that such emerging technologies will be inevitably weaponized. . . . The PLA intends to achieve an operational advantage through seizing the initiative in the course of this transformation.[96]

Western military leaders are increasingly attuned to these matters.[97] In 2023, an Australian team demonstrated that it was feasible for an individual to remotely "drive" a robotic vehicle through a series of waypoints through mental visualization while wearing a Microsoft HoloLens augmented/mixed-reality headset.[98] A more invasive example can be seen in a controversial effort by the US Defense Advanced Research Projects Agency to create an implantable neural interface that directly links the brain to the digital world. The result from this and other pathbreaking initiatives has the potential to allow troops to control weapons and equipment directly from thoughts rather than through a physical interface.[99] As far back as 2015, a paralyzed woman implanted with such a chip was able, by means of neural signals alone, to pilot a virtual model of the F-35 Lightning II stealth fighter.[100]

It might even be possible to use devices like this to deactivate thoughts and emotions related to fear when doing so might provide a battlefield advantage.[101] The ethical dimensions of such an idea would, of course, require a great deal of consideration. Indeed, arguments have been made that artificially limiting a soldier's ability to engage in moral reasoning on the battlefield raises the likelihood of war crime incidents.[102] Significant thought would also be required on the practical issue of whether warfighter performance might unintentionally be hampered rather than helped. Fear—or at least a controlled version of it—for example, plays a useful role in preventing soldiers from being reckless.[103]

On the opposite end of the spectrum, technologies have begun to emerge that allow personnel to assess physiological signs pertaining to an individual's mental and emotional state. When combined with data-driven learning via artificial intelligence systems, a ship's captain might, for example, be given forewarning when an officer under his or her command was likely to make a stress-induced mistake.[104] The chief technology officer of the organization that developed one such tool described the results of a late 2019 evaluation: "We put a test guy up there [and] asked him some questions that made him fairly uncomfortable. Now, getting 'intent' is hard; but we could tell: 'Hey, this person is nervous when you ask this question. Their countenance changed.' Micro-expressions is another way to look at this."[105]

Early 2023 reports of hopeless charges by (allegedly) drug-addled Russian forces during the country's war in Ukraine deserve thought from this perspective. "They're like zombies. You can shoot them and more come constantly," said one Ukrainian fighter. In describing his part in fending off four such waves of attacks by individuals who seemingly had little ability to think coherently, another soldier asserted, "The orcs [Russian troops] have some serious drugs . . . they have no motivation to fight so they are given some euphoric shit. . . . Truly, they are under some kind of influence."[106] Usage of some drugs also may inhibit other areas of cognitive function. A captured Russian soldier told reporters in the summer that followed that Russian commanders were giving irrational orders while on painkillers.[107]

* * *

The American military is at risk of falling behind on potentially pivotal new paradigms of performance enhancement.[108] The ethical and operational implications of such methods will require significant study before much of significance is done on the matter. The downsides of focusing on pharmacological means of performance may in the end weigh decisively against their implementation. The following chapter will examine digital gaming technologies and unmanned platforms as potential alternatives in the race to boost combatant performance.

Digital Gaming Technologies, Unmanned Systems, and the American Military Body

Military innovators have for centuries been interested in gaming technologies as tools for war planning and operational and strategic decision-making.[1] The rise of Peter the Great has been described as greatly enabled by his extensive playing of war games as a child.[2] The first modern appearance of such an instrument occurred in 1824 with the invention of "The War Game of Prussia" by Lieutenant Georg von Reisswitz of the Prussian Guard Artillery.[3] This was followed a half century later by the first modern naval game—one created by British Navy Captain Philip Howard Colomb.[4] In 1886, retired US Navy Lieutenant William McCarty Little delivered a lecture on the topic at the US Naval War College.[5]

Building on these early developments, the mid-twentieth century witnessed new forms of interest in gaming within US military and civilian circles. During the 1950s, two projects (one conducted by the RAND Corporation and the other by Charles Roberts, an entrepreneur who would eventually start the Avalon Hill commercial entertainment firm) created rigorous board-based war-gaming systems.[6] During this period, military and civilian strategists also began to employ early forms of computers to test out various war scenarios in a more mathematically sophisticated fashion. It was around this time as well that the military's first computerized battle network was deployed in the form of an airspace command-and-control system called the Semi-Automatic Ground Environment (SAGE).[7] By the 1960s, these types of technologies had evolved to the point that digital systems could offer virtually instantaneous, synchronized updates of a simulated battlespace. Examples could be found across the country's network of military strategy and training offices, including in its service war colleges and external think tanks.[8] The ongoing miniaturization of computers alongside advancements in communications technologies allowed increasingly advanced forms of war-gaming over the next several decades.[9]

While advancements were made in the following decades, it was not until the 1990s that the true battlefield potential of digital technologies was realized.[10] Early in the decade—during Operation Desert Storm—the United States employed a number of seemingly groundbreaking Information Age tools within its command-and-control and targeting networks.[11] These capabilities helped American and allied forces quickly overwhelm an Iraqi army that was, at the time, one of the largest in the world.[12] Observers of the conflict argued that the US military's performance demonstrated no less than a revolution in the way war could be conducted.[13] Others saw Desert Storm as simply the first event in a longer evolution in military technology that was taking place. In the aftermath of the conflict, for example, the head of the Defense Department's internal think tank, the Office of Net Assessment, Andrew W. Marshall, said, "It's probable that we are near the beginning of the real revolution in military affairs. The Gulf War needs to be seen as something like [the World War I Battle of] Cambrai. A first trial of new technology and new ways of operating was undertaken. Because we are at the beginning, perhaps in 1922 in the analogy with the 20's and 30's, we cannot fully foresee how things are going to work out."[14] There was much to be said for this point of view. Indeed, and despite the obviousness of their effectiveness, many of the technologies employed by America and its allies were far from state-of-the-art.[15]

Given these lessons, US officials became ever more creative in their efforts to take advantage of digital tools. It was in this atmosphere that, a few years later, US Marine Corps (a service that had a long history of innovation in war-gaming) Commandant General Charles Krulak issued an order on the issue of digital gaming technologies.[16] The key section of the directive asserted that "the use of technological innovations, such as personal computer (PC)-based wargames, provide great potential for Marines to develop decision making skills, particularly when live training time and opportunities are limited. Policy contained herein authorizes Marines to use Government computers for approved PC-based wargames."[17] In the aftermath of the order's issuance, the service modified a popular off-the-shelf computer game called *Doom II* into a virtual training tool for small-unit tactics.[18]

Using such a game for tactical training remained relatively rare, however. More representative of how the military saw such innovations at the time was a recruiting initiative sponsored by the US Army. In 1999, Army Colonel Casey Wardynski began working on a project with the goal of creating a digital platform to provide civilian participants a virtual military experience. The release three years later of the first-person shooter video game *America's Army* gave the service a powerful new recruiting tool aimed at the country's tech-minded youth. By February 2009, the game had more than 9.7 million users, and the military employment of digital gaming technologies has accelerated ever since.[19] Many of the games that have since been released reflect the characteristics of what in 2003 defense analyst Max Boot dubbed "The New American Way of War"—one founded on special operations forces, precision-guided munitions, and integrated targeting networks.[20]

As the post-9/11 wars appeared to draw to a close during the 2010s, American military leaders confronted a pressing concern: how to maintain realistic training for their forces once they transitioned from the real-world battlefields of Iraq and Afghanistan. In-depth analyses showed that the latest in video game–based simulations was less effective than had been hoped. For instance, an Army exercise conducted in January 2012 revealed that soldiers on the ground required around 30 percent more time to accomplish identical tasks compared to their avatars in supposedly representative virtual training scenarios. "We were honestly quite shocked at the difference," said Colonel Anthony Krogh as the then head of the military's National Simulation Center. Rather than proving to be a true difference maker for "real-world" skills, the military experience so far suggested that digital training might simply make gamers better at gaming. As Krogh continued, "If your thumb-eye dexterity as an X-box player is better than mine, you're going to appear to be a better soldier than I am." Hope was not lost, however. Indeed, Krogh himself planned to alter course in an attempt to avoid this outcome. The avatars featured in the games that his command employed would, for example, be given skill trait scores (including ones for physical fitness and weapons proficiency) designed to more accurately reflect the real-world attributes of their military users.[21]

Man-Machine Teaming

The middle point of the decade also featured a major rethinking of what it would take to win the wars of the future in light of the ongoing digital revolution within modern society. In a December 2015 speech sponsored by the Washington, D.C.-based Center for a New American Security, US Deputy Secretary of Defense Robert Work described the results of an eighteen-month study on the topic—one he believed reflected the fact that a new great power competition was at hand. "The theme that came out over and over and over again," he said, "is what we call human-machine collaboration and combat teaming. . . . . and our work suggests that artificial intelligence and autonomy will allow entirely new levels of what we refer to as man-machine symbiosis on the battlefield." The emerging new performance-enhancing technologies, in his opinion, necessitated considerable social debate. "Our adversaries," he said, "quite frankly are pursuing enhanced human operations. And it scares the crap out of us, really. We're going to have to have a big, big decision on whether or not we are comfortable going that way."[22] It was soon announced that the Defense Department would spend some $18 billion on autonomous and semi-autonomous warfighting technologies.[23]

The ideas that Work expressed had been in development for some time. Indeed, over the course of the various conflicts which emerged after the September 11, 2001, terrorist attacks on America, unmanned aerial vehicles (UAVs, also referred to as remotely piloted aircraft and popularly known as drones) became the most prominent early manifestation of the man-machine interface in military affairs.[24] With these systems, human pilots "flew" their drones from remote stations that could be located literally on the other side of the world. The concept offered a perfect tool for long-duration flights that would protect the personnel operating the drones from harm. While most UAV time in the air was dedicated to reconnaissance and surveillance, they over time became increasingly important close air support and strike platforms.[25] The enhanced role for UAVs in the post-9/11 period is demonstrated by the fact that by 2010 the US Air Force was training more drone operators than it was pilots of manned aircraft.[26] Reflective as well on the subject is the fact that, as noted in a congressional report, unmanned aircraft went from being only 5 percent of US military platforms in 2005 to 31 percent in 2012.[27]

Moreover, digital gaming technologies became important recruiting and training tools for the growing fleet of unmanned platforms. A declassified 2009 Air Force report noted that the service planned in time to conduct some 75 percent of UAV training through digital simulation and virtual instruction.[28] In addition, a 2014 documentary titled *Drone* revealed that specially designed video games were being used at US military recruiting fairs to attract prospective pilots for unmanned aerial vehicles.[29] Whether intentional or not, the decision was made in accordance with research that suggested that gamers possessed a set of skills that made them better candidates for the position than traditional pilots.[30] The documentary focused on the complex ethical dilemmas faced by those flying the drones. Former drone operator Michael Haas succinctly captured this sentiment in reflecting on his transition from civilian gaming to real-life military operations: "I thought it was the coolest damn thing in the world . . . play video games all day. . . . And then the reality hits you . . . that then you may have to kill somebody."[31] While disconnecting the human from the battlefield offered far greater safety to individual operators, such a means of war thus also posed considerable psychological risks for the drones' human operators.[32] Deputy Secretary of Defense Work articulated an important additional potential dilemma: By lowering the stakes of combat operations, political leaders might well be more likely to lead a nation into war (and keep it there) than otherwise.[33]

Over time, the United States and its allies have become ever more ambitious in their ideas on gaming and the man-machine interface. In December 2016, Air Marshall Greg Bagwell, who had recently retired as the British Royal Air Force's deputy commander of operations, suggested that "we need to test harder whether we can take a young 18- or 19-year-old out of their PlayStation bedroom and put them into a Reaper [unmanned aircraft] cabin and say: 'Right, you have never flown an aircraft before [but] that does not matter, you can operate this.'"[34] In May 2018, US Air Force Lieutenant General Steven Kwast, then head of the service's Air Education and Training Command, announced that the service was beginning development work on a flight simulation game that would appeal to high school students. The IP addresses of players would be tracked so that recruiters could spot particularly promising future aviators. "The game gives me insight into their insight, their skill, their knowledge, their attributes and their characteristics," Kwast said. "And I don't need to know their name, but I know that 'IP address so-and-so' has really got unique skills to be a helicopter pilot, or an F-35 pilot or whatever it may be, and then I can start recruiting that person . . . marrying up their propensity, their passion and their talents with my requirement." Going into more detail, he went on to say that "if they're interested in flying and they see this game the Air Force has . . . if I find a 15-year-old kid that's just brilliant, I'll probably send a message to that IP address saying, 'Go tell your mom and dad that you are special and that I will offer you a $100,000 signing bonus and I will send you to Harvard for four years for free.'"[35]

Data for the design of the game came from the Air Force's "Pilot Training Next" initiative. The program aspired to take advantage of the latest advancements in virtual reality and digital simulation to provide a personalized learning experience for each student aviator. In the inaugural class, students were provided access to flight simulators both in the classroom and within their living quarters. Lieutenant Nate Lewis, a member of the initial cohort, highlighted the system's benefits, stating, "We get our schedule and I can plan with my roommate or other students to practice. . . . We can practice and work out the details of what we want to do together and what we are expecting for the actual flights or the graded simulations. If I'm not doing a loop right, I can go home and practice my visual cues and my scanning. That is where it has really helped out in the long run by helping my thought process and being very adaptive."[36] The Air Force was not alone in moving toward these types of new training technologies. The Navy's "Project Avenger" resulted in similar changes. Students undergoing primary flight training (meaning for brand-new pilots) would receive personalized instruction integrating traditional methods with a suite of virtual and mixed-reality simulators and tablet- and watch-based aviation apps.[37]

These advancements were made possible by the fact that flight simulators developed for the civilian world had become so accurate as to allow military units to adopted them into their formal training programs. The Air Force in 2020 collaborated with Google in designing a new aviation learning platform called the Joint Immersive Training System. "This enables them [student pilots] to use commercial off-the-shelf equipment, virtual reality headsets, joysticks, gaming chairs, so the Air Force doesn't have to build its own immersive training devices," said Google Public Sector Vice President Mike Daniels.[38] In terms of real-world incorporation of such ideas, the Air Force's 355th Training Squadron had by May 2021, for example, begun to employ a video game called *Digital Combat Simulator (DCS)* for A-10 Thunderbolt II (popularly known as the Warthog) attack pilots at Davis-Monthan Air Force Base in Arizona. Employed in conjunction with commercially available virtual reality headsets and gaming joysticks, throttles, and rudder pedals, the game offered a low-cost and efficient means to supplement more traditional training methods.[39] Major Drew Glowa, one of the 355th instructor pilots, described the initiative thusly:

> This prepares them for flying events better than any other training tool we have at a higher availability. . . . Now we can review specific tasks in a

real world environment by supplementing our four traditional simulator bays and giving students 22 more cockpits to practice on their own or with an instructor. We can expedite ground training to prepare students for flights at a pace we have never seen. We are transforming the way we train pilots and will be able to actually increase quality faster and add remedial training at almost any time. This will expedite the time it takes to get students ready for their flights. . . . [40]

355th Commanding Officer Lieutenant Colonel Tim Manning added that "we are not cutting [physical] flights or flying hours in the syllabus, but instead we are using this technology to produce more lethal, efficient and safer A-10 pilots. Better ground training methods will result in making the most of every flight hour. It also has the potential to be more cost efficient for the Air Force, and maybe even a faster way to accomplish training."[41]

Although their efforts are less widely known, American ground units have also utilized gaming systems to enhance their operations. Even old-fashioned contests of chance and commercial board games have been found to have utility in this regard.[42] In 2017, a class of officer candidates for the Illinois National Guard found that playing the board game Space Cadets: Dice Duel helped build communication and planning skills alongside ones related to managing complex actions under pressure.[43] The relatively new hex-and-counter board game Littoral Commander: Indo-Pacific, which was intentionally designed as a tool for educating junior leaders, has also received praise by military practitioners.[44] So too has the well-known fantasy role-playing game Dungeons & Dragons by a pair of US military intelligence officers.[45]

In terms of the employment of commercially available digital technologies, US Army tank crews used the publicly available online combat game *War Thunder* as a training mechanism in reaction to the reduced training opportunities caused by the 2020–23 Covid-19 pandemic. While *War Thunder* only gave them access to simulated Second World War and Cold War–era machinery, the troops found the game useful for understanding the tactical concepts needed on the modern battlefield. "We are able use the game as a teaching tool for each crew member," said Staff

Sergeant Tommy Huynh, a platoon section leader in the First Calvary Division's 3rd Armored Brigade Combat Team. "For example, drivers can train on maneuver formations and change formation drills. Of course online games have their limitations, but for young Soldiers it helps them to just understand the basics of their job."[46] While real tanks and formal simulators offered far better training on things like control and weapons interfaces, brigade spokesperson Captain Scott Kuhn asserted that *War Thunder* provided complementary utility from a broader perspective. "If you are a driver and you're inside a tank for real," he said, "you don't get to see what it looks like from above. You don't always understand that bigger picture because you're just focused on the role of driving the tank. This kind of broadens that. It provides a training opportunity to teach younger soldiers how what they do impacts the bigger picture for the platoon or the company."[47]

There is, of course, a potential risk if the military services lean too heavily on virtual training. This concern surfaced when a series of US military aircraft mishaps unfolded in 2019–20. According to an Air Force fighter pilot interviewed for a story on the situation:

> Students receive less total flight experience in their initial flight screening, in UPT [Undergraduate Pilot Training], and IFF [introduction to fighter fundamentals], with increased emphasis on virtual training and efficiency on the horizon. . . . While FTUs [Formal Training Units] and undergraduate training programs can get student pilots up to par for a few flights in each phase [enough for them to progress along the timeline], the lack of experience is beginning to show in basic flight discipline. . . .
>
> While a student may reach the level of expertise to graduate a fighter basic course, they can still find themselves in situations where experience would be of great value.[48]

The advantages of using virtual training technologies as a supplement seem clear, however. *DCS*, for example, can be used anywhere and at any time so long as a personal computer and requisite peripherals are in place. In November 2022, the Air Force announced that it was planning to remove an entire step of live training in its pipeline for air mobility pilots (i.e. those whom it slotted for cargo, tanker, and other non-fighter or bomber aircraft). In that track, pilots have historically followed their primary flight training in small aircraft with five months in the twin-engine T-1 Jayhawk before finishing in their final operational aircraft. Under the new plan, the intermediate step in the Jayhawk would be replaced entirely by time in simulators. According to Colonel Robert Moschella, the head of pilot training reform in the 19th Air Force, the initiative was based on rigorous study. "We started from the target back," he said. "We established what [the Air Force] would be looking for. Then we built a syllabus using some of our new technology and what we've learned over the years." While the training was far cheaper, Moschella explained, the quality of its trainees showed "no discernible difference" from those who had gone through the preexisting traditional route of training.[49]

The Lockheed Martin defense corporation has employed an open-source 3D game engine built by Epic Games to create cutting-edge training environments.[50] Plans are also in the works to collect physiological data from personnel undergoing training in simulators so that the systems can be optimized in terms of human-performance outcomes.[51] These advancements highlight a commitment to remain at the forefront of training methodologies, ensuring that military personnel are equipped with the necessary skills to excel in dynamic environments. As these initiatives continue to evolve, they have the potential to revolutionize military training, fostering agility, adaptability, and readiness among military members.

Digital Gaming and Cultural Change in Modern Military Affairs

Managing the many dimensions of modern warfare will be complicated and difficult.[52] And despite a set of inherent limitations, this is where digital gaming has the potential to be truly revolutionary. To reach this full capacity, however, a cultural and intellectual change needs to happen in military thinking about the subject.[53] Former Navy captain and current US Naval War College Professor Emeritus Robert C. Rubel emphasizes the need for this change, advocating for a return to a learning ethos within the service. He asserts, "Wargaming will be one powerful instrument for transforming the Navy's culture. . . . But to achieve the necessary power and influence for wargaming, the Navy's approach to it must change; the practice must be woven more tightly into the fabric of the Navy's professional culture."[54]

Digital gaming does indeed offer a powerful overarching paradigm for how the warfare of the future might be conducted. In an October 2016 issue of the US Army's *Military Review* professional journal, intelligence officer Mark Van Horn even went so far as to assert that "advanced war games and not advanced weapon systems will be the most promising technological investment for the future force."[55] As a practical example of this perspective, today the Marine Corps uses gaming to "help create critical thinkers to hone adaptive warfighting.[56] The service sees its new 100,000-square-foot Marine Corps Wargaming and Analysis Center as a key catalyst for force design and operational doctrine in this regard.[57] The Air Force also lets its personnel play the commercial war game *Command* on its secure internal networks as a means of testing out various scenarios.[58] Foreign policy and defense-oriented think tanks additionally use war games to flesh out strategic lessons for policymakers.[59] And American efforts on the matter are expanding. In May 2020, the US Joint Chiefs of Staff directed that "curricula should leverage live, virtual, constructive and gaming methodologies with wargames and exercises involving multiple sets and repetitions to develop deeper insight and ingenuity. We must resource and develop a library of case studies, colloquia, games and exercises for use across the PME enterprise."[60] Moreover, emerging artificial intelligence tools will in short order allow military organizations to play out wartime scenarios literally millions of times so that successful strategies can be identified and incorporated into practice.[61]

Even so, some observers have questioned whether the military too often sees war games solely as a strategic learning device for higher-level officers rather than one that has utility for lower-level tactical operators as well.[62] And, despite a set of limitations inherent to the genre, this is where training devices built on what the public sees as video games might come into play.[63] An example is provided in the form of the Mobile-Navy Enhanced Simulation for Training (Mobile-NEST), which employs a small, commercially available gaming device for training at sea. After working with the tool, submarine missile technician Jolley Gowipy declared that "the handhelds directly relate to the younger generation of operators and technicians, allowing them to continuously train and develop into 21st-century warfighters. . . . I highly recommend continuing this effort and expanding its functionality across all career and rating fields."[64] Given the space restrictions aboard ships and submarines, the compactness of the device serves as one of its most appealing characteristics. "This is the only feasible way to provide a realistic simulation of all manner of casualties [maritime incidents that cause damage]," Navy Lieutenant Commander Jacob L. Christiansen said in agreement. "In an effort to prepare our military for combat, they need this level of training."[65]

However, a hurdle exists to widespread adoption of such ideas as service leaders often depict video games as a challenge to maintaining a physically prepared force. A February 2022 release by the US Military Health System provides a representative example of such an outlook. In it, Army Major Jon-Marc Thibodeau, a clinical coordinator at Fort Leonard Wood in Missouri, was cited as saying, "The 'Nintendo Generation' soldier skeleton is not toughened by activity prior to arrival, so some of them break more easily."[66] But the fact is that the value of video gaming goes well beyond just that of a recruiting and training tool. Research sponsored through the Office of Naval Research's (ONR) Warfighter Performance Department found that it offers significant benefits as a cognitive performance enhancer. "People who play video games are quicker at processing information," explained program officer Dr. Ray Perez. "Ten hours of video games can change the structure and organization of a person's brain. In the past few years we have gathered data through research that backs that up. The data will eventually be applied for training to enhance warfighter performance."[67]

And yet digital gaming has so far been limited by the lack of a professional learning pathway in the military.[68] Moreover, the US military services have until recently largely failed to employ live, virtual, constructive gaming systems at the higher strategic and operational levels.[69] In a June 2022 essay penned for the Modern War Institute at the US Military Academy, a pair of British and American field-grade Army officers (Lieutenant Colonel Nicholas Moran of the former and Colonel Arnel P. David of the latter) stressed this point as a lesson driven home by their experiences as members of the NATO Allied Rapid Reaction Corps (a quickly deployable headquarters unit for the NATO alliance's military and crisis response units). The title of their essay, "Why Gamers Will Win the Next War," gives a strong sense of their degree of conviction on the matter.[70] Moran's and David's eyes had been opened, they said, during an experiment with a digital war-gaming simulation produced by the Battlefront.com gaming company called *Combat Mission*. In that event, the officers recalled, an experienced collection of military combat leaders had been bested by an amateur assortment of civilian gamers. "The military gamers were constrained in their thinking and clung dogmatically to doctrine," Moran and David wrote. "They discovered, to their frustration, that their speed of decision-making was lacking against gamers with greater intuition and skill."[71]

What was needed, the essay argued, was a way of looking at the world that reflected something of writer Orson Scott Card's classic science fiction tale *Ender's Game*.[72] The US and allied military services are slow to change, however. Moran and David thus conclude their essay with a cautionary passage on the likely result of ignoring the possibilities afforded by digital gaming technologies and processes:

The world's leading military academies contain the portraits of history's most storied leaders—to use Theodore Roosevelt's words, those that were "actually in the arena." But what if the future, with an emphasis on rapid judgment rather than physical example, demands that our leaders remain above the arena, not actually in it. That they exploit the cold, dispassionate vantage point of a spectator to orchestrate a clear strategy, rather than an emotive calculation influenced by the "sweat and dust and blood" of battle. . . .

The idea that the generals of the future are the gamers of today will be anathema to institutions built upon practical example. Yet, if we cling to the past and remain fixed in the present, we will inevitably mortgage our future.[73]

Digital gaming, in short, deserves the attention of military leaders.

Competition from Abroad

The United States is not alone in thinking about the military implications of gaming technologies. Indeed, gaming practices have been (or are being) adopted by both state and non-state actors around the world. In 2003, Hezbollah, the Lebanese political party with a militia arm, published a computer game called *Special Force*, another entry into the first-person shooter genre. The game placed players in direct action scenarios against the group's chief enemy, Israel. In describing the group's motivation for the development effort, Hezbollah media official Sheikh Ali Daher asserted, "This game presents the culture of the resistance to children: that occupation must be resisted and that land and the nation must be guarded. . . . Through this game the child can build an idea of some of . . . the most prominent battles and the idea that this enemy can be defeated."[74] A later version of the game was released to show Hezbollah's efforts to fight ISIS extremists in Syria.[75] Following in those footsteps, the Global Islamic Media Front (affiliated with al-Qaeda) modified a popular Western game called *Quest for Saddam*, calling its version, *Quest for Bush*. The game, which was offered free online, provided virtual terrorists with six levels of play that culminated in an attack on US President George W. Bush.[76] And rebels in Syria took a cue from US defense contractors' integration of Xbox components, designing a homemade tank with a machine gun operated via a Sony PlayStation controller.[77]

At the state level, China boasts a long-standing tradition of wargaming.[78] In a manner akin to the types of non-state uses of video games above, the Chinese People's Liberation Army released its game *Glorious*

Mission in 2011.[79] The product serves as both a propaganda tool and a training device, placing Chinese soldiers in large-scale battle scenarios involving the United States. Moreover, China's military leaders have in recent years demonstrated conviction of the ability of advanced war games to reveal lessons at both the strategic and operational levels of armed conflict. They see such tools as especially useful in helping overcome the relative lack of recent real-world experience on the part of the Chinese military compared to that of US forces.[80]

*   *   *

It would be a mistake to read this chapter as asserting that war has somehow become something of a video game in today's world. To the contrary, warfare will forever remain a fundamentally human-centric endeavor. The chapter instead argues that the adoption of advanced war-gaming technologies represents a fundamental transformation in how military organizations prepare their personnel for the complexities of modern battlefield environments.[81] By embracing their potential, US forces will be better equipped to cultivate a new generation of warriors capable of navigating uncertainty and defeating rapidly evolving threats. And physical conditioning drills can be built into this type of training. Although he went beyond digital gaming, Marine officer Brian A. Kerg proposed in 2024 a framework for doing so. "Achieving this goal, he argued, requires a *think-and-fight* drill. Such drills integrate physical fatigue, complex decision-making, and an opponent acting against you in real time."[82]

Moreover, the thoughtful integration of digital gaming tools also has the advantage of building on the everyday habits of the young Americans who dominate the personnel ranks of the country's military forces. For this reason, digital gaming technologies possess significant potential in addressing the interrelated issues of recruiting and operational effectiveness. Even so, the challenges that the US armed forces are having in their attempts to attract new recruits are likely to remain endemic. Given the scale of the problem, leaders would do well to identify those population groups with the most potential to ameliorate the problem. Such an analysis is likely to identify women as being the largest such collection of individuals. Indeed, although women constitute over half of the national population, they represent only around 17 percent of the country's military personnel. With this reality in mind, the next section of this book focuses on how the place of women in the armed services should be reflected upon within the context of the ongoing revolution in military affairs.

CHAPTER 4

WOMEN AND THE AMERICAN MILITARY BODY

Women have contributed to America's combat forces since the days of the Continental Army and the War for Independence from Great Britain.[1] During the American Civil War of the next century, a number of female soldiers disguised themselves as men in order to join combat units and take part in action against the enemy.[2] Alongside others who contributed to the struggle as civilians, the Great War of 1914 to 1918 saw women serve openly in the country's military units for the first time.[3] In the US Army Signal Corps, a group of bilingual women nicknamed the "Hello Girls" played a key role in connecting phone calls between French and American military forces.[4] The Second World War saw women make myriad contributions both at home and in foreign lands.[5] During the Cold War that ensued in the following decades, women's involvement in military affairs became broader still.[6] And yet up until very recently, the US military leadership remained staunchly against the integration of women into the bulk of its combat arms branches.[7]

As for the contemporary context of the American military body, the ongoing technological transformations that are the focus of this book also deserve consideration from the perspective of their connections to gender and biological sex. To a considerable degree, debates about women in combat have generally evolved around concepts of physicality. Indeed, and as scholar Christina S. Jarvis has compellingly argued, American notions of physicality and war are very much the product of a set of culturally constructed ideals in which the male form holds primary position.[8] Moreover, each of the country's military services has attempted to define and then ultimately leverage that masculine body to its benefit.[9] The lives of wartime women, on the other hand, have been carefully regulated in order to relieve cultural anxieties about the safety of America's conventional gender norms.[10]

Recent critiques about the integration of women into direct combat arms branches have largely been couched in the language of unwarranted risks to battlefield effectiveness.[11] In this manner, critics have argued that physical standards for women are often made easier in order to increase their odds of success in training. This, they say, only heightens the risk of substandard performance in the real field of battle. Moreover, critics assert that the small number of success stories for women in ground combat units should be placed in the context of a much larger number of failures. Women in the military, it has also been argued, have higher rates of musculoskeletal injuries than their male counterparts and have performed poorer in military marksmanship evaluations.[12] Wholesale integration of women into all combat arms units, the critics assert, thus constitutes an inefficient use of resources that places both themselves and others at risk.[13]

Some of these arguments can be rebutted with countervailing data. In high-level rifle shooting competitions, for example, women have performed just as well as men.[14] Evidence also exists that optimized military training programs can significantly narrow any physical performance gap that might exist between men and women.[15] And historical examples abound of admirable battlefield performances by women.[16] Nevertheless, physiological differences do exist between the two sexes. Indeed, the best athletes among males typically outperform those of females by some 10 to 30 percent in events that require strength and endurance.[17] It is thus at least a reasonable question to ask whether certain occupational specialties that require very specific physical abilities should remain exclusive to men. The direct-action components of special operations forces (SOF) constitute the most obvious such personnel areas. (Other SOF positions have skill priorities that are less physical in focus.)[18] And even here for the most arduous of such roles, it should be acknowledged that women very much exist who can meet their physical standards. As of September 2022, for example, women have achieved an approximate 10 percent passage rate in the assessment and selection program for US Army Special Forces. This number, though, should be reflected upon in comparison to the close to 36 percent historical average for men.[19] With the caveat that further study may result in a different set of conclusions, it thus seems at least logical to ask whether the trade-off of significantly higher failure rates for women (on average) in the training pipelines for direct-action special operations forces—and consequent reductions in yearly numbers of graduates for crucial combat billets in them—may make opening these areas to women imprudent.

Perhaps outside of only this narrow personnel range, any analysis of the female body in combat should begin with an appreciation for the fact that much of the historically dominant mode of thinking on the subject has been greatly influenced by the way in which military physicality is customarily framed as being based largely on speed, strength, and endurance. Such a belief played a role in the implementation in 2025 of "sex-neutral" fitness tests requiring women to meet the same standards as men in direct combat positions.[20] However, the physical demands of contemporary war go beyond speed, strength, and endurance. To take one example, military aviation requires excellent vision among aircrews. While surgical techniques have for some time been able to correct for nearsightedness and farsightedness, they cannot do so for color blindness. It is therefore noteworthy that women are considerably less likely to be colorblind than men in all population groups.[21] Women also, on average, have better hearing at high frequencies, and studies have shown that women may also have slightly better tolerance to decompression sickness as well as potentially a better ability to multitask (one of the most critical traits for combat personnel).[22] In terms of the real-world impact of these characteristics—and although other studies provide conflicting data—at least some evidence indicates that female military helicopter pilots have lower accident rates than male military helicopter pilots.[23] Given the mental health struggles faced by US veterans of the decades-long Global

War on Terror, it is perhaps noteworthy as well that women are less likely than men to develop substance-abuse problems (although this gap is narrowing) as well as lower rates of death by suicide (a gap that is widening).[24] Moreover, and of special relevance to the focus of this book, adult American women have a lower obesity rate than their male counterparts.[25] Thus, while the average male possesses combat-related physical advantages over the average female, the average female holds others.[26]

The masculine norms that exist within the military have, of course, their own set of downsides, including heightened risks of prisoner mistreatment and sexual assault.[27] In addition, policymakers would do well to recognize that physicality constitutes only one dimension alongside an assortment of physical and non-physical variables that denote a unit's effectiveness.[28] Studies, for example, show that a group's "collective intelligence" is positively affected by the percentage of female inclusion in it.[29] When US military forces encounter foreign cultures that frown upon contact between strangers of the opposite sex, female soldiers often play critical roles in interacting with local populations.[30] This point became especially salient during the post-9/11 interventions abroad by US forces. While women were officially excluded from American ground combat units for a number of years during this period, they served approximately 170,000 tours of duty in Afghanistan and Iraq within "support" units in the three-year stretch of 2002 to 2005.[31] Eventually, some 300,000 women deployed to the two countries.[32] As a workaround from the preclusion on "assigning" women to combat arms units, commanders would temporarily "attach" them as members of a Cultural Support Team.[33] In addition, the nature of the guerrilla-type battlefields meant that women were in many places essential to American success. As put by US Navy Captain Lory Manning, "with the cultural differences [in] Iraq and Afghanistan, if you were going to search a [local] woman, it had to be done by another woman. If you were going to get the sort of intelligence you could get by talking to the local people, our men couldn't talk to the local women. Our women could talk to both the men and women."[34]

Despite these contributions, the military has spent less effort recruiting women than it has for men despite the fact that women are outpacing men in college and postgraduate educational programs.[35] Moreover, and despite bipartisan calls for reform on the issue, women remain exempt from the draft under the Selective Service Act.[36] In addition, the services haven't been able to find ways to retain women in their force structures on the same levels that they have for men.[37] The American military remains an environment in which instances of gender discrimination, sexual assault, and other forms of harassment too often take place.[38] All of this despite the fact that women have performed spectacularly in a number of combat roles. Women have been instructors at the Navy's famous TOPGUN Strike Fighter Weapons School, commanded aircraft carriers, and been nominated to head two of the nation's military services in the Coast Guard and the Navy.[39]

To their credit, the military services appear aware that the female portion of the American population constitutes an important group that can help overcome their recruiting challenges. Noting the particular need for more cyber-warfare specialists in the force alongside the fact that women receive 52 percent of the technical degrees in the country, Navy Chief of Personnel Vice Admiral Robert Burke said in a 2018 congressional hearing, "That's where the talent is. We're very aggressively going after them."[40]

Women in Direct Combat Units

In 2015, US Secretary of Defense Ash Carter mandated that all American combat arms specialties be opened to women.[41] While significant doubts existed at some levels regarding the policy change, Captain Kristen Griest—also one of the first women to successfully complete the branch's backbreaking Ranger training—became the Army's first female infantry officer in 2016.[42] That same year also saw thirteen females within the group who graduated from the service's Armor Basic Officer Leader Course.[43] In 2017, a woman passed the Marine Corps's demanding infantry officer course for the first time.[44] This was followed a few months later with the first occasion in which a woman was accepted into one of the nation's elite special operations units—in this case, the Army's 75th Ranger Regiment.[45] July 2021 witnessed the first woman to successfully complete the training necessary to become a Naval Special Warfare combatant-craft crewman.[46] And in November 2023, a woman graduated from the US Army Sniper course.[47]

Each of these events marked an astonishing physical accomplishment. At the same time, the 2015 memorandum that opened all combat roles to women recognized that the order was, "by itself, not sufficient for their [females'] full integration."[48] Retired Army Major General Robert H. Scales has pointed out that the true indicator of success will be the degree of acceptance women receive within the "culture of the rank and file" of the military.[49] The experiences of Afghanistan and Iraq indicate that the bonds forged by professional competence will in time lead to this result.[50] To further facilitate progress in this way, the services have much to do as the hypermasculine norms of the armed forces are deeply entrenched.[51]

At the material level, combat vehicle and equipment design constitute two additional areas in which the physical differences between the sexes should be taken into greater consideration. For many years, military aviators who were women had access only to flight suits sized for men.[52] When women began to be integrated into ground combat units in the years following the 2015 order, many found the body armor issued to them to be too ill-fitting to be of use. Those in the initial ground combat personnel pipelines listed above, for example, had to wear equipment designed for male bodies. This restricted their arm movements to a degree that inhibited the firing of weapons in the anatomically prescribed fashion.[53] They were therefore forced to alter the equipment, often in dangerous ways— by removing their protective side panels, for instance. Female recruits also had higher rates of stress fractures than men due in part to the fact that their rucksacks and hip belts were designed for larger male bodies.[54]

Research shows that these issues can be ameliorated by appropriately designing gear for female bodies as well as by implementing physical training programs that are optimized to prepare women for the physical tasks of combat.[55] In line with a general pattern of excluding women from biomechanical and physiological studies, however, the armed forces failed going back to 2011 in repeated attempts to field body armor that fit the female figure.[56] Among the observations made in the 2018 report from the Defense Advisory Committee on Women was the following: "Although [the committee] is encouraged by the scientific and technological advances the Services are employing to accommodate the variety of body types present across the Armed Forces, including the Reserve and Guard, servicewomen continue to lack accessibility to this new equipment for both training and deployment."[57] Chairman of the Joint Chiefs of Staff General Joseph Dunford pushed in that same year for accelerated effort on resolving the problem.[58] Congress even introduced legislation on the matter in the form of the Body Armor for Females Modernization Act.[59]

The war in Ukraine has reflected a similar need for the provision of appropriate uniforms, body armor, and other forms of tactical gear for the approximately 42,000 Ukrainian female military personnel, including some 5,000 serving on the front lines. As of summer 2023, it was difficult for female combatants to find even combat boots in appropriate sizes.[60] It was thus a significant accomplishment when Defense Minister Oleksii Reznikov announced that same month: "Another big achievement for Ukraine and for the #UAarmy: field military uniforms for women have been approved! We are also working on developing women's body armor. It's an honor to work for Ukraine, a trailblazer in meeting women's needs for comfort while carrying out their duties. We value each and every member of our Armed Forces."[61] And to their credit, the American military service branches have similarly worked toward a range of solutions.

In April 2019, Air Force Chief of Staff General David L. Goldfein outlined the matter for US forces succinctly in saying, "We have women performing in every combat mission, and we owe it to them to have gear that fits, is suited for a woman's frame and (one) can be in for hours on end."[62] Over the next few years, the US Army's Program Executive Office Soldier helped design a new modular scalable vest with an eye toward providing an adaptable piece of equipment that would fit across the range of body shapes found in the force—including those of women.[63] Another initiative focused on the creation of a fireproof "tactical brassiere" that would provide better support and protection to female soldiers by integrating directly with their body armor.[64] The US Navy has gone a step further in working toward implementation of a "smart bra" and undershirt designed to monitor the breathing, heart functioning, and acceleration of female aircrew members. Among the aspirations is to preempt a physiological emergency such as g-force-induced loss of consciousness.[65] Sports bras helped usher in a new age of women's athletic performance.[66] A combat analogue may just offer the same potential for the battlefield.

Despite these initiatives, a 2001 internal report issued by US Army Special Operations Command found that some 44 percent of its women soldiers continued to experience problems with the fit of their equipment. Some even resorted to buying their own armor in lieu of the approved gear provided to them.[67] And these aren't the only areas where additional help is needed. Unlike men, women face challenges in long-duration missions due to a scarcity of appropriate options to urinate in such situations. Current combat uniforms for women have proven much more difficult than those for men in terms of ease of use on the matter. And cultural norms within some societies constitute another issue that women must navigate while deployed. Moreover, women hoping to avoid the awkwardness of disrobing in front of their colleagues by wandering away in search of a modicum of privacy face potentially grave security risks.[68]

According to the Air Force's Air Combat Command, "Past systems have been more geared toward male aircrew, making it difficult for female aircrew on longer sorties. The current options are adult diapers, holding it, or using a bag system to take the urine away [a method that can take up to forty-five minutes and which requires unstrapping from one's seat and gear]." Intentional dehydration prior to a flight is mentioned by some aviators as the current preference for dealing with the issue. The Air Force's perspective, however, is that this constitutes the least "safe" of the available possibilities. More specifically, according to the service, "This method can cause fatal errors and health issues, such as lowering the aircrew's ability to withstand high-g-forces by 50% and increase headaches." Fortunately, progress has been made on new technologies to redress the long-standing problem. In 2023, the Air Force began testing a new cup-liner-based system developed by the Airion Health company called the Advanced Inflight Relief Universal System (AIRUS). "We brought in a lot of female experts to really take a look at what this is and how to make a system for women from women versus being a male-driven design," said Airion manager Cam Chidiac.[69]

To take another issue in need of redress, as a result of a set of standards put in place after a 1967 survey of male pilots, the cockpits of military fighter aircraft have historically been fabricated based on the body sizes of average to larger males.[70] In 2020, it was reported that only around 9 percent of women in the Air Force met the body anthropometry requisites for piloting the F-15C Eagle, which has served as the service's primary air superiority platform since the 1980s. The Air Force's chief acquisition executive, Will Roper, described the rationale for changing course in an August 2020 media interview. "All well and good when you're a country that's going to face a country with a population that's four times your own by the end of this decade," he said. "But if we begin with a recruitment population that we've artificially halved because of how we design our cockpits and workstations, we've just doubled our work, and now we make every operator in the seat have to be eight times better than the counterpart they will face in a nation like China."[71]

The F-15 was far from the only combat aircraft from which female aviators were excluded as a result of their anthropometric design. Beginning in 2015, pilots weighing less than 135 pounds (which disproportionally affected female aviators) were barred from flying all versions of the country's most advanced tactical aircraft, the stealthy F-35 Lightning II, after tests showed that an ejection from its cockpit at even relatively slow speeds was highly likely (98 percent) to be fatal for individuals below that limit. Even those who weighed between 136 and 165 pounds faced a "serious risk" (23 percent) of death.[72] In May 2023, the Naval Air Warfare Center Aircraft Division (NAWCAD) announced an initiative to better understand the physical characteristics of contemporary Navy and Marine aviators. Data on everything from forehead size to hand length was to be included with the idea of better informing equipment acquisition and eligibility for flying various platforms. "It's just to keep in mind that the population is always changing," asserted Lieutenant Jennifer Knapp as the study's military liaison. "We have evolved. Humans have evolved since the 1960s. And we want to capture the current picture." Women are likely to experience considerable improvements in their equipment as a result. "It's not just to affect industry. It's not just to affect our partnerships, but it goes to the end user," said Knapp. "It's to ensure that everyone who walks through the door into their squadron has access to survival gear that fits, access to uniforms that fit, boots that fit. And they're able to get those uniform pieces readily available and get out and do their job effectively."[73]

Problems remain. The weight loads carried by US infantry forces have grown dramatically in recent decades (which constitutes a problem for both men and women in such units).[74] With this in mind, the Marine Corps Infantry Officer Course has for some years included a requirement that students successfully ruck a 152-pound load at a minimum three-mile-per-hour pace for over nine miles.[75] This constituted a difficult barrier for otherwise well-qualified female attendees, including ones who finished in the top 10 percent of the course's entry fitness screening, the Combat Endurance Test. The test has been criticized as overly difficult regardless of the gender integration issue.[76] Proposals have been made to reevaluate body armor in a way that compares the protection advantages of added weight against their costs. As put in a 2018 report issued by the Center for a New American Security, "The benefit of additional armor should be balanced against its effect on mobility, survivability, and mission performance."[77] Moreover, by promising to ameliorate the problem of manual weight loads to an even greater extent, the ongoing revolution in robotics and autonomous technologies may well constitute a true game changer for the place of women in combat.[78] To take one example, motorized exoskeletons (first proposed in the 1960s) have been cited by some observers as potentially leveling the physical differences in strength of men and women on the battlefield.[79]

Other initiatives have included the development of robotic "mules" to ease the load-carrying requirements for ground combatants. The fielding of such ideas has, of course, encountered challenges. Indeed, such was the case for early versions of the US Marine Corps's Legged Squad Support System, which was developed through a partnership between the US Defense Advanced Research Projects Agency and the private robotics firm Boston Dynamics. In testing the equipment at the 2014 Rim of the Pacific Exercise, the Marines found it to be too noisy for their needs. As Kyle Olson described as a spokesperson for the Marine Corps Warfighting Lab, "As Marines were using it, there was the challenge of seeing the potential possibility because of the limitations of the robot itself. They took it as it was: a loud robot that's going to give away their position."[80]

Horses, mules, and other such animals have as a result been suggested as possibilities for this role. "One potential answer to the mobility ashore question is pack animals. Pack animals can reliably carry hundreds of pounds of water, fuel, munitions, or equipment over terrain that is too difficult for wheeled or tracked vehicles. In many situations they can survive off local water and forage, and they have much lower signatures—optically, on radar, and infrared—than vehicles."[81] This said, Ukrainian forces have begun to demonstrate the combat utility of robotic methods of getting wounded troops off the battlefield.[82] And unmanned resupply aircraft constitute another technologically advanced idea.[83]

In fact, considerable progress has been made in these areas. In 2019, both the US Marines and the Army took receipt of four potential contenders for what the services called a prospective Squad Multipurpose Equipment Transport.[84] After further refinement, late 2022 saw the first batch of some 409 "robotic mules" (officially designated Small Multipurpose Equipment Transports) delivered to the US Army for deployment with the service's ground combat units. With the ability to run for three days over a sixty-mile range, the devices can carry between 1,000- and 2,500-pound loads. Considering the expanding collection of digital devices at the small-unit level, it is noteworthy as well that the mules can deliver up to three kilowatts of internally generated power.[85] So too is the fact that the Chinese have demonstrated the ability to deploy robotic ground vehicles via insertion by drone aircraft.[86]

Other ideas for reducing combat loads that cannot be divested are in the works. April 2023 saw the unveiling of a lighter and more flexible type of Kevlar body armor whose creation was enabled by recent advancements in chemistry and manufacturing processes.[87] After previewing a model at a June 2023 exposition, a Marine special operations specialist also expressed enthusiasm for a new piece of equipment called the rapid accessory integration device (RAID) plate.[88] Built by the Tomahawk Robotics company as a modular system for employing multiple tools, including communications devices and controllers for unmanned vehicles, the device has the potential to lessen a combatant's load by up to twenty-five pounds compared to legacy gear. Also of note is the fact that the plate comes with artificial intelligence technologies driven by significant computing power, which together promise to reduce the training timetable for learning how to operate robotic devices to a mere three to five days.[89]

*   *   *

Clear anthropometric and biological differences exist between the average man and the average woman. Some of these advantage women in battle, and others advantage men. While a few legitimate questions may perhaps exist as to the physical readiness of women for combat, most worries on the subject reflect outdated views as to what is possible. As American military leaders consider how future combat will be waged, it will be important to distinguish between the two and adjust where possible to allow for the largest and most talented pool of potential recruits.

The fact is that the nature of warfare itself has changed. In some areas, physical strength and endurance offer little benefit. The military services have for many decades worked hard to overcome any physical limitations on the part of men. It seems reasonable to take the same approach to women at this point in the nation's history—particularly considering the ongoing recruiting crisis as well as the need for wider readiness in the American population going forward. As has been apparent for some time now, women constitute the largest group available for filling the gap in US military personnel needs.[90] As the following chapter will show, an important lesson of the ongoing war in Ukraine is that numbers will matter should the United States ever finds itself engaged in large-scale combat with a peer or near-peer adversary.

LESSONS OF THE WAR IN UKRAINE

In terms of the focus of this book, the initial years of the 2020s have been noteworthy in several respects. The war in Ukraine, triggered by Russia's invasion in February 2021, constitutes the first "major power" conflict to take place on the European continent in decades and represents a pivotal event in shaping contemporary warfare and international relations.[1] Amid the chaos and tensions, it serves as a reminder of the evolving nature of conflict in the twenty-first century. Despite the difficulty of ascertaining the long-term lessons of any such event, the combat taking place in Ukraine nevertheless serves as a prism through which to contemplate the battlefields on which the United States and its Western allies may someday find themselves.[2] In doing so, the war may allow the militaries of these countries to break their long-standing pattern of constructing forces to fight the last war in which they found themselves.[3] "We are witnessing the ways wars will be fought, and won, for years to come," said US Chairman of the Joint Chiefs of Staff General Mark A. Milley.[4] In an appeal for more advanced weaponry to employ against Russian forces, Ukrainian Minister of Defense Oleksii Reznikov echoed this point in a July 2022 event organized by The Eurasia Center of the Washington, D.C.-based Atlantic Council think tank:

I often say that Ukraine is now essentially a testing ground. Many weapons are now getting tested in the field in the real conditions of the battle against the Russian Army which has plenty of modern systems of its own . . . We are sharing all the information and experience with our partners. We are interested in testing modern systems in the fight against the enemy and we are inviting arms manufacturers to test their new products here. So, as you know, for example, some kind of different equipment just are starting [for the first time] in our battlefield. For example, Polish Krab artillery systems. It's a really organic unit, but they are distinct in this Russian-Ukrainian war. So, I think for our partners in Poland, in the United States, France, or Germany, it's a good chance to test the equipment. So, give us the tools. We will finish the job and you will have all the new information.[5]

In the months after the invasion began, the Ukrainian military and associated units engaged in around-the-clock, desperate efforts to hold off an ostensibly superior Russian force.[6] In the immediate term, stemming the tide required significant cunning alongside a willingness to rapidly train with unfamiliar equipment supplied in intermittent waves by external supporters. These requisites were eased by the fact that Western military advisors had been engaged with the country ever since a Russian intervention in 2014 into the Crimean region of southern Ukraine.[7] Over the ensuing years, resourcefulness and a national "will to fight" became enmeshed as defining values within Ukrainian military culture at virtually every level (including in its volunteer militias).[8] Indeed, once it became clear that the country would withstand the initial assault, national military leaders began to engage with ideas on a longer time horizon— among them ones that might never have occurred in countries in less dire straits.[9] Moreover, a nimble acquisition structure was put in place to quickly field systems based on these concepts.[10] So too was an adaptive, distributed, and decentralized concept of military operations.[11] "Do you see these innovations now?" asked University of New Haven defense intellectual Matthew Schmidt. "I think it will be even better [moving forward]. Now it has to be done quickly. You have to improvise. But in a future stabilised conflict, we will see many long-term innovations."[12]

The significance of the digital battlefield in Ukraine is unmistakable in these developments.[13] A March 2023 *Wall Street Journal* editorial noted the substantial impact that some 300,000 Ukrainian computer scientists were having on the front lines.[14] Indeed, a popular podcast focusing on the evolution of the internet dubbed the conflict "the Most Online War of All Time Until the Next One."[15] Others labeled the fighting as "the First TikTok War," while to former Google CEO Eric Schmidt it constituted "the First Broadband War."[16] A more sober depiction was provided by Ukraine's minister of digital transformation, Mykhailo Fedorov. "This is the most technologically advanced war in human history," he said. "It's quite different from everything that has been seen before."[17]

Technologies related to remotely operated systems have been particularly important to Ukraine's successes—so much so that questions have been raised as to the future survivability of manned combat platforms.[18] For example, in 2024 the US Army canceled its manned Future Attack Recon Aircraft program in large part due to what was happening in Ukraine.[19] Indeed, the war is famous for video images of Ukrainian military personnel remotely engaging their Russian counterparts by means of unmanned aerial vehicles.[20] "Drones are critically important for us. They're critical to our combat advantage, and that's why we're scaling this," said Fedorov.[21] Helping illustrate this reality is the fact that by February 2025 some 60 to 70 percent of all Russian losses were taking place at the hands of such systems.[22] Unmanned maritime vessels have additionally exerted considerable damage to Russian ships operating in the Black Sea.[23] Artificial intelligence technologies have been incorporated into some of these systems.[24] Ukrainian developers can now build unmanned vehicles with such features that allow them to continue to target an enemy combatant even after radio or satellite connections to their human operators is lost.[25] In June 2025, AI-enabled drones were used to carry out a coordinated, long-range attack on strategic bombers at several Russian air bases, the most distant of which was located nearly 5,000 miles away.[26] The success aside, such equipment remains far from a panacea on the battlefields of Eastern Europe. Ukraine's head of military intelligence has estimated that Russian forces destroy between 60 and 70 percent of the unmanned sea vessels that threaten their ships.[27] And a

May 2023 report issued by a British defense establishment estimated that some 10,000 Ukrainian drones are lost every month.[28]

Despite these challenges, and as reflected in the creation of independent service branches for such systems by both Ukraine and Russia, unmanned platforms have undeniably played a fundamental role in the war.[29] Moreover, for the former, their impact has been amplified by the temporal and material efficiencies offered by the types of digital gaming technologies that are discussed in chapter 3 of this text. According to a May 2022 CNN report, a significant portion of Ukraine's drone operators were previously young video gamers. A deputy commander of one such unit named Volodymyr Demchenko described this development in the following manner:

> Imagine you are having your infantry . . . and they need to go into the village, imagine how much easier it is if there is a drone in front of them telling them, "Hey guys, there is a tank, there is a vehicle, there is a group of enemies on the left" . . . Now in each platoon we have a guy with a drone. It's funny to say but most of these drone guys are just young nerds who used to play video games. . . . Of course if they are the youngest guys, like 18 years-old, nobody will send them to liberate the villages [on the ground], but they can do their job sitting from five kilometers away. It's kind of fun for them.[30]

Two years later in the war, the *Wall Street Journal* admiringly declared these individuals "the nerdy gamers who became Ukraine's deadliest drone pilots." In covering their efforts, the article argued that traditional conceptions of military physicality should be revised. It pointed to the fact that by then several of the Ukrainian video gamers had more than doubled the number of kills achieved by the celebrated American sniper Chris Kyle in the Global War on Terror. "The movie image of elite soldiers as macho hulks has fueled concerns that today's flabby and screen-addicted youths couldn't cut it in a real fight," said the article. "But piloting drones demands quick thinking, sharp eyes and nimble thumbs, the kind of prowess more readily associated with computer games than military combat."[31]

A key to the effective transference of skills from the gaming world to the battlefield in Ukraine was an innovative set of training programs, such as one code-named Raven, which were designed to generate capable operators within a matter of days.[32] In the summer of 2022, Ukraine's military forces teamed up with video game developers to establish virtual learning environments at a collection of training centers. Unlike traditional training complexes, these facilities provide trainees with realistic battlefield scenarios without requiring the expenditure of precious ammunition—stocks of which have been continuously depleted over the course of the conflict.[33] Such was the potential of virtual training tools that a Ukrainian fighter pilot (who asked to remain anonymous) suggested that he and his colleagues could learn to fly advanced Western 4th-Generation fighter aircraft in just a few weeks (a feat that would ordinarily take years in the US military)—especially if they prepared ahead of time by using commercially available simulation software in conjunction with the flight manuals of the aircraft.[34]

Another example occurred when a forty-six-year-old, mid-ranked Ukrainian infantry officer helped initiate a training program utilizing open-source digital gaming software and commercial virtual reality components. This program aimed to prepare pilots for scenarios involving the potential acquisition of A-10 Thunderbolt II attack jets from the Unites States, which were rumored to soon be divested from American service. One of the primary inspirations for the initiative was a 2020 You-Tube video showing the US Air Force 355th Training Squadron's use of the *Digital Combat Simulator* (*DCS*) flight simulation game (which was discussed in chapter 3 of this book).[35] Prior to beginning formal training with American instructors following the summer 2023 announcement that F-16 fighter aircraft would be provided to the country, Ukraine also turned to *DCS* to give its aircrews a head start in preparation for their deployment.[36] In view of this early integration of virtual training methods, it is not surprising that a subsequent US Air Force study found that Ukrainian pilots could learn to fly the type much faster than had been previously projected. "Given the skillset demonstrated by the UKR AF [Ukrainian Air Force] pilots, and the requirement to develop a specialized syllabus only focused on min [minimum] required tasks," the report asserted, "~4 months is a realistic training timeline."[37]

Moreover, such forms of digital ingenuity have extended beyond training to very real combat environments. Following the 2014 Russian incursion into Crimea, crowdsourcing began in the hopes of funding a remote gun turret for Ukraine dubbed the "Sabre."[38] Years later (in April 2023), a release by Ukraine's TPO Media news outlet featured soldiers remotely employing the platform by means of a video game controller offered by an American game developer and distributor known as the Valve Corporation.[39] Given the scale of the modern gaming culture, for training personnel to use such systems offers considerable temporal and financial efficiency. In describing the lesson for US military forces, defense intellectual Peter Singer puts the point insightfully: "The gaming companies spent millions of dollars developing an optimal, intuitive, easy-to-learn user interface, and then they went and spent years training up the user base for the U.S. military on how to use that interface. . . . These designs aren't happenstance, and the same pool they're pulling from for their customer base, the military is pulling from . . . and the training is basically already done."[40]

An instructive case on Singer's point was provided in late 2022 and early 2023 by a Ukrainian drone operator named Mykhailo (last name omitted in the original source for security reasons), who gained his skills through video games while growing up. "When I put on my glasses and grab the remote, I remember my mom saying that these games won't do any good," he said. "But here's the use of them."[41] Another occurred just over a year later when a commander/gunner of a Ukrainian Bradley Fighting Vehicle named Serhiy (last name withheld for security reasons) used his experience as a video gamer to defeat a Russian T-90M main battle tank that possessed significantly thicker armor and heavier weapons. After initially failing to knock out the tank, Serhiy recognized as a result of his gaming experiences that he needed to hit his Russian foe where it was most vulnerable—a weak point in its armor or sensor systems. "But as I played video games," he said, "I remembered everything. Both how to hit and where." In the end, the Ukrainian commander was able to obtain a "mission kill" by disabling the optical devices that allowed the crew of the tank to visualize its surroundings. "Due to issues with anti-armor, I started blinding him so he couldn't leave," said Serhiy.[42]

Vulnerabilities and Attributes

Operational security risks are of course associated with the employment of virtually any electronic device on the battlefield—including ones that pertain to drone operations. On numerous occasions, troops on both sides of the conflict have accidentally given away their (and/or their fellow troops') positions and made them available for targeting through the use of radios, cell phones, and other personal electronic devices.[43] Equipment called cell-site simulators (often deployed via unmanned aerial vehicles or trucks) can gain access to a smartphone's internal GPS unit. They can alternatively be used to "trick" mobile devices into reporting signals from cell phone masts in the region. Once analyzed, the returns from these simulators can be used to pinpoint cell phone locations for attack. Turning off a personal device doesn't entirely eliminate the risk of such an event. If "hacked," a device operating system might allow the piece of equipment to remain "on" (and emitting) while appearing "off" to the owner.[44] As put by Roman Horbyk, a Ukrainian academician at Örebro University in Sweden, "The mobile phone is among the soldier's first non-human allies alongside their weapon, but it is not a trusted one. If it is still a friend, it is the kind of friend you would not let in on the most important secrets."[45]

These recognized digital capabilities and technical expertise at the small-unit level have undoubtedly been of crucial importance on the battlefields in Eastern Europe. It has helped that Ukraine is home to one of the most technologically educated populations in the world.[46] As put by retired US Army general Wesley Clark, "They got more educated people than we do in our armed forces, to be honest with you. Much better science and technology in the schools in Ukraine than we have in the United States."[47] And yet Western observers have noted shortcomings in small-unit equipment. Indeed, they have remarked that better night vision goggles, body armor, communication tools, and other tactical gear would be of great benefit to Ukrainian infantrymen. "What we need is a whole kit for X number of troops," said an American Marine veteran embedded in the country who asked to remain anonymous. "Pick a brigade and fully supply that whole brigade: So, give them NVGs [night vision goggles]; give them good weapons; give them lasers, drone jammers, all of that, like fully kit them out—and that brigade will be fully functional now."[48]

The war in Ukraine has also demonstrated that any future great power war will likely be protracted and prone to produce mass casualties.[49] Indeed, both sides have lost troops in the many tens of thousands—over 150,000 in the case of Ukraine and over 400,000 in the case of Russia.[50] The pre-invasion militaries of the two countries have, in fact, been so decimated as to require virtually complete rebuilds.[51] To do so, both sides have brought in massive numbers of new recruits and conscripts into their respective armies.[52] For reasons ranging from poor morale and high casualty rates to distrust in superior officers, Russia especially has experienced significant challenges in sustaining its military induction numbers.[53] At the same time, the country's forces have also suffered from high desertion rates.[54] Regardless, a key challenge for the Ukrainian and Russian militaries alike has been to identify efficient measures by which to turn inexperienced troops into effective fighting forces.[55] Aided by a large set of international partners such as the United States and a number of countries in Western Europe, Ukraine has, despite many setbacks, clearly been more successful on the matter than its more rigid foe.[56] In reflecting upon these points, retired US Navy Rear Admiral James Robb asserts that "the current conflict highlights not only the need for cutting-edge technology that is easy to use, but more importantly in my view, a fighting force confident in their leadership and trained to fight for its country and that is educated enough to assimilate technology quickly and smart enough to employ to maximum effect."[57]

Ukraine has also shown that continued service for wounded personnel offers another potential area of advantage for Western forces over their potential foes. When the country's recruiting system for the armed forces showed signs of exploitation in the summer of 2023, the nation's leadership replaced its personnel with qualified individuals from the wounded ranks. "Corruption in military recruiting will be eliminated," announced President Zelenskyy on the Twitter social media platform. "The heads of all regional recruitment centers will be fired and replaced by brave warriors who have lost their health on the frontlines but have maintained their dignity."[58] The *New York Times* reported that same month that Ukrainian combat units weakened in the country's 2023 counteroffensive were being bolstered in large part by previously wounded troops.[59] An innovative initiative titled Combat Path Debriefing has been particularly successful in keeping Ukrainian troops mentally resilient despite multiple frontline deployments.[60]

On the other hand, due to deficiencies in medical care for wounded personnel in the Russian military forces, injured soldiers from that country are far less likely to return to fight compared to their Ukrainian opponents.[61] In contrast, many individuals among the latter who suffer from even the most grievous of wounds have the potential to resume a combat role.[62] An important battlefield advantage is the fact that some 80 percent of Ukrainian troops receive medical care within sixty minutes (often called the "golden hour") of sustaining a battlefield injury.[63] Citing the director of a combat medical training division in an intelligence update, the UK Department of Defence estimated in July 2023 that some half of Russia's dead could have been prevented by appropriate use of first aid treatment on the battlefield. For those of Russia's wounded lucky enough to make it out of the combat zone, the country's medical facilities have simply been unable to keep up with the numbers.[64]

At a broader level, among the many reasons for Russia's failures in its ongoing operations in Ukraine is the fact that its population is relatively unhealthy. This became especially apparent after Russian President Vladimir Putin announced in 2022 the country's first partial mobilization since World War II.[65] In outlining what he saw as the likely consequences of the move, retired US Army General Jack Keane pointed to this fact in asserting that "when they do join the battlefield, some of them are medically unfit, some of them are physically unfit, a lot of them are mentally unfit in the sense that they really don't want to be there. And then they're going to join a military force that's already displayed a significant amount of incompetence, I don't believe they're going to be decisive."[66] The combined result, according to RAND think tank analyst Dara Massicot, is likely to be a two-pronged predicament for Russia: a significant challenge in the future recruitment and retainment of military personnel and a generational crisis regarding the mental health of its veterans.[67] Beyond the battlefields of Eastern Europe, the conflict in Ukraine underscores the physical and psychological complexities of modern warfare. Moreover, it is evident that the war in Ukraine reflects significant shifts in the dynamics of warfare.

*   *   *

Columbia University scholar Stephen Biddle has persuasively argued that the war in Ukraine demonstrates the enduringly physical nature of war. "But in many ways, this war seems quite familiar," he writes. "It features foot soldiers slogging through muddy trenches in scenes that look more like World War I than *Star Wars*. Its battlegrounds are littered with minefields that resemble those from World War II and feature moonscapes of shell holes that could be mistaken for Flanders in 1917."[68] Biddle is of course correct in asserting that war exists as a fundamentally human endeavor—one in which skill and personal innovation matter alongside numbers and technological advancement.[69] "The soldier is the Army. No army is better than its soldiers," claimed George S. Patton Jr. in this regard.[70]

At the same time, it also seems clear that the combat taking place in Ukraine reflects a shift in the relationship between the human body and war. The conflict, for example, helps illuminate the diminished battlefield importance of traditional notions of speed, strength, and endurance. For some time now, neither Ukraine now Russia has had meaningful standards for new recruits. Moreover, the tasks of troop maneuver and concealment are rendered far more complex when the battlefield is littered with unmanned weapons systems. At the risk of oversimplification, little hope exists for a human combatant to outrun a drone on the hunt in the skies above Ukraine.[71] Even heavily armored formations face grave risks in such an environment.[72]

In short, a new form of combat physicality has emerged by means of linking the human form to a network of digital tools and robotic equipment. In this way, according to Schmidt, "Ukraine offers a preview of future conflicts: wars that will be waged and won by humans and machines working together."[73] Indeed, echoed columnist David Ignatius in a 2022 issue of the *Washington Post*, "That's the central fact of the extraordinary drama the world has been watching since Russia invaded so recklessly in February. This is a triumph of man and machine, together."[74] In this way, the conflict in Ukraine helps illuminate what will likely take place in the future, where an interplay between human ingenuity and cutting-edge technology will redefine the landscape of war. As the following chapter will show, however, significant challenges will confront American leaders as they attempt to transform their forces with these points in mind.

CHAPTER 6

American Challenges and the Evolving Physicality of Combat

While the war in Ukraine has helped weaken one of America's most long-standing rivals, it has also served to highlight a significant set of ongoing challenges facing US and other Western military leaders. By the 2021–22 time frame, US armed services had entered into a dire recruiting position—the worst such state of affairs in half a century.[1] By early summer 2022 worries were mounting that each of the military's six branches might fail to reach its recruiting goals for the year.[2] The reality was that the different forces were in direct competition with one another for the approximately 412,000 (out of 32.8 million) Americans in the primary age group for the military who were both interested in and eligible for such service.[3] In the context of a September 2022 hearing before the US Senate Armed Services Personnel Subcommittee, policymakers expressed gloom on the matter. "There is no sunlight on the horizon," said North Carolina Senator Thom Tillis, the ranking Republican on the body. "It's becoming clear the all-volunteer force that has served our country well over the last 50 years is at an inflection point."[4] For retired Navy Admiral (and former Supreme Allied Commander of NATO) James Stavridis, the situation constituted no less than "a National Security Crisis . . . [which] puts the future of the all-volunteer force in doubt." Moreover, he said, "The Pentagon needs to reverse these trends or there will be grave risk to national security in an era of great-power competition."[5]

Making matters worse was that many of those who were already in the armed forces had also demonstrated a reluctance to enter an array of much-needed lines of work. In October 2022, the US Navy's Personnel Command announced that the maritime service was facing some 9,000 empty at-sea billets across a considerable number of rating types.[6] And this was despite the fact that the force had enacted several "up-and-return" provisions designed to ease the path for a resumption of service when an enlisted individual needed a break from the military.[7] The Coast Guard and the National Guard experienced similar problems.[8] A bright spot occurred in 2024, when each of the services met its recruiting numbers. The long-term trends remain worrisome, however.[9] In fact, the 2024 results were more the product of cutting service personnel levels than any sort of triumph in attracting significant numbers of new troops.[10]

The military forces of one of America's closest allies, the United Kingdom, have also for some years been beset by recruiting challenges—most worryingly in key personnel areas.[11] Of interest to the subject matter of this book, the British Army designed an advertising campaign that it hoped would convince potential recruits that, despite the trends toward automation in society, people would remain the primary drivers of military strength. Colonel Nick Mackenzie, the service's assistant director of recruitment, said in describing the campaign that "we want to tell future recruits that no matter what technological advancements we make, it is the judgment, intelligence and even the wit of our soldiers that is indispensable to the future of the army."[12] Germany has experienced similar problems, with some 18 percent of all military positions listed as higher than junior rank going unfilled as of 2020.[13] A 2016 report issued by the Canadian government stated that three-quarters of the country's military personnel were categorized as overweight or obese.[14] Not surprisingly, six years later the country's forces likewise experienced a shortfall of some 7,600 members; by 2024 this had grown to 12,000.[15] Australia has also admitted problems, which were especially alarming given the country's aspirations to grow its forces significantly in the near-term future.[16]

Any solution to the recruitment problems will by necessity be multi-pronged and creative.[17] This is especially true given the wish among some of the services to expand their personnel numbers in the coming years.[18] While recruiting problems are nothing new to the US armed services, insightful observers recognize the occurrence of a "generational shift" in the population of potential American military entrants.[19] Short of a return to some form of the draft, today's combination of both declining eligibility and decreasing interest means that only around 1 percent of the country's young people are likely prospects for the armed forces.[20] This is far from ideal should the United States ever find itself at true risk of entrance into a major war. "It's time for a national security strategy for military recruiting," said retired Lieutenant General Thomas W. Spoehr of the Heritage Foundation think tank.[21] As exemplified by the presence of Army and Marine recruiters at a November 2023 anime pop culture convention in New York, considerable attention is being given to the unique traits and interests of today's youth as well as to what those might mean for military recruitment and retention of talent.[22]

Such innovative thinking is to be encouraged, as the time-worn strategy of offering higher and higher monetary incentives for enlistment simply isn't producing the results needed.[23] Many of the young Americans in Generation Z worry about death or injury should they enlist, and they are disinclined toward the types of demands they associate with life in the military.[24] Given the circumstances, a potentially effective idea would be to reopen the doors for a significant number of foreign nationals to receive citizenship in return for military service.[25] Yet another is to raise the upper age limit for military service when needed—something that has been done repeatedly in recent years.[26] It should be noted at this point that quality has not been sacrificed by doing so. If anything, the available evidence indicates that older recruits tend to outperform younger inductees.[27] In fact, if a draft is needed, the country might do better by bringing in older, more technically proficient individuals than has been the norm in past uses of the Selective Service System.[28]

This all aside, the fact remains the physical activity and obesity trends, which were the focus of chapter 1 of this book, show no signs of abating.[29] In fact, the interrelated problems became even more acute over the course of the Covid-19 pandemic, which spread across the globe beginning in the spring of 2020.[30] Hope remains, however. Former naval officer (and noted author) David Poyer suggests a dual-track means to achieve a larger proportion of young Americans ready for service: universal military training and universal health care.[31] In addition, personnel decisions should be made with an eye toward the sorts of skills that will be most needed on the battlefields of the future. More specifically, and as articulated by scholars Brandon Leshchinskiy and Andrew Bowne, deep cultural change is needed in terms of the military's approach to emerging technologies.[32]

On this point, it seems clear that the wars of tomorrow will require a new type of warrior physicality—one that operates in linkage to an assortment of digital tools and robotic equipment.[33] A future conflict with a peer or near-peer rival such as China or Russia, says the Center for a New American Security, will revolve around "techno-cognitive confrontation."[34] Not surprisingly given the broader challenges outlined above, however, the services have experienced significant difficultly in their attempts to attract recruits with relevant technical skills.[35] Even just recently, for example, nearly a quarter of the Pentagon's cyber positions remained unfilled.[36] Fortunately, the nation's military leaders seem to understand the reality of the situation. In an October 2022 editorial penned for the *Wall Street Journal* titled "Uncle Sam Wants You for a Job That Matters," the service secretaries of the US Army, Air Force, and Navy jointly expressed that "since the end of the draft gave way to the all-volunteer military in 1973, new technologies have emerged that shape how we engage with those who seek to do us harm. Today more than ever, the armed forces need data scientists, coders and engineers as much as we need pilots, submariners and infantry. If you join, you'll get the chance to change lives, use technology and develop skills that the private sector can't match."[37]

Too often missed as well is the point that even small-unit combatants will need expertise in the employment of electronic devices and other emerging forms of technology. As then–Secretary of Defense James

Mattis said in February 2018 regarding the establishment of a new Close Combat Lethality Task Force, "These [small, tactical-level] formations have historically accounted for almost 90 percent of our casualties and yet our personnel policies, advances in training methods, and equipment have not kept pace with changes in available technology, human factors science, and talent-management best practices."[38] As part of a broader retooling of the service so that it is equipped for dealing with a future conflict with China in the Pacific, the US Marine Corps has been particularly aggressive in implementing reforms in recognition of this reality.[39] In 2018, for example, the service announced that it had begun to reorganize its rifle squads with this issue in mind. Among the changes would be the implantation in every squad of a common handheld tablet as well as the creation of a new position called a "squad systems operator" that would oversee the employment of small drones and other digital interfaces.[40] Marine Commandant Robert Neller described such individuals as needing to be proficient at a level usually associated with the use of electronic consumer goods.[41] Even so, a November 2022 article in the maritime services' leading professional journal argued that, while recent change "introduces much needed modernization to the infantry, it does not do enough to take the infantry into the next generation. To dominate the 21st-century battlefield, Marine Corps infantry must integrate drones and remote sensors at every level, field affordable loitering munitions, and equip every unit to counter small aerial targets, such as drones and loitering munitions."[42] Even in the vast maritime spaces of the Pacific, Marine ground patrols enabled by such tools will possess the ability to act as "sensor webs" for the larger force.[43]

By September of the following year, the service was considering the incorporation of autonomous systems across all its combat formations. "If you take the [person] out of it, things get simpler and they typically get more efficient, and they get less expensive. You should try to go after everything," said Lieutenant General Karsten Heckl in his role as Marine deputy commandant for combat development and integration.[44] "Marines must fight at machine speed or face defeat at machine speed," asserted the service's 2023 update to its overarching Force Design 2030 planning guidance.[45] During the following year's Modern Day Marine Conference, Lieutenant Colonel Keenan Chirhart declared as the service's lead officer in the field of robotics systems integration that a new occupational specialty was needed with such expertise. "We need to have robot Marines, IRAS [Intelligent Robotics and Autonomous Systems] Marines, integrated in every formation so we can fully leverage these capabilities."[46] Moreover, the breadth of data available to contemporary forces has led to calls for Marine information technology specialists to be integrated into the service's tactical combat units. From his position as deputy director for the Intelligent Robotics and Autonomous Systems office, Lieutenant Colonel Scott Humr asserts in this manner that "Marines embedded across the Marine Air Ground Task Force (MAGTF) with skill sets that range from data curation and application development to data science can provide a hedge against the fluctuating dynamics of the modern battlefield while supporting data-fusion requirements for better decision-making."[47]

As sensors and devices driven by artificial intelligence eventually move down to the small-unit level, the need for these types of skills will only grow.[48] In order to prepare for this eventuality in the context of Great Power competition with China, Marine leaders have additionally regeared their recruitment and retention priorities. Whereas the Marines have traditionally relied on a continuous stream of young, single-tour recruits to keep up the force's personnel levels, the service announced in 2023 that its focus going forward would be to become a more "mature" organization by means of retaining a larger portion of troops who are already serving every year.[49] At the same time, newcomers to the force would enter into an environment of revamped training courses that emphasize cognitive development and technological adeptness.[50]

While the Marines have been particularly aggressive in enacting reform over the past few years, their maritime partners in the US Coast Guard and Navy have been comparatively sluggish.[51] Only very recently has the latter brought forth a vision of the service that incorporates autonomous systems on a truly significant level.[52] Moreover, it took until November 2022 for production to begin of an unmanned aerial vehicle for the service's aircraft carriers.[53] In contrast to this has been the Air Force's longtime utilization of remotely operated aircraft. Regardless, the entirety of the American military—including the Army—seems finally to have bought into the necessity of large-scale employment of uncrewed systems.[54] In October 2024, for example, the Army announced that it was creating a new occupational specialty for robotics technicians.[55] As this book was heading to press in summer 2025, the Defense Department moreover announced plans to begin "Top Gun"–style training events for first-person-view drone operators.[56] A prime motivator for these developments has been lessons learned from the battlefields of Ukraine discussed in the previous chapter.

It will be particularly difficult for the military services to find, develop, and retain experts in such critical cutting-edge areas as quantum computing and artificial intelligence.[57] Importantly, only around 25 percent of US Air Force officers with graduate degrees have them in STEM subjects. The Army, Navy, and Marine Corps also feature officer corps with insufficient science and technology expertise.[58] The cost and time needed to attain degrees in such areas are also steep, leading a growing set of commentators to call for reforms in American higher education with implications for national security in mind.[59] For their part, the services unfortunately too often disincentivize science and technology expertise compared to other fields.[60] Such skills are to a considerable degree relegated to personnel in acquisition areas rather than being allowed to permeate throughout the operational force.[61]

In the long term, the United States should also implement national, state, and local policies designed to instill scientific and technological literacy within the population from childhood. A 2022 essay published by the US Military Academy's Modern War Institute analogized this need to the pre–World War II nutrition reforms put in place to create a more militarily prepared citizenry.[62] The Pentagon has additionally been behind the times in focusing its technological efforts on hardware rather than software.[63] So too in cyber operations it has been focused on reacting to threats rather than preventive defense.[64]

If given sufficient equipment and time to train, a new National Digital Reserve Corps and US Digital Service Academy would help to address this issue.[65] Another step would be a proposed paramilitary cyber reserve or auxiliary modeled along the lines of the Public Health Service Ready Reserve Corps.[66] As perhaps best expressed in a November 2023 strategy document issued by the US Navy, the reality is that technological adeptness has become at least equal to physical ability as a martial trait. "The next fight against our major adversary will be like no other," it says. "The use of non-kinetic effects and defense against those effects prior to and during kinetic exchanges will likely be the deciding factor in who prevails."[67] And this means that the United States must find a way to fill its personnel ranks with individuals in possession of such non-kinetic technological skills. With this in mind, the Army announced in 2023 that mandatory training on handheld counter-drone devices would be included in boot camp.[68] In addition, the service made clear that in the future, electronic warfare would be studied by every single member of its force, regardless of occupational specialty.[69] Also needed, according to technology entrepreneur Abdul Subhani, is a concrete software development career path within the armed forces. "The service needs to value coding the same way as shooting," he said.[70]

American Fitness and Obesity in the Context of the New Physicality of Warfare

At the same time, bureaucratic inertia for many years has hindered novel thinking on ways to improve the physical fitness of American military personnel. As biomedical physicist Jan De Backer writes, "Former officials . . . have emphasized the need for the Department of Defense to enhance its agility, particularly in adopting new technologies to maintain military dominance. This need for speed and efficiency in embracing innovation is not just a matter of maintaining a technological edge, but also crucial for the physical and mental preparedness of the troops."[71] In a March 2022 bit of progress on the subject, the Pentagon issued an order allowing each service branch to develop its own fitness and body composition standards based on its respective personnel and mission needs.[72] The US Space Force announced just a few days later that a yearly fitness test would be put aside in favor of a "holistic" approach to health. The fact that the service had itself only recently been created may have played a role in its openness to the idea.[73] Key features of the initiative included an emphasis on access to nutritional and physical activity instruction as well as the incorporation of wearable devices that track health data.[74] Moreover, service leaders mentioned that they believed that gamification would be a natural component of the program. "It's impossible not to start gamifying," asserted Roger Towberman, the service's highest-ranking enlisted member. "I'll say, sir, what was your sleep score last night? And he wants to know mine.... So we're building this community, this culture, of fitness where in a fun way people just immediately start taking care of each other. They're connected."[75]

Of course, many wearable devices fail to deliver all that they promise. And data-management and collection hurdles exist to effectively using them in military environments. A ship's motion across the waves might be mistaken for the movement of an individual wearer, and a risk exists that reams of useless information might needlessly be collected.[76] Moreover, and as discussed in the preceding chapter on the operational lessons of combat in Ukraine, security risks associated with the data collected by cell phones, radios, and wearable devices will need to be appropriately addressed.[77] The 2023 assassination of a Russian submarine captain is especially instructive in this regard, as Ukrainian operatives purportedly traced the officer's whereabouts by means of a smartphone fitness tracking application he used called Strava.[78] Indeed, the episode should resonate strongly with members of the US military in light of the fact that the very same app had several years earlier revealed the location and movement patterns of a significant number of American troops, including ones based overseas as well as in combat zones within Syria and Afghanistan.[79] Notably, Deputy Secretary of Defense Patrick M. Shanahan issued an order in August 2018 prohibiting the use of geolocating services on both governmentally owned and personal electronic devices in military operational areas.[80] The degree of compliance with the order is difficult to ascertain, however. Efforts are also underway to develop technological solutions that will help obscure the use of wireless networks from enemy combatants.[81]

By enabling the tracking of manned and unmanned military aircraft by publicizing the transmissions of their Automatic Dependent Surveillance–Broadcast transponders (which themselves also constitute defense risks), X (formerly known as Twitter) and other social media platforms constitute another point of weakness in informational security—and one that national legislators are trying to ameliorate.[82] The fact is that contemporary society is saturated with open-source information that can be exploited by intelligence operators.[83] This includes data that can compromise US military units operating under even the most stringent of electronic signature controls. A cell phone picture uploaded by a friendly local observer of a Marine amphibious force's landing might, for example, be accessed and then used as a targeting tool by a hostile actor.[84] As put in a December 2016 US Army evaluation of Russian military tactics, "The newest cadre of Soldiers, known as the first generation of truly 'digital natives,' is often culturally resistant to regulations imposed on their online social interactions. . . . A simple, innocuous-seeming post on Twitter or Facebook can now give away location, movement, and military capabilities in the stroke of a key. . . . Never before has the actions of one lone individual been so visible and prone to manipulation by the adversary."[85]

One member of a US special operations force described the importance of reducing his unit's electronic signature in the field accordingly: "The ability to hide in the noise is the No. 1 thing I think we need to attack, from my perspective."[86] Naval and Marine experts also express the need for regular emissions control exercises within the fleet so that such practices can be implemented seamlessly during wartime.[87] In the worst of cases, US forces may be forced to fight with their digital networks severely reduced or disabled.[88] Summarizing these points in a December 2022 event put on by the Defense Writers Group, US Marine Corps Commandant David Berger asserted that the great majority of American troops remained unaware as to just how much information their electronic devices emit. "They don't think anything about pressing a button. This is what they do all day long. Now we have to completely undo 18 years of communicating all day long and tell them that's bad. That will get you killed, so turn your cell phone off," Berger said. "They're like, 'I won't touch it. It just stays on.' No, there's parts of the cell phone you don't understand." Berger continued, "We have to be distributed. You have to have enough mobility that you can relocate your unit pretty often. You have to learn all about—like some of us learned 30 years ago—camouflage, decoys, deception. . . . What we didn't worry so much about 30 years ago now is every time you press a button, you're emitting."[89]

With the caveat that these issues must be successfully overcome, and while the results may ultimately prove disappointing, those who hope to address the military's fitness and obesity issues would do well to reflect on the Space Force's efforts to leverage the everyday technologies that Americans use. Even the more traditional military services have experimented with potential innovations that might offer a way around the problems posed by America's fitness and obesity epidemic. The Coast Guard requires annual fitness testing for only around a quarter of its members—such as search-and-rescue swimmers and divers.[90] As far back as 2012, Colonel Anthony Krogh of the Army's National Simulation Center proposed that soldiers might be motivated to improve their real-world physical fitness if doing so would improve their digital gaming traits. "If I were to tell you that the only way to improve your avatar in [the popular first-person shooter video game] *Call of Duty 3* is to go out and run in real life, what would you do?" Krogh remembers asking his son (who was then a member of his university's Reserve Officer Training Corps program). "He said, 'I wouldn't sit here playing—I'd be out running.'"[91]

More recently (in July 2022), the Army announced the creation of a pilot initiative it called the Future Soldier Preparatory Course.[92] Within the program, recruits who failed to meet physical or mental test requirements for enlistment could receive up to ninety days of guided instruction and drilling. The early results were promising.[93] In December 2022, the Army decided to expand the operation after some 93 percent of the 3,206 individuals who took part were able to meet Army entrance requirements for basic training.[94] Early data indicated that once basic training began, those who completed the preparatory course were outpacing regular Army entrants in terms of receiving leadership positions.[95] By September 2023, the service had decided to cement the initiative as a permanent option.[96] The military branches have also provided informal guidance on how to avoid injuries that might delay or prevent entrance as a recruit, including suggestions that potential inductees avoid underwater swimming, combat-related physical activities such as boxing, long-distance running, outdoor biking, and other types of competitive sports.[97] Given the reality of the future battlefield, military leaders in the other services are likewise considering the adoption of such ideas.[98] In the spring of 2023, for instance, the Navy announced a new Future Sailor Preparatory Course that mirrored many of the ideas found in its Army forebearer. Here again, out-of-shape or academically unqualified recruits would receive training designed to improve their prospects of successfully meeting basic training requirements.[99] More recently, the service announced that the program could potentially add around a thousand new recruits per year.[100]

Musculoskeletal injuries remain a risk well beyond an individual's initial time in the military, of course. Indeed, some 800,000 US military personnel suffer such injuries every year.[101] They especially remain a major concern for members of the nation's ground combat units. Such individuals routinely carry heavy loads over long distances, and airborne troops experience intense shocks to knees and ankles when they land by parachute. Given the pressing need to boost retainment numbers, physical therapists have been directly embedded into military units.[102] Alternative fitness assessments have also been considered for those personnel who have suffered permanent injuries.[103]

More cutting-edge still is service-sponsored research on the use of body-worn sensors as a potential means to predict when an injury is about to occur so that measures can be taken to prevent a problem from arising. With this goal in mind, the Uniformed Services University has, for example, employed such devices with volunteers from the Army's

82nd Airborne Division. As the unit's holistic health and fitness director and chief physical therapist, Major Matthew Helton asserted that "these types of tests give us insight on every day movement patterns performed by paratroopers during physical training and during combat operations, both critically important to not only enhance performance but also to potentially prevent injuries as well."[104] Should the project prove effective, the next step would be to partner at-risk troops with athletic trainers and physical therapists so that they can strengthen problem areas.[105] Even so, more can be done. As Defense Intellectual Katherine Kuzminski argues, "The military needs to make human-performance optimization part of daily ops."[106]

Moreover, an assortment of innovative drugs offers the potential to significantly curb the obesity crisis. In 2018, the Defense Health Agency opened the door to greater access to these drugs for military personnel.[107] Military leaders might also consider separate operational and strategic fitness and health "tracks" for tech specialists. As Texas A&M historian Brian McAllister Linn has noted, "There has always been inherent tension between recruiting soldiers for immediate readiness—usually in the combat arms—and recruiting and retaining soldiers with essential technical-administrative skills to sustain the force."[108] As part of its vision for readying the force for the modern battlefield, the Marine Corps has emphasized physical fitness as a core component of a new Expeditionary Communications Course. "There is Fleet [Marine Force] buy-in—especially on those we support—on us being physically fit. We're going to sustain that," said Major Toby Pope as one of the course designers. "Really, for us, this is part of being communicators . . . being able to keep up with those that we are supporting. If we can't keep up, then we are useless. Period. . . . So throughout the course, we did rucks, we did hikes, we did water PT, we swam. All this to make the Marines better and to be a complete communicator."[109] The US Congress has implied its support for such an approach. The National Defense Authorization Act for Fiscal Year 2024 tasks the Army to increase fitness testing standards only for "close combat force military occupational specialties," for example.[110]

Digital Gaming Technologies

Western militaries have already begun to incorporate ideas gleaned from the ongoing hostilities in Eastern Europe into their combat simulations.[111] "A major game changer is what we learn from Ukraine," said John-Mikal Størdal in his position as director of the NATO Science and Technology Organization's collaboration support office. Among the most important lessons gleaned so far, he continued, is "the need for speed and multi-domain operations, new ways to operate, how to use technology."[112] In learning how to implement such tools going forward, the merging of live and virtual training methodologies will be of particular importance.[113] Doing so will allow American and allied forces to do things that might otherwise be impossible given physical and temporal realities.

As mentioned in chapter 3 of this book, simulations can be run on a given scenario millions of times thanks to artificial intelligence technologies.[114] Moreover, virtual simulations will give military leaders insights into the very different tactics and strategies that enemy forces might employ. "We envision our adversary thinking and operating and fighting like we do, which is a big mistake," explained former Marine Commandant David Berger at a November 2021 professional conference. "In other words, modeling and simulation can help bake in how we think the adversary is going to fight, how they are going to operate, which is not necessarily how we would. . . . We need help in creating the software, the simulation, that doesn't replicate us, it replicates the adversary, whether it's Russia or Iran or the [People's Republic of China]."[115] In a recent example of such an undertaking, the US Army's 2nd Multi-Domain Effects Battalion used software developed in Ukraine as the basis for a simulator system for drone operators and supporting electronic warfare specialists. "Essentially, it's a video game," explained the unit's commander, Lieutenant Colonel Aaron Ritzema.[116]

Moreover, the potential of digital gaming technologies extends beyond recruitment and training to operational scenarios as well. To that end, the demonstrated utility of employing already-familiar gaming controllers and interfaces should cause defense planners to consider bringing gamers directly into equipment design and acquisition processes. What better way to optimize future vehicles and equipment in terms of the prevailing talents and weaknesses of the next generation of operators? Some early efforts have already taken place in this manner. Indeed, Western defense contractors have on several occasions drawn inspiration from gaming culture in their product designs to create interfaces familiar to today's gaming youth. In 2014, the Boeing Company employed an Xbox control pad as the command mechanism for its High Energy Laser Mobile Demonstrator, a weapons system designed to shoot down mortar rounds and unmanned aerial vehicles.[117] Around this time period, similar control devices were used to operate robotic ground vehicles (particularly ones designed to neutralize improvised explosive devices), unmanned airframes, and even US Navy periscopes.[118]

Going forward, intuitive user interfaces designed with today's warriors in mind (such as ones developed from the gaming world) will enable better decision-making on the battlefield. Having worked with the US Defense Department closely, Matt McElvogue, the vice president of the Teague global design consultancy, put the matter thusly:

> We're at a moment in time when a new generation is coming into the military and entering command structures. They've grown up post-Internet, with smartphones being something they had in their pockets from an early age. They're used to using products that have had millions of dollars put into their user experience to make them easy to use.
>
> Whether they realize it or not, they're bringing that perspective to the tools that they use at work. They're asking for better UX [user experience] as they're brought into the process of evaluating new equipment and interfaces.[119]

Operational personnel once played important roles in the processes by which military equipment was designed, tested, and improved over time.[120] The monetary and temporal efficiency of digital gaming systems can help in current efforts to return to this paradigm. Indeed, war-gaming and digital simulation have been described as crucial components of the process by which the military should conduct "technology roadmapping" of the battlefields of the future.[121] In 2017, the US Army announced its aspirations to receive soldiers' feedback on fresh innovations via an online gaming platform (dubbed "Project Overmatch") within a broader initiative it called Early Synthetic Prototyping (ESP). As the program lead of ESP, Lieutenant Colonel Brian Vogt asserted, "What we want is two-way communication, and what better medium to use than video games." Through the system, Vogt explained, equipment designers could try out literally thousands of permutations without incurring the cost and time penalties of actual hardware fabrication. "In a game environment," he said, "we can change the parameters or the abilities of a vehicle by keystrokes. . . . We can change the engine in a game environment and it could accelerate faster, consume more fuel or carry more fuel. All these things are options within the game—we just select it, and that capability will be available for use. Of course, Army engineers will determine if the change is plausible before we put it in the scenarios."[122] The result, it was hoped, would be faster and cheaper implementation of new tactics and equipment. "The great thing about this is we're not building an actual model here. We're not actually putting dollars to build something on the ground," said Major General Robert "Bo" Dyess from his office as acting director of the Army Capabilities Integration Center at Training and Doctrine Command. "This has the potential not only to guide future science and technology research, but [also] inform the way we fight."[123]

Soldiers in battle are already using technologies that began as virtual training tools.[124] It is thus no accident that a number of Western weapons systems have been conceptualized with remote controllers borrowed from the gaming world.[125] In 2017, US and British military researchers entered into a collaborative relationship with the goal of exploring autonomous resupply vehicles. Here again, Xbox-type controllers were utilized as the devices for operating the platforms—including hoverbikes and four-wheel-drive all-terrain vehicles.[126] According to Pentagon planners, distributed forces in the Pacific will constitute one key to winning any potential war with the People's Republic of China in the future. With this in mind, the US Marine Corps developed a new strategy for working in the region that it calls Expeditionary Advanced Base Operations.[127] Under the concept, the Marines have asked to purchase anti-ship missile systems for the first time in the service's history. The initial proposal, called the Navy Marine Expeditionary Ship Interdiction System (NMESIS), includes two naval strike missiles mounted on a remotely controlled Joint Light Tactical Vehicle.[128] Images released by the service make it clear that the system is operated through a handheld controller like those found in the gaming world.[129] The Army has similarly begun development of what it calls a Maneuver-Short Range Air Defense (M-SHORAD) system to be mounted on its M1126 Stryker Combat vehicles. Here again, a video game–style controller is employed.[130] Given the ongoing fielding of such systems, a key going forward will be the development of a universal set of standards for control devices.[131]

An even more robust example of the possibilities afforded through such ideas can be found in the Israel Aerospace Industries' (IAI's) Carmel armored fighting vehicle, which was recently selected for incorporation into the Israel Defense Forces.[132] In order to leverage the skills of Israelis who would soon be reaching military service age, IAI designers brought in young gamers for consultations. The original design called for complex control mechanisms like ones found in advanced fighter aircraft. After the consultations, the plan was changed in favor of video game–like handsets for everything from driving the vehicle to weapons employment. In describing the effort from his perspective as the general manager of IAI's robotics systems operations, Meir Shabtai said, "From teenagers up to pre-military guys, and guys who are after their service, we let each one play with the [Carmel simulator] and define what kind of skills and what kind of accessories we should use, and according to that we developed the whole system." The result, according to Shabtai, should be greater effectiveness on the battlefield. "They know exactly the position of those buttons," he said, "and they can reach much better performances with that system. . . . The controller is just the interface, the whole idea is to present a sophisticated technology in a way they can deal with."[133]

The 3D software engines through which game designers create and manage their products have for some time possessed incredible potential in this regard. As the leader of defense contractor Lockheed Martin's Collaborative Human Immersive Laboratory, Darin Bolthouse has seen what such platforms offer: "You start by creating your virtual world in a 3-D game engine, put on a virtual reality headset, add full body motion tracking and haptic feedback, and your virtual world starts to look and feel real. Now you're a virtual avatar within your own personal holodeck capable of assembling a virtual spacecraft or aircraft with virtual tools."[134] As an outcome of its aspiration to become the "world's first fully Digital Service," the Space Force has expressed interested in an immersive, service-wide virtual reality platform modeled on Mark Zuckerberg's Metaverse.[135]

From a long-term perspective, national security officials should contemplate the creation of games designed to be early training for the next generation of cyberwarriors and electronic attack operators.[136] The need for such skills has been shown time and again in the Ukrainian war zone. For instance, Russia has demonstrated an ability to "jam" Western munitions that use GPS for navigating to their targets.[137] Gamified learning systems also have potential in non-combat-related areas. For example, they have been shown to be effective at improving the skills of defense acquisition personncl.[138] The broader point, however, is that digital tools have the potential to boost the performance levels of military personnel in a variety of areas going forward.

*   *   *

The US armed forces are likely to continue to face challenges in their attempts to attract and retain sufficient numbers of tech-savvy personnel. Young Americans appear increasingly disinclined toward military service. At the same time, shortcomings in physical fitness within the national population alongside obesity-related health problems will render ineligible many of those who do have an interest in joining. For this reason, modified standards for personnel in certain high-demand areas may need to be considered. In addition, policymakers should give thought to how to increase combatant performance most effectively. The digital technologies at the heart of the ongoing revolution in military affairs will demand a new type of warrior—one who is equally adept at working with a machine as he or she is at navigating an obstacle course. The Pentagon would do well to reflect on this reality as it prepares for the wars of the future.

Conclusion

For much of human history, debates about physical preparedness for combat have centered around the corporeal traits of speed, strength, and endurance. With their perceptions rooted in this framework, a growing number of national leaders have understandably depicted the deteriorating state of American physical fitness and accompanying obesity levels as constituting nothing less than a national security crisis. Although considerable thought has been devoted to the problem over the past decades, for many reasons an effective set of policy solutions has yet to be implemented, and none are likely forthcoming anytime soon. Indeed, approximately 260 million Americans are projected to be overweight or obese by the year 2050.[1]

The situation should be reflected upon within its larger context, however. Chairman of the US Joint Chiefs of Staff General Mark Milley predicted in June 2023 that in ten to fifteen years unmanned systems would comprise a full third of the battle forces wielded by the world's most powerful militaries.[2] To take a narrower example, the US Navy expects its air wings of the future to be 60 percent composed of uncrewed aircraft.[3] Such widespread adoption of robotic equipment threatens to overturn much of the historically dominate modes of thinking about the nature of the human body in military affairs. Individual physical capacity has diminished in importance across multiple (though certainly not all) dimensions of modern combat. Within the ongoing revolution in military affairs, it is instead the performance of the human and machine operating together as a team that is most important—even in those aspects of warfare requiring high levels of physical capacity such as infantry and special operations.[4] For this reason, ideas like the embedding of familiar gaming technologies into military systems have greater potential in most situations than ones that pertain nearly exclusively to the physical performance of a human operator (such as the employment of pharmacological methods to enhance speed, strength, and/or endurance). In short, enhancing the pairing of human and machine should be a driving force going forward in military affairs. As put by Army Futures Command Gen James E. Rainey, "We're never going to replace humans with machines, it's about optimizing them."[5]

Of course, a host of legal, ethical, and operational questions pertain to the evolving relationship between humans and machines on the battlefield.[6] In point of fact, such questions will become ever more complex as autonomous weapons systems come online.[7] It has been suggested, for example, that public confidence in the ability of robots to act as America's frontline troops might at some point unduly lower the risk calculation for entering the country into combat operations.[8] Others argue that autonomous platforms designed to kill rather than disable may by their very definition violate a protocol in the Geneva Convention that states that "it is prohibited to employ weapons, projectiles and material and methods of warfare of a nature to cause superfluous injury or unnecessary suffering."[9] However, most of the debates on the subject focus on to the degree of necessary human control of such technologies. In 2012, the US Department of Defense enacted specific guidance on the subject that mandated that "autonomous and semi-autonomous weapon systems shall be designed to allow commanders and operators to exercise *appropriate levels of human judgment over the use of force* [emphasis added]."[10]

In an operational context, one suggestion has been that, in order to avoid criminal culpability for a flawed attack by an autonomous system, a commander must show that the decision to strike was based on "a narrow, extensively tested algorithm with an extremely high level of certainly (for example, 99 percent or higher)."[11] Regardless, it has been apparent for some time that serious consideration will need to be given to the ethical dimensions of military employment of artificial intelligence and autonomous weapons technologies—especially given the likelihood that America's rivals will pay less attention to international norms on the matter.[12] In 2024, the Defense Advanced Research Projects Agency launched an initiative to ascertain the ability of autonomous systems to remain in compliance with human ethical standards on the battlefield. Called the Autonomy Standards and Ideals with Military Operational Values (ASIMOV) program, its goal is to establish an "ethical common language" by which to appraise autonomous platforms on the matter.[13]

The worlds of fiction and cinema have given us an idea as to where these types of technologies might lead if handled in the wrong way. *The Terminator* film franchise constitutes a particularly terrifying such example.[14] In the real world of 2023, Colonel Tucker Hamilton, then the head of US Air Force AI test and operations, described a simulated combat mission in which an autonomous aircraft had eliminated its human teammates in a "friendly fire" incident.[15] Although Hamilton later clarified his remarks as a "thought experiment" rather than a depiction of an actual occurrence, the episode received considerable attention as an ominous sign of the future of war.[16] Although the pace of future warfare may well render such promises impossible to keep, Western militaries for now remain dedicated to keeping humans in direct operational control of unmanned systems.[17] As said regarding a prospective new optionally manned fighting vehicle by Army Major General Glenn Dean from his position as program executive officer for ground combat systems, "Now, we know we're never going to put lethality fully autonomous without a human being on the loop."[18]

Going forward, it will remain important for defense leaders to keep in mind the primary role played by humans in their relationships with machines. Indeed, and in contrast to what might be initially expected, the growing employment of robotics systems will likely require greater rather than fewer numbers of military personnel.[19] In October 2024, for example, it was reported that the US Army had prior to that point largely underestimated how many crewed vehicles would be needed to control the service's unmanned ground platforms. The updated projection asserted that for every four of the latter in a formation, three of the former would be required.[20] In addition, the armed services might well benefit from adding seats (despite the higher cost of doing so) to their crewed tactical aircraft so that more effective command and control can be exerted over unmanned wingmen.[21] The reality is that the cutting-edge technologies of today (and tomorrow) demand an ability to deal with ever greater mental workloads.[22] Speaking at a spring 2024 military conference, US Air Force Lieutenant General Tony Bauernfeind noted in this regard, "Cognitively, it will require us to train our air crews in a new way. . . . You got to figure out how to handle an epic level of multitasking."[23]

Though a full transition may at some point take place, human operators will remain essential even once advances in artificial intelligence make fully autonomous systems possible.[24] Given the ongoing need for personnel with relevant tech skills, the continuing recruitment challenges faced by the military services constitute a real threat to the future force. Among the set of potential solutions for the problem, one of the most promising centers on making better use of women in the armed services. Collectively they remain by far the largest underutilized population group in the American military.[25] While US Secretary of Defense Pete Hegseth hinted before entering office about the possibility of cutting women from close combat roles, a major war may render doing so impossible.[26] It is instructive in this regard that when the Israeli military had manpower shortages while engaged on multiple fronts in 2025, a large number of women were transferred to frontline units, where they served effectively.[27] Given the scale of a potential conflict with a peer or near-peer adversary, it seems prudent for American forces to remain committed to including women in combat formations. At the same time, the US armed services should consider whether eligibility standards may need to be revised in certain military occupational specialties. A precedent certainly exists for doing so. While the services set their own guidelines for new recruits, the US Congress determines by law the maximum and minimum ages for entering troops. In January 1968, males could be no older than thirty-five years old.[28] In response to lobbying from the Army, the 2006 National Defense Authorization Act changed the maximum age to forty-two.[29] It has also been suggested that the US Selective Service System should be modified so that gender and age restrictions are removed for individuals possessing certain technical and/or scientific expertise.[30]

Although it has been compellingly argued that the US military should move toward a smaller and more technically proficient force, the war in Ukraine has shown that an ability to maintain significant troop levels will matter in a future conflict with a peer enemy.[31] The best path forward may be to ensure that today's smaller American military can be scaled to much larger levels in the case of such an event.[32] This is rendered particularly meaningful given that exercises have shown that US casualty rates in a major future war are likely to reach or exceed 50 percent.[33] The prospect of a conflict with China in the vast expanses of the Indo-Pacific region is especially ominous in terms of the time it will take to get care to wounded troops.[34] Fortunately, American combat medicine is already starting to adapt to this reality.[35] The Navy has designed a new class of Expeditionary Fast Transport ships with expanded hospital facilities.[36] Work has also started on forging an agreement that would allow blood supplies for wounded American troops to come from allied nations.[37] Concepts involving the use of virtual tools and unmanned vehicles for delivering medical help to operational forces have additionally begun to take shape.[38] The employment of artificial intelligence is also being studied for what it might bring to the table on the issue of combat medicine.[39] Medical researchers are also looking at other promising developments such as cold atmospheric pressure plasmas (CAPPs) as a means to extend the "golden hour" of effective combat casualty care.[40] Even so, the scale of a future Great Power war may lead to casualties of such magnitude as to render American efforts to regenerate forces exceedingly difficult.[41]

It is also worth our attention that the battlegrounds of Ukraine demonstrate the advantages of prioritizing numerical capacity (as opposed to individual platform sophistication) of a force's unmanned weapons platforms.[42] Until very recently, the United States has mainly focused on the development of large, high-end systems in this area. Given the lessons of Ukraine, however, the US military services likely need to make greater investment in larger numbers of more affordable units.[43]

Within this paradigm, new swarm tactics and training methodologies will need to be developed to make best use of the voluminous number of drones that will operate in collaborative partnership with human operators on the battlefield.[44] The results will need to be written into service combat manuals going all the way down to the squad level. The employment of unmanned systems should be as familiar to small units as firing a rifle.[45] Prior to their deployment to combat, of course, the services will require clear experiential evidence that unmanned platforms perform as they are tasked.[46] Indeed, absent considerable testing and operational experience in the field, human actors are unlikely to truly trust the myriad technologies discussed in the preceding chapters.[47]

This will especially be the case for military forces that feature particularly human-centric organizational cultures.[48] Infantry units constitute one such example.[49] The 2007 version of the US Army's field manual *The Infantry Rifle Platoon and Squad* asserts this viewpoint on the very first page of its very first chapter:

> Of all branches in the U.S. Army, the Infantry is unique because its core competency is founded on the individual Soldier—the Infantry rifleman. While other branches tend to focus on weapon systems and platforms to accomplish their mission, the infantry alone relies almost exclusively on the human dimension of the individual rifleman to close with and destroy the enemy.[50]

A more succinct expression along the same lines can be found in the Marine Corps's slogan "Every Marine a rifleman."[51] It goes against everything in such a culture to allow a reliance to form on something that might eventually break down. And yet, as Marine officer Nick Brunetti-Lihach pointed out in an award-winning 2018 article, while "every Marine is a rifleman for very real life and death reasons. . . . Now, with military operations within both the physical and cyber domains, the definition of martial strength must expand."[52] In agreement, Captain Karl Flynn argues that the service should "make every Marine a drone killer" while fellow officer Ken Hampshire puts forth another potential motto: "Every Marine a data scientist?"[53]

Linking the various emerging technologies on the battlefield (and beyond) will take considerable time and effort.[54] In order to accelerate the pace of connecting humans to machine, Western militaries increasingly include operational combat personnel in the development of tactics and procedures related to new technologies. The US armed services have gone to considerable effort in recent years to bring its fighting forces into such discussions.[55] This is necessary as the various cultures of military organizations will need to be grappled with as new technologies undergo the process of adoption.[56] An example is an initiative described by US Army Chief of Staff General Randy George as "Transformation in Contact."[57] Under this framework, American combat troops undergo exercises with the types of technologies and tactics seen in Ukraine and other contemporary conflict zones around the world.[58] The results are meant to inform much faster integration of new technologies into operational forces. General Rainey described the urgency of the situation thusly: "The amount of technical disruption in the character of war is unprecedented, and it just continues to go faster and faster . . . Whatever you think you know this year, come back in 90 days, and you'll know something different."[59]

Failures will no doubt occur as the military services incorporate new technologies and develop the tactics, techniques, and procedures for using them. In some cases, technical hurdles may be so steep as to prevent some seemingly promising robotic solutions from being implemented. Those hoping for a true "Iron Man" suit, for instance, may be disappointed given the limits of power supply and mechanical articulation. Robotic devices are simply unable to replicate today—with a precise degree of fidelity anyway—the complex movement patterns of flexible human joints and appendages.[60] Moreover, the novel power sources such as lithium-ion batteries that would be needed to power this type of equipment exert a considerable weight cost; they are also constructed out of flammable materials that are at risk of "thermal runaways" associated with catastrophic explosions.[61] In light of the danger, it doesn't take much to question the sanity of anyone willing to encase themselves in an exoskeleton that incorporates this type of power source. Happily, and with a caveat that the People's Republic of China so far dominates the supply chains of such minerals, advances in batteries are on the horizon that promise to moderate these problems.[62] Solid-state configurations are being developed, for instance, that will be lighter, safer, and more energy-dense than today's cutting-edge lithium-ion batteries.[63]

In any event, the broader point is that a profound shift is taking place in the relationship of the human body to war. Given the complex security environment facing the country, policymakers would do well to consider several actions in the near-term future. A good first step would be for the US armed services to begin a set of studies regarding potential organizational changes that might better align with the newly emergent tools of war.[64] As put by Army General Charles A. Flynn, "These organizational adjustments . . . are things that we in the Army are having to have a good, healthy debate about. . . . I believe that the organizational adjustments are maybe more important or equally as important as, I'll just say, the bent metal things that are showing up."[65] An example is a 2024 proposal by the US Congress to create a new "Drone Corps" within the Army.[66] While the idea may well turn out to be inappropriate to the situation, it is the type of innovative thinking that deserves consideration on such issues.[67] Indeed, the conflict in Ukraine has clearly demonstrated the critical role that unmanned systems will play on the battlefields of the future. Policymakers have fortunately begun to recognize this truth, and they are currently at work enacting reforms designed to greatly increase the number and quality of unmanned systems in operational units. Guidance issued in 2025 by Secretary of Defense Pete Hegseth offers a strong step in the right direction. Among its directives is a requirement that unmanned weapons systems should be rapidly adopted across the entirety of war-fighting units in the US armed services. To facilitate this outcome, the defense secretary ordered the recategorization of small drones as "consumables" in procurement policies; even lower-level commanders can now order such devices directly without going through traditional acquisition pipelines.[68]

Growing the industrial capacity to produce such platforms will be an essential part of such an undertaking—especially when considered alongside similar programs within the nation's other armed services.[69] Here again, policymakers are making progress. In August 2023, for example, Deputy Secretary of Defense Kathleen Hicks announced that a new program dubbed the "Replicator Initiative" aimed to field thousands of affordable autonomous systems within a relatively short time frame.[70] Moreover, the effort offered a more efficient model for the adoption of new technologies compared to what had historically been possible within the United States' vast and complex defense acquisition structure—one that appears increasingly ill-suited to the rapid pace of change in today's world.[71]

At the same time, the Defense Department should update its administrative processes to conform with modern digital best practices.[72] Its talent management systems constitute but one example of how such systems can be improved. Outside a relatively small group of military occupational specialty codes, the current system fails to provide a coordinated mechanism by which to track and make best use of numerous tech skills. These include traits like the ability to write software code and/or pilot small personal drones.[73] Moreover, appropriate resourcing should be devoted to such proven initiatives as the Army's Future Soldier Preparatory Course (which was discussed in chapter 6). A 2025 advisory issued by the Inspector General's office of the US Department of Defense asserted that troops enrolled in the program had been given insufficient medical supervision and that a significant number of participants eventually entered service despite falling short of Army fitness and/or health guidelines.[74] The Air Force, Space Force, Coast Guard, and Marine Corps have so far shown little interest in starting their own analogous programs. Should they experience the same recruiting problems as their larger sister services, such disinclination might merit reconsideration.[75]

Finally, the armed services should mandate modern data and robotic literacy as fundamental traits among tactical and operational leaders.[76] As put by US Army Colonel Kevin Bradley in a July 2025 essay penned for West Point's Modern War Institute, "To effectively lead a human-machine integrated formation, soldiers must develop technical and digital fluency that matches their physical endurance and tactical knowledge."[77] It seems increasingly clear, after all, that the future of war will center on digital networks that tie together sensors, weapons, platforms, and command and control systems—all of which will be enabled by artificial intelligence tools.[78]

*   *   *

The stakes could not be higher for the country. In a period increasingly defined by rapid technological change and heightened potential for war between major powers, US policymakers should reimagine the nature of combat physicality and embrace new tools of war. Traditional conceptions of war-fighting power appear increasingly outdated in the face of autonomous weapons platforms, cyber warfare, and AI-enabled combat systems. Future conflicts will hinge to a large degree on data dominance and human-machine teaming. To stay ahead of adversaries, military and civilian leaders alike should prioritize innovation and a forward-thinking doctrine that reflects the technological realities of an emerging revolution in military affairs.[79]

NOTES

Introduction

1. On this story, see Hanna Arhirova, "Ukraine Hails Teen Drone Operator Who Spied Russian Armor," AP News, June 12, 2022, https://apnews.com/article/russia -ukraine-kyiv-politics-1115558b2d4db5a1146a2bc65ec8a275.

2. Thomas M. Hunt, *Drug Games: The International Olympic Committee and the Politics of Doping, 1960–2008* (University of Texas Press, 2011).

3. Robert Work, "Deputy Secretary of Defense Speech, CNAS Defense Forum," US Department of Defense, December 14, 2015, https://www.defense.gov/News /Speeches/Speech/Article/634214/cnas-defense-forum/.

4. Work, "Deputy Secretary of Defense Speech."

5. For an overarching overview on this fact, consult Martin Van Creveld, *Technology and War: From 2000 B.C. to the Present* (Free Press, 1989).

6. On the elements of such revolutions, see James R. Fitzsimonds and Jan Van Tol, "Revolutions in Military Affairs," *Joint Force Quarterly*, Spring 1994, 24–31.

7. Andrew F. Krepinevich, *The Origins of Victory: How Disruptive Military Innovation Determines the Fates of Great Powers* (Yale University Press, 2023), 4. For an excellent overview of how this revolution will impact armed conflict going forward, see John F. Antal, *Next War: Reimagining How We Fight* (Casemate Publishers, 2023). For an argument that artificial intelligence is driving the current revolution in military affairs, see Michael Raska, "The Sixth RMA Wave: Disruption in Military Affairs?," *Journal of Strategic Studies* 44, no. 4 (2021): 456–79, https://doi.org/10 .1080/01402390.2020.1848818.

8. See Centers for Disease Control and Prevention, "New CDC Data Show Adult Obesity Prevalence Remains High," CDC Newsroom, September 12, 2024, https:// www.cdc.gov/media/releases/2024/p0912-adult-obesity.html.

9. Centers for Disease Control and Prevention, "Childhood Obesity Facts," Obesity, April 2, 2024, https://www.cdc.gov/obesity/php/data-research/childhood-obesity -facts.html.

10. See US Department of Defense, "2020 Qualified Military Available Study," 2020, https://prod-media.asvabprogram.com/CEP_PDF_Contents/Qualified_Military _Available.pdf.

11. See Pew Research Center, *From Businesses and Banks to Colleges and Churches: Americans' Views of U.S. Institutions* (Pew Research Center, 2024), 11, https://www.pewresearch.org/politics/2024/02/01/the-u-s-military/.

12. Vazirani's statement can be found at David Vergun, "DOD Addresses Recruiting Shortfall Challenges," US Department of Defense, December 13, 2023, https://www.defense.gov/News/News-Stories/Article/Article/3616786/dod-addresses-recruiting-shortfall-challenges/.

13. For a useful primer on how to evaluate combat effectiveness, consult T. J. Holland, "Decoding Lethality: Measuring What Matters," *Military Review: The Professional Journal of the U.S. Army*, Online Exclusive (October 2024): 1–8.

14. On this point, see Michael L. Hefti, *Training the Professional Soldier: Bridging Inexperience and Sophisticated Warfighting Technologies*, School of Advanced Military Studies Monograph (School of Advanced Military Studies, US Army Command and General Staff College, 2021), 58.

15. For an interesting overview of the relationship between new military technologies and human performance research, see Daniel C. Billing et al., "The Implications of Emerging Technology on Military Human Performance Research Priorities," *Journal of Science and Medicine in Sport* 24, no. 10 (2021): 947–53, https://doi.org/10.1016/j.jsams.2020.10.007. On the optimization of the human fighter within this paradigm, consult the various contributions in Nicholas Wright et al., eds., *Human, Machine, War: How the Mind-Tech Nexus Will Win Future Wars* (Air University Press, 2025).

16. On this point, see Travis Sharp and Tyler Hacker, *Evaluate Like We Operate: Why DOD Should Evaluate Weapons Systems as Networked Force Packages, Not Individual Platforms* (Center for Strategic and Budgetary Assessments, 2023), https://csbaonline.org/research/publications/evaluate-like-we-operate-why-dod-should-evaluate-weapons-systems-as-networked-force-packages-not-individual-platforms.

17. Noel Williams, "Military Force Design in an Age of Accelerating Technologic Change," *Marine Corps Gazette* 108, no. 2 (2024): 11.

18. See Daniel Pfaltzgraf and Gary S. Insch, "Digitally Native, yet Technologically Illiterate: Methods to Prepare Business Students to Create Versus Consume," *Journal of Applied Business and Economics* 23, no. 2 (2021): 25–34, https://articlearchives.co/index.php/JABE/article/view/2181.

19. See Consumer Technology Association, "CTA Research: Exploring Gen Z Views and Preferences in Technology," February 20, 2024, https://www.cta.tech/Resources/Newsroom/Media-Releases/2024/February/CTA-Research-Exploring-Gen-Z-Views-and-Preferences. These types of technologies can impact the relationship between humans and war. In the case of the smartphone and app-based tools, see Matthew Ford, *War in the Smartphone Age: Conflict, Connectivity and the Crises at Our Fingertips* (Oxford University Press, 2025).

20. For a very brief overview of the subject, see Aimee Pearcy, "Meet Gen Alpha, the 'Mini-Millennials' Who Are Poised to Take Over the Internet," *Business Insider*, November 7, 2023, https://www.businessinsider.com/gen-alpha-explained-technology-views-mental-health-2023-10.

21. See US Department of Defense, "Defense Department Report Shows Decline in Armed Forces Population While Percentage of Military Women Rises Slightly," November 6, 2023, https://www.defense.gov/News/Releases/Release/Article/3580676 /defense-department-report-shows-decline-in-armed-forces-population-while -percen/.

22. See US Department of Defense, *Department of Defense Annual Report on Sexual Assault in the Military: Fiscal Year 2022* (US Department of Defense, 2023), https:// www.sapr.mil/sites/default/files/public/docs/reports/AR/FY22/DOD_Annual _Report_on_Sexual_Assault_in_the_Military_FY2022.pdf.

23. See Ernesto Londoño, "A Decade into War, Body Armor Gets Curves," *Washington Post*, September 20, 2012, https://www.washingtonpost.com/world/national -security/a-decade-into-war-body-armor-gets-curves/2012/09/20/ffaa06c4-0353 -11e2-91e7-2962c74e7738_story.html.

24. To give one example, it took until 2024 for the US Navy to field a submarine designed for both men and women. See Diana Stancy, "Navy to Commission First Sub Designed for Both Men and Women Sailors," *Navy Times*, September 12, 2024, https://www.navytimes.com/news/your-navy/2024/09/12/navy-to-commission -first-sub-designed-for-both-men-and-women-sailors/.

25. This phrase is borrowed from T. J. Holland, "Technology at the Point of Contact: Shaping the Future of Warfighting," *NCO Journal*, Muddy Boots Online Collection (November 2024): 1–4.

26. Dwight D. Eisenhower, "Executive Order 10673—Fitness of American Youth," *Federal Register* 21, no. 138 (1956): 5341. For greater context on the body, see Matthew T. Bowers and Thomas M. Hunt, "The President's Council on Physical Fitness and the Systematisation of Children's Play in America," *International Journal of the History of Sport* 28, no. 11 (2011): 1496–511, https://doi.org/10.1080/09523367.2011.586789.

27. See Thomas Spoehr and Bridget Handy, *The Looming National Security Crisis: Young Americans Unable to Serve in the Military*, no. 3282, Backgrounder (The Heritage Foundation, 2018), 10, https://www.heritage.org/defense/report/the-looming -national-security-crisis-young-americans-unable-serve-the-military.

28. For context, see Joseph J. Knapik and Whitfield B. East, "History of United States Army Physical Fitness and Physical Readiness Testing," *U.S. Army Medical Department Journal*, US Army Medical Department Center & School, April 1, 2014, 5–20; and Bradley C. Nindl, *Strategies for Enhancing Military Physical Readiness in the 21st Century* (US Army War College, 2012), https://apps.dtic.mil /sti/citations/ADA561612.

29. For one example, see Spoehr and Handy, *The Looming National Security Crisis*.

30. See Norman Ohler, *Blitzed: Drugs in Nazi Germany*, trans. Shaun Whiteside (Houghton Mifflin, 2017).

31. See Nicolas Rasmussen, "Medical Science and the Military: The Allies' Use of Amphetamine during World War II," *Journal of Interdisciplinary History* 42, no. 2 (2011): 205–33, https://doi.org/10.1162/JINH_a_00212.

32. David L. Emonson and Rodger D. Vanderbeek, "The Use of Amphetamines in U.S. Air Force Tactical Operations During Desert Shield and Storm," *Aviation, Space, and Environmental Medicine* (US) 66, no. 3 (1995): 260–63.

33. See David Larter, "'Performance Enhancing Drugs' Considered for Special Operations Soldiers," *Defense News*, May 16, 2017, https://www.defensenews.com/digital-show-dailies/sofic/2017/05/16/performance-enhancing-drugs-considered-for-special-operations-soldiers/.

34. The incident is discussed in Richard Martin, "It's Wake-Up Time," Tags, *Wired*, November 1, 2003, https://www.wired.com/2003/11/sleep/.

35. See chapter 4 of Benjamin M. Jensen et al., *Information in War: Military Innovation, Battle Networks, and the Future of Artificial Intelligence* (Georgetown University Press, 2022).

36. The war is depicted in detail in Rick Atkinson, *Crusade: The Untold Story of the Persian Gulf War* (Houghton Mifflin, 1993).

37. A representative statement can be found in Andrew W. Marshall, "Some Thoughts on Military Revolutions—Second Version," ONA Memorandum for the Record, August 23, 1993, https://stacks.stanford.edu/file/druid:yx275qm3713/yx275qm3713.pdf.

38. See Rob Riddell, "Doom Goes to War," Tags, *Wired*, April 1, 1997, https://www.wired.com/1997/04/ff-doom/; and Michael J. Jernigan, "Marine Doom," *Marine Corps Gazette* (Quantico, United States) 81, no. 8 (1997): 19–20.

39. For an in-depth investigation of the game (and others), see Robertson Allen, *America's Digital Army: Games at Work and War* (University of Nebraska Press, 2017).

40. *Drone*, directed by Tonje Hessen Schei, with Brandon Bryant et al. (Flimmer Film, Radiator Film, Volt Film, 2014), 1h18m.

41. See Jacqueline M. Wheatcroft et al., "Unmanned Aerial Systems (UAS) Operators' Accuracy and Confidence of Decisions: Professional Pilots or Video Game Players?," *Cogent Psychology* 4, no. 1 (2017), https://doi.org/10.1080/23311908.2017.1327628.

42. See Oriana Pawlyk, "Here's How the Air Force Plans to Recruit Teenage Gamers," Military.com, May 25, 2018, https://www.military.com/defensetech/2018/05/25/heres-how-air-force-plans-recruit-teenage-gamers.html.

43. See Colin Schultz, "A Military Contractor Just Went Ahead and Used an Xbox Controller For Their New Giant Laser Cannon," Smart News, Smart News Science, Articles, *Smithsonian Magazine*, September 9, 2014, https://www.smithsonianmag.com/smart-news/military-contractor-just-went-ahead-and-used-xbox-controller-their-new-giant-laser-cannon-180952647/.

44. See David Hambling, "Game Controllers Driving Drones, Nukes," Tags, *Wired*, July 19, 2008, https://www.wired.com/2008/07/wargames/; Nathan Hodge, "Future Warbot Powered by Xbox Controller," Tags, *Wired*, June 12, 2009, https://www.wired.com/2009/06/future-warbot-powered-by-xbox-controller/; Leore Dayan and Noah Smith, "A New Israeli Tank Features Xbox Controllers, AI Honed by 'StarCraft II' and 'Doom,'" *Washington Post*, July 28, 2020, https://www.washingtonpost.com/video-games/2020/07/28/new-israeli-tank-features-xbox-controllers-ai-honed-by-starcraft-ii-doom/; and Travis M. Andrews, "The Navy's Adding a New Piece of Equipment to Nuclear Submarines: Xbox Controllers," *Washington Post*, September 25, 2017, https://www.washingtonpost.com/news

/morning-mix/wp/2017/09/25/the-navys-adding-a-new-piece-of-a-equipment-to
-nuclear-submarines-xbox-controllers/.

45. This policy has come under question recently. In December 2024, then secretary
of defense candidate (he was shortly thereafter nominated and later confirmed by
the Senate in that position) Pete Hegseth discussed ending the practice. Hegseth's
remarks on the subject can be found in *Pete Hegseth—Secretary of Defense Nomi-
nee | SRS #143*, Shawn Ryan Show, 2024, YouTube video, 0:15, https://www.youtube
.com/watch?v=DoN5ovwB8s4.

46. Linell Letendre, "Women Warriors: Why the Robotics Revolution Changes the
Combat Equation," *PRISM: The Journal of Complex Operations* 6, no. 1 (2016):
91–103.

47. For one such argument, see Leanne K. Simpson, "Eight Myths About Women on
the Military Frontline—and Why We Shouldn't Believe Them," *The Conversation*, April 1, 2016, http://theconversation.com/eight-myths-about-women-on-the
-military-frontline-and-why-we-shouldnt-believe-them-55594. On the 1960s roots
of military exoskeletons, see "News Notes: Future Soldier-Human Tank," *Armor*
70, no. 3 (1961): 54–55.

48. For a useful overview of why the first few days of the war turned out poorly for
Russia, see Zack Beauchamp, "Why the First Few Days of War in Ukraine Went
Badly for Russia," *Vox*, February 28, 2022, https://www.vox.com/22954833/russia
-ukraine-invasion-strategy-putin-kyiv. A penetrating analysis of perhaps the
most important battle during this early period in the war can be found in Liam
Collins et al., "The Battle of Hostomel Airport: A Key Moment in Russia's Defeat
in Kyiv," *War on the Rocks,* August 10, 2023, https://warontherocks.com/2023/08
/the-battle-of-hostomel-airport-a-key-moment-in-russias-defeat-in-kyiv/.

49. See Lucian Kim, "How U.S. Military Aid Has Helped Ukraine Since 2014," World,
NPR, December 18, 2019, https://www.npr.org/2019/12/18/788874844/how-u-s
-military-aid-has-helped-ukraine-since-2014; and Daniel Michaels, "The Secret of
Ukraine's Military Success: Years of NATO Training," World, *Wall Street Journal*,
April 13, 2022, https://www.wsj.com/articles/ukraine-military-success-years-of-nato
-training-11649861339.

50. For an overview on this matter, see Seth G. Jones et al., *Ukrainian Innovation in
a War of Attrition*, CSIS Briefs (Center for Strategic and International Studies
[CSIS], 2023), https://www.csis.org/analysis/ukrainian-innovation-war-attrition.

51. Eric Schmidt, "Innovation Power: Why Technology Will Define the Future of
Geopolitics," *Foreign Affairs* 102, no. 2 (2023): 43.

52. David Ignatius, "How the Algorithm Tipped the Balance in Ukraine," Opinion,
Washington Post, December 19, 2022, https://www.washingtonpost.com/opinions
/2022/12/19/palantir-algorithm-data-ukraine-war/.

53. See Dave Philipps, "With Few Able and Fewer Willing, U.S. Military Can't Find
Recruits," *New York Times*, July 14, 2022, https://www.nytimes.com/2022/07/14
/us/us-military-recruiting-enlistment.html. For a useful overview of the national
security implications of the matter, see the following report by an American think
tank: Thomas W. Spoehr, *The Administration and Congress Must Act Now to
Counter the Worsening Military Recruiting Crisis*, no. 5283, Issue Brief (Center for

National Defense, The Heritage Foundation, 2022), 8, http://report.heritage.org/ib5283.

54. See Courtney Kube and Molly Boigon, "Every Branch of the Military Is Struggling to Make Its 2022 Recruiting Goals, Officials Say," *NBC News*, June 27, 2022, https://www.nbcnews.com/news/military/every-branch-us-military-struggling-meet-2022-recruiting-goals-officia-rcna35078.

55. For a thoughtful set of ideas on the subject, see the following essay by a US Marine Corps commandant: David H. Berger, "Recruiting Requires Bold Changes," *U.S. Naval Institute Proceedings* 148, no. 11 (2022), https://www.usni.org/magazines/proceedings/2022/november/recruiting-requires-bold-changes. See also Christopher McMahon and Colin Bernard, "Storm Clouds on the Horizon—Challenges and Recommendations for Military Recruiting and Retention," *Naval War College Review* 72, no. 3 (2019): 84–100.

56. The United States has a history of this policy going back at least to 1950. See Brad Hardy, "Citizen Candidates: Cold War Naturalization, Military Service, and the Lodge Act of 1950," *Journal of Military History* 87, no. 1 (2023): 169–88. For an argument to reinstate and expand the policy after it was halted during the Trump administration, see Collin Fox, "Restore MAVNI for Legal Aliens to Enter the Military," *U.S. Naval Institute Proceedings* 147, no. 8 (2021), https://www.usni.org/magazines/proceedings/2021/august/restore-mavni-legal-aliens-enter-military.

57. See Stephen A. Cheney and Stephen N. Xenakis, *Perspective—Obesity's Increasing Threat to Military Readiness: The Challenge to U.S. National Security* (American Security Project, 2022), 1–15, https://www.jstor.org/stable/resrep46869; and Matthew Wallin, "Briefing Note—The Military Recruiting Crisis: Obesity's Impact on the Shortfall," American Security Project, March 2023, https://www.americansecurityproject.org/briefing-note-the-military-recruiting-crisis-obesitys-impact-on-the-shortfall/.

58. See Kirsten Weir, "The Extra Weight of COVID-19,"American Psychological Association, July 1, 2021, https://www.apa.org/monitor/2021/07/extra-weight-covid.

59. Klaus Schwab, "The Fourth Industrial Revolution," *Foreign Affairs*, December 12, 2015, https://www.foreignaffairs.com/world/fourth-industrial-revolution.

60. For interesting overviews of how fiction has been employed in this way, see Meredith Roaten, "Telling Tales: Startup Helps Officers Explain Future Warfare Through Storytelling," *National Defense* 107, no. 829 (2022): 36–39; and Nicholas Sarantakes, "The Future-War Literature of the Reagan Era—Winning World War III in Fiction," *Naval War College Review* 76, no. 3 (2023): 1–20.

61. Orson Scott Card, *Ender's Game* (Century Publishing, 1985); Robert Anson Heinlein, *Starship Troopers* (Putnam, 1959); Joe Haldeman, *The Forever War* (St. Martin's Press, 1974); and John Scalzi, *Old Man's War* (Tor Publishing, 2005).

62. Craig Thomas, *Firefox* (Holt, Rinehart and Winston, 1977); and *Firefox*, produced by Clint Eastwood, directed by Clint Eastwood (Warner Bros., 1982), https://www.imdb.com/title/tt0083943/.

63. General Sir John Hackett, *The Third World War: A Future History* (Sidgwick and Jackson, Limited, 1978); and Tom Clancy, *Red Storm Rising* (G.P. Putnam Sons, 1986). The real-world impact of these and other fictional accounts is outlined in Sarantakes, "The Future-War Literature of the Reagan Era."

64. See, for example, Walker Mills and Timothy Heck, "War Books—Preparing for Great-Power Competition: A Fiction Reading List," Modern War Institute (at West Point), May 13, 2020, https://mwi.westpoint.edu/war-books-preparing-great -power-competition-fiction-reading-list/; and Modern War Institute staff, "War Books: Science Fiction and Modern War," Modern War Institute (at West Point), May 12, 2023, https://mwi.westpoint.edu/war-books-science-fiction-and-modern -war/.

65. For an award-winning fictional work in the journal, see Dale Rielage, "How We Lost the Great Pacific War," *U.S. Naval Institute Proceedings* 144, no. 5 (2018), https:// www.usni.org/magazines/proceedings/2018/may/how-we-lost-great-pacific-war. Another excellent work of fiction in the journal is James A. Winnefeld, "AI Surprise in the Black Ditch," *U.S. Naval Institute Proceedings* 147, no. 1 (2021), https://www .usni.org/magazines/proceedings/2021/january/ai-surprise-black-ditch.

66. See Atlantic Council, "Atlantic Council Announces New Art of Future Warfare Project," *Atlantic Council*, November 19, 2014, https://www.atlanticcouncil.org /news/press-releases/atlantic-council-announces-new-art-of-future-warfare -project/.

67. August Cole, ed., *The Atlantic Council Art of Future Warfare Project: War Stories from the Future*, ebook (Atlantic Council, 2015), https://www.atlanticcouncil.org /in-depth-research-reports/books/war-stories-from-the-future/.

68. See Allison Kuntzman, "Mad Scientist FY17: A Retrospective," Mad Scientist Laboratory, January 8, 2018, https://madsciblog.tradoc.army.mil/18-mad-scientist -fy17-a-retrospective/.

69. Jasper Jeffers, "AN41," Modern War Institute (at West Point), May 6, 2019, https:// mwi.westpoint.edu/an41-2/.

70. P. W. Singer and August Cole, *Ghost Fleet: A Novel of the Next World War* (Houghton Mifflin Harcourt, 2015); Although it is not specifically focused on a future war, Singer and Cole have written a second fictional account that is also relevant to the sorts of things that this book concerns: *Burn-In: A Novel of the Real Robotic Revolution* (Houghton Mifflin Harcourt, 2020).

71. For an admiring essay on Poyer's broader set of fictional works on modern naval warfare, see C. Herbert Gilliland, "Dan Lenson Comes Home: An Appreciation of David Poyer's The Academy," *U.S. Naval Institute Proceedings* 150, no. 3 (2024), https://www.usni.org/magazines/proceedings/2024/march/dan-lenson-comes -home-appreciation-david-poyers-academy.

72. For an overview of his intent for the series to be a sort of warning, see David Poyer, "The War with China, as I Saw It," Commentary, *Air Force Times*, November 27, 2020, https://www.airforcetimes.com/opinion/commentary/2020/11/27/the-war -with-china-as-i-saw-it/.

73. Elliot Ackerman and James Stavridis, *2034: A Novel of the Next World War* (Penguin, 2021). See also a follow-up by the same authors: Elliot Ackerman and James Stavridis, *2054: A Novel* (Penguin Press, 2024).

74. Mick Ryan, *White Sun War: The Campaign for Taiwan* (Casemate Publishers, 2023).

75. Driscoll's quote can be found in Jennifer Hlad, "In the Pacific, Army Leaders Expect Today's Fiction to Be Near-Term Reality," *Defense One*, July 25, 2025,

https://www.defenseone.com/technology/2025/07/pacific-army-leaders-expect
-todays-fiction-be-near-term-reality/407001/.

76. Garrett Gatzemeyer, *Bodies for Battle: US Army Physical Culture and Systematic Training, 1885–1957* (University Press of Kansas, 2021).

77. Gatzemeyer, *Bodies for Battle*, 398.

78. James D. Campbell, *"The Army Isn't All Work": Physical Culture in the Evolution of the British Army, 1860–1920* (Ashgate Publishing Company, 2012).

79. Whitfield B. East, *Fit to Serve: A History of US Army Physical Readiness* (Army University Press, US Army Combined Arms Center, 2024). See also an earlier report by the same author: *A Historical Review and Analysis of Army Physical Readiness Training and Assessment* (Combat Studies Institute, US Army Command and General Staff College, 2013), https://apps.dtic.mil/sti/citations/ADA622014.

80. Christina S. Jarvis, *The Male Body at War: American Masculinity During World War II* (Northern Illinois University Press, 2004).

81. Rachel Louise Moran, *Governing Bodies: American Politics and the Shaping of the Modern Physique* (University of Pennsylvania Press, 2018).

82. Jon Robert Adams, *Male Armor: The Soldier-Hero in Contemporary American Culture* (University of Virginia Press, 2008).

83. For a useful anthology of essays on the subject, see Robert Egnell and Mayesha Alam, *Women and Gender Perspectives in the Military: An International Comparison* (Georgetown University Press, 2019).

84. Kate Germano and Kelly Kennedy, *Fight Like a Girl: The Truth Behind How Female Marines Are Trained* (Prometheus Books, 2018).

85. To take just a few examples, see Thomas G. Mahnken, *Technology and the American Way of War* (Columbia University Press, 2008); Keith L. Shimko, *The Iraq Wars and America's Military Revolution* (Cambridge University Press, 2010); Ian G. R. Shaw, *Predator Empire: Drone Warfare and Full Spectrum Dominance* (University of Minnesota Press, 2016); Michael P. Kreuzer, *Drones and the Future of Air Warfare: The Evolution of Remotely Piloted Aircraft* (Routledge, 2016); Daniel R. Lake, *The Pursuit of Technological Superiority and the Shrinking American Military* (Springer, 2019); Christian Brose, *The Kill Chain: Defending America in the Future of High-Tech Warfare* (Hachette Books, 2020); Sam J. Tangredi and George Galdorisi, eds., *AI at War: How Big Data, Artificial Intelligence, and Machine Learning Are Changing Naval Warfare* (Naval Institute Press, 2021); and Mick Ryan, *War Transformed: The Future of Twenty-First-Century Great Power Competition and Conflict* (Naval Institute Press, 2022). For a thought-provoking text on the subject, see the various contributions (some of which focus on non-technological matters) in Artur Kaempf and Sebastian Gruszczak, eds., *Routledge Handbook of the Future of Warfare* (Routledge, 2023), https://doi.org/10.4324/9781003299011.

86. Lukasz Kamienski, *Shooting Up: A Short History of Drugs and War* (Oxford University Press, 2016).

87. Jai Galliott and Mianna Lotz, *Super Soldiers: The Ethical, Legal and Social Implications* (Routledge, 2016); and Jean-François Caron, *A Theory of the Super Soldier: The Morality of Capacity-Increasing Technologies in the Military* (Manchester University Press, 2018). See also Andrew Bickford, "The 'Superman' Solution: 'Super Soldiers' and 'Superheroes' in the United States Military," *Anthropology*

Today 36, no. 5 (2020): 14–17, https://doi.org/10.1111/1467-8322.12605; Andrew Bickford, *Chemical Heroes: Pharmacological Supersoldiers in the US Military* (Duke University Press, 2020); Dave Shunk, "Ethics and the Enhanced Soldier of the Near Future," *Military Review: The Professional Journal of the U.S. Army* 95, no. 1 (2015): 91–98; and Luke J. Matthews et al., *Plagues, Cyborgs, and Supersoldiers: The Human Domain of War*, Research Report RR-A2520-1 (RAND Corporation, 2024), https://www.rand.org/pubs/research_reports/RRA2520-1.html.

88. Bickford, *Chemical Heroes*.

89. Jean M. Twenge, *iGen: Why Today's Super-Connected Kids Are Growing Up Less Rebellious, More Tolerant, Less Happy—and Completely Unprepared for Adulthood—and What That Means for the Rest of Us* (Simon and Schuster, 2017).

90. John Wills, *Gamer Nation: Video Games and American Culture* (Johns Hopkins University Press, 2019).

91. Matthew B. Caffrey Jr., *On Wargaming: How Wargames Have Shaped History and How They May Shape the Future*, Naval War College Newport Papers 43 (US Naval War College Press, 2019).

Chapter 1

1. For an excellent overview of how the US Army handled the subject from the mid-nineteenth century through the 1950s, consult Garrett Gatzemeyer, *Bodies for Battle: US Army Physical Culture and Systematic Training, 1885–1957* (University Press of Kansas, 2021).

2. Theodore Roosevelt, *Theodore Roosevelt: An Autobiography* (Charles Scribner's Sons, 1913), 46. For more on Roosevelt's ideas on military fitness, consult André Sobocinski, "Teddy Roosevelt, Navy Medicine and the Birth of Physical Readiness," US Navy website, July 18, 2022, https://www.navy.mil/Press-Office/News-Stories/Article/3097256/teddy-roosevelt-navy-medicine-and-the-birth-of-physical-readiness/.

3. See James Rainey, "The Questionable Training of the AEF in World War I," *The US Army War College Quarterly: Parameters* 22, no. 1 (1992): 89–103, https://doi.org/10.55540/0031-1723.1638; Stephen J. Lofgren, "Unready for War: The Army and World War I," *Army History*, no. 22 (1992): 11–19; and Gregory C. Hope, *"Army Training, Sir": The Impact of the World War I Experience on the Evolution of Training Doctrine in the US Army*, Art of War Papers (Army University Press, 2021). US Marine Corps recruits seemed better prepared than their Army brethren. See Peter F. Owen, *To the Limit of Endurance: A Battalion of Marines in the Great War* (Texas A&M University Press, 2007), 1–25. The Marine Corps, as a result, forged a unique form of American masculinity within and without the service. See Mark Ryland Folse, *The Globe and Anchor Men: U.S. Marines and American Manhood in the Great War Era*, Modern War Studies (University Press of Kansas, 2024). For a more concise work on the subject by the same author, see Mark R. Folse, "'The Cleanest and Strongest of Our Young Manhood': Marines, Belleau Wood, and the Test of American Manliness," *Marine Corps History* 4, no. 1 (2018): 6–22. For an additional useful text, see Heather P. Venable, *How the Few Became the Proud: Crafting the Marine Corps Mystique, 1874–1918* (Naval Institute Press, 2019).

4. Miranda Summers Lowe, "'Swat the Kaiser' and Stork Stands: The History of Army Physical Fitness," *New York Times Magazine*, March 28, 2019, https://www.nytimes.com/2019/03/28/magazine/army-physical-fitness-test.html.

5. A helpful overview of these issues can be found in David Woodward, *The American Army and the First World War*, Armies of the Great War (Cambridge University Press, 2014).

6. The last name of the soldier was Hersch (no given name was listed). The summary of the statements can be found in Intelligence Section, the General Staff of the American Expeditionary Forces, "Candid Comment on the American Soldier of 1917–1918 and Kindred Topics by the Germans: Soldiers, Priests, Women, Village Notables, Politicians, and Statesmen" (Headquarters, General Staff of the American Expeditionary Forces, 1919), 8, https://irp.fas.org/agency/army/wwi-soldiers.pdf.

7. On the interwar Army and the lessons learned from World War I, consult Brian McAllister Linn, *Real Soldiering: The US Army in the Aftermath of War, 1815–1980*, Modern War Studies (University Press of Kansas, 2023), 74–114.

8. Marshall quote from Lofgren, "Unready for War," 15–16.

9. Jonas H. Ingram, "Physical Fitness: A Necessary Requisite for War-Time Leadership and Longer Life," *U.S. Naval Institute Proceedings* 68, no. 1 (1942): 10.

10. Omar Nelson Bradley, and Clay Blair. *A General's Life: An Autobiography* (Simon and Schuster, 1983), 106.

11. David J. Grisdale, "Unfit to Fight: Medical Readiness Lessons Learned from the Draft from 1940 to 1947, and Why the United States Should Classify Registrants of the Selective Service" (master's thesis, Fort Leavenworth, KS, US Army Command and General Staff College, 2019), https://apps.dtic.mil/sti/citations/AD1106237.

12. Donald William Rominger, "The Impact of the United States Government's Sports and Physical Training Policy on Organized Athletics During World War II" (Ph.D. diss., United States—Oklahoma, Oklahoma State University, 1976), https://www.proquest.com/docview/302843272/citation/D5F8FC91514C4E38PQ/1; Donald W. Rominger, "From Playing Field to Battleground: The United States Navy V-5 Preflight Program in World War II," *Journal of Sport History* 12, no. 3 (1985): 252–64.

13. US War Department, *FM 21–20 Physical Training 1941* (US Government Printing Office, 1941), 7.

14. On Army training leading up to D-Day itself, see Stephen A. Bourque, "'An Incredible Degree of Rugged and Realistic Training': The 4th Infantry Division's Preparation for D-Day," *Military Review: The Professional Journal of the U.S. Army* 104, no. 3 (2024): 20–32.

15. Jessica Anderson-Colon, "Marine Corps Boot Camp During World War II: The Gateway to the Corps' Success at Iwo Jima," *Marine Corps History* 7, no. 1 (2021): 46–63, https://doi.org/10.35318/mch.2021070103. A 2023 monograph argues that the Marine Corps Basic School for Officers also played a key role in the service's wartime effectiveness. See Jennifer L. Mazzara, *Shared Experience: Organizational Culture and Ethos at the U.S. Marine Corps Basic School, 1924–1941* (Marine Corps University Press, 2023), https://doi.org/10.56686/9798986259406.

16. A large body of literature exists on these matters. To take one important work within them, see Paul A. C. Koistinen, *Arsenal of World War II: The Political Economy of American Warfare, 1940–1945* (University Press of Kansas, 2004).

17. See Jeffrey Montez De Oca, "The 'Muscle Gap': Physical Education and US Fears of a Depleted Masculinity, 1954–1963," in *East Plays West: Sport and the Cold War*, ed. Stephen Wagg and David Andrews (Routledge, 2012), 137–62, http://www.taylorfrancis.com/books/e/9781134241682/chapters/10.4324%2F97 80203007112-13.

18. Faye Flam, "Obese Kids: How Did We Get Here?," *Washington Post*, September 27, 2013, https://www.washingtonpost.com/postlive/how-childhood-obesity-became -a-crisis/2013/09/26/b2f87652-1708-11e3-804b-d3a1a3a18f2c_story.html.

19. US War Department, *FM 21–20 Physical Training 1946*.

20. See Bradley J. Warr et al., "Testing and Evaluation of Tactical Populations," in *NSCA'S Essentials of Tactical Strength and Conditioning*, ed. Brent A. Alvar, Katie Sell, and Patricia A. Deuster (Human Kinetics, 2017), 136.

21. Dwight D. Eisenhower, "Executive Order 10673—Fitness of American Youth," *Federal Register* 21, no. 138 (1956): 5341. For greater context, see Matthew T. Bowers and Thomas M. Hunt, "The President's Council on Physical Fitness and the Systematisation of Children's Play in America," *International Journal of the History of Sport* 28, no. 11 (2011): 1496–511, https://doi.org/10.1080/09523367.2011.586789.

22. The test results were published in Hans Kraus and Ruth P. Hirschland, "Minimum Muscular Fitness Tests in School Children," *Research Quarterly. American Association for Health, Physical Education and Recreation* 25, no. 2 (1954): 178–88, https://doi.org/10.1080/10671188.1954.10624957.

23. Headquarters Marine Corps, "1956 US Marine Corps PFT Order" (Headquarters, US Marine Corps, August 9, 1956), http://archive.org/details/1956UsMarine CorpsPftOrder.

24. John F. Kennedy, "The Soft American," *Sports Illustrated* 13, no. 26 (1960), 16. For more on Kennedy's outlook on physical activity, consult Hendrik W. Ohnesorge, "Profile in Vigor: John F. Kennedy and the Quest for Athletic Excellence," in *Sports and the American Presidency: From Theodore Roosevelt to Donald Trump*, ed. Adam Burns and Rivers Gambrell (Edinburgh University Press, 2022), 35–55, https://www.cambridge.org/core/books/sports-and-the-american-presidency /profile-in-vigor-john-f-kennedy-and-the-quest-for-athletic-excellence /2177B6913CE98991AB3048EBB57566B6.

25. Kennedy, "The Soft American," 16.

26. See Whitfield B. East, *Fit to Serve: A History of US Army Physical Readiness* (Army University Press, US Army Combined Arms Center, 2024), 134; and Whitfield B. East, *A Historical Review and Analysis of Army Physical Readiness Training and Assessment* (Combat Studies Institute, US Army Command and General Staff College, 2013), 130, https://apps.dtic.mil/sti/citations/ADA622014.

27. See James A. Hodgdon, *A History of the U.S. Navy Physical Readiness Program From 1976 to 1999*, Technical Document No. 99–6F (Naval Health Research Center, 1999), 3, https://doi.org/10.21236/ADA375322.

28. See East, *Fit to Serve*, 155–57; and East, *A Historical Review and Analysis*, 150–51.

29. Appendix E, US Army, *Army Field Manual FM 21–20: Physical Readiness Training* (Headquarters, US Department of the Army, 1980), https://www.google.com /books/edition/Physical_Readiness_Training/bU5jQXqmTVIC?hl=en&gbpv=1.

30. See Joseph J. Knapik and Whitfield B. East, "History of United States Army Physical Fitness and Physical Readiness Testing," *U.S. Army Medical Department Journal*, April 1, 2014, 14.

31. For the broader context of the American military during this period, consult David Fitzgerald, *Uncertain Warriors: The United States Army between the Cold War and the War on Terror* (Cambridge University Press, 2023).

32. Rieger quoted in Steven Lee Myers, "The Old Army, It Turns Out, Was the Fitter One," *New York Times*, June 25, 2000, https://www.nytimes.com/2000/06/25 /health/the-old-army-it-turns-out-was-the-fitter-one.html.

33. Thomas quoted in Myers.

34. Thomas Spoehr and Bridget Handy, *The Looming National Security Crisis: Young Americans Unable to Serve in the Military*, Backgrounder (The Heritage Foundation, February 13, 2018), https://www.heritage.org/defense/report/the-looming -national-security-crisis-young-americans-unable-serve-the-military.

35. Muth's explanation is quoted in Meghann Myers, "America's Obesity Is Threatening National Security, According to This Study," *Army Times*, October 10, 2018, https://www.armytimes.com/news/your-army/2018/10/10/americas-obesity-is -threatening-national-security-according-to-this-study/.

36. Thomas Novelly, "Even More Young Americans Are Unfit to Serve, a New Study Finds. Here's Why," Military.com, September 28, 2022, https://www.military.com /daily-news/2022/09/28/new-pentagon-study-shows-77-of-young-americans-are -ineligible-military-service.html.

37. Peter T. Katzmarzyk et al., "Results from the United States 2018 Report Card on Physical Activity for Children and Youth," *Journal of Physical Activity and Health* 15, no. s2 (2018): S422–24, https://doi.org/10.1123/jpah.2018-0476.

38. JoEllen M. Sefton, K. R. Lohse, and J. S. McAdam, "Prediction of Injuries and Injury Types in Army Basic Training, Infantry, Armor, and Cavalry Trainees Using a Common Fitness Screen," *Journal of Athletic Training* 51, no. 11 (2016): 849, https://doi.org/10.4085/1062-6050-51.9.09.

39. Palkoska quoted in Thomas E. Ricks, "Military Physical Training: It's a Problem Bigger than Obesity, with No Easy Solutions," *Foreign Policy*, August 10, 2015, https://foreignpolicy.com/2015/08/10/military-physical-training-its-a-problem -bigger-than-obesity-with-no-easy-solutions-2/.

40. James Benjamin, "The Hidden Battlefield: Diet and Lifestyle in Military Readiness," *U.S. Naval Institute Proceedings* 151, no. 2 (2025), https://www.usni.org /magazines/proceedings/2025/february/hidden-battlefield-diet-and-lifestyle -military-readiness.

41. Charles Blake, Christopher W. Boyer, and David R. Hourani, "The Musculoskeletal Imperative: Enhancing Combat Capability through Effective Injury Management," *Military Review: The Professional Journal of the U.S. Army* 104, no. 5 (2024): 129.

42. Lionel Beehner and John Spencer, "To Win at War, Make America Fit Again," *The Hill* (blog), January 12, 2017, https://thehill.com/blogs/congress-blog/healthcare

/313866-to-win-at-war-make-america-fit-again/. See also John Spencer, "The Softer American," *Modern War Institute* (at West Point), July 5, 2016, https://mwi .westpoint.edu/the-softer-american/.

43. An overview of states with high enlistment rates can be found in Dave Philipps and Tim Arango, "Who Signs Up to Fight? Makeup of U.S. Recruits Shows Glaring Disparity," *New York Times*, January 10, 2020, https://www.nytimes.com/2020 /01/10/us/military-enlistment.html.

44. These states were Alabama, Arkansas, Florida, Georgia, Louisiana, Mississippi, North Carolina, South Carolina, Tennessee, and Texas. Daniel B. Bornstein et al., "Which US States Pose the Greatest Threats to Military Readiness and Public Health? Public Health Policy Implications for a Cross-Sectional Investigation of Cardiorespiratory Fitness, Body Mass Index, and Injuries Among US Army Recruits," *Journal of Public Health Management and Practice* 25, no. 1 (2019): 36–44, https://doi.org/10.1097/PHH.0000000000000778. According to a 2023 article, southern recruits received 50 percent of the injuries in Army basic training despite being only 41 percent of recruits. See Thomas Spoehr and Isaac Tang, "Obesity Epidemic Threatens Not Just Public Health, but Also National Security," *Washington Examiner*, August 8, 2023, https://www.washingtonexaminer.com /restoring-america/courage-strength-optimism/obesity-epidemic-threatens-not -just-public-health-but-also-national-security.

45. Quote from Bornstein et al., "Which US States Pose the Greatest Threats to Military Readiness and Public Health?," 40–41.

46. Robert L. Wilkie, "DoD Policy for Non-Deployable Service Members," Office of the Under Secretary of Defense for Personnel and Readiness Memorandum, February 14, 2018, https://www.usar.army.mil/Portals/98/Users/151/87/1687/MEMO%20 20180218%20DoD%20Policy%20for%20Non-deployable%20Service%20Members .pdf.

47. See Jonel Aleccia, "Pandemic Pounds Push 10,000 U.S. Army Soldiers into Obesity," AP News, April 2, 2023, https://apnews.com/article/military-obesity -pandemic-army-covid-404bbc1a67408d390a7d462f1ecbef75.

48. See Doug G. Ware, "Nearly 70% of US Troops Are Overweight or Obese, Research Report Says," *Stars and Stripes*, October 17, 2023, https://www.stripes.com/theaters /us/2023-10-17/military-troops-obese-overweight-11738212.html.

49. Centers for Disease Control and Prevention, "Unfit to Serve: Obesity and Physical Inactivity Are Impacting National Security," July 2022, https://www.cdc.gov /physicalactivity/downloads/unfit-to-serve-062322-508.pdf.

50. Matt Clark, "The Army Has a Physical Fitness Problem, Part 1: Eight Myths That Weaken Combat Readiness," *Modern War Institute* (at West Point), January 25, 2020, https://mwi.usma.edu/army-physical-fitness-problem-part-1-eight-myths -weaken-combat-readiness/.

51. Courtney Manning, "Combating Military Obesity: Stigma's Persistent Impact on Operational Readiness," White Paper (American Security Project, October 2023), introductory summary, https://www.americansecurityproject.org/wp-content /uploads/2023/10/Ref-0286-Combating-Military-Obesity.pdf.

52. "MARADMIN 601/12: Announcement of High Intensity Tactical Training Program" (US Marine Corps, October 17, 2012), https://www.marines.mil/News

/Messages/Messages-Display/Article/895087/announcement-of-high-intensity
-tactical-training-program/.

53. US Army, *Field Manual FM 7–22: Army Physical Readiness Training October 2012* (Washington, D.C.: Headquarters, Department of the Army, 2012), 1–4, https://www.google.com/books/edition/FM_7_22_Army_Physical_Readiness_Training/9tZvDwAAQBAJ?hl=en&gbpv=0.

54. The pilot testing is covered in Meghann Myers and Charlsy Panzino, "The Army Is Testing a New Combat Fitness Test," *Army Times*, August 15, 2017, https://www.armytimes.com/news/your-army/2017/08/15/the-army-is-testing-a-new-combat-fitness-test/. On the goal of reducing injuries, see Steve Beynon, "Recruits, Especially from the South, Are Getting Injured at Alarming Rates in Basic Training," Military.com, April 4, 2023, https://www.military.com/daily-news/2023/04/04/recruits-especially-south-are-getting-injured-alarming-rates-basic-training.html. On the equivalent men/women standards, see Jeff Schogol, "Here's What the Army's Proposed Gender-Neutral Combat Test Really Looks Like," *Task & Purpose*, February 22, 2018, https://taskandpurpose.com/news/army-gender-neutral-combat-test/.

55. On the failure rates, see Matthew Cox, "Men and Women Seeing Different Failure Rates on Army's Gender-Neutral Fitness Test," Military.com, October 15, 2019, https://www.military.com/daily-news/2019/10/15/men-and-women-seeing-different-failure-rates-armys-gender-neutral-fitness-test.html.

56. On the permanent adoption of the test, see Davis Winkie, "Army Combat Fitness Test Debuts with Major Changes to Scoring April 1," *Army Times*, March 23, 2022, https://www.armytimes.com/news/your-army/2022/03/23/army-combat-fitness-test-debuts-with-major-changes-to-scoring-april-1/.

57. See Steve Beynon, "Debate on Army Fitness Test Takes New Turn with Chief Nominee Testimony," Military.com, July 12, 2023, https://www.military.com/daily-news/2023/07/12/armys-next-top-officer-wants-higher-gender-neutral-fitness-standards.html.

58. Milley quote from Lowe, "'Swat the Kaiser' and Stork Stands."

59. For such criticism, see David Barno and Nora Bensahel, "Dumb and Dumber: The Army's New PT Test," *War on the Rocks*, October 16, 2018, https://warontherocks.com/2018/10/dumb-and-dumber-the-armys-new-pt-test/.

60. For the independent assessment, see Chaitra M. Hardison et al., "Independent Review of the Army Combat Fitness Test: Summary of Key Findings and Recommendations" (RAND Corporation, 2022), https://www.rand.org/pubs/research_reports/RRA1825-1.html. On the deletion of the claimed linkage, see Winkie, "Army Combat Fitness Test Debuts."

61. These remarks are from Rick Blank and Tyler Patterson, "Relearning the Timeless Lessons of Land Operations in Asia," Modern War Institute (at West Point), February 28, 2025, https://mwi.westpoint.edu/relearning-the-timeless-lessons-of-land-operations-in-asia/.

62. US Secretary of Defense, "Combat Arms Standards" (US Department of Defense, Office of the Secretary of Defense, March 30, 2025).

63. US Secretary of Defense, "Combat Arms Standards."

64. US Government Publishing Office, "Cyber Strategy, Policy, and Organization" (2017), https://www.govinfo.gov/content/pkg/CHRG-115shrg28907/html/CHRG-115shrg28907.htm.

65. While its recommendations go beyond physical fitness and obesity standards, see, on this point, Crispin Burke, "The Pentagon Should Adjust Standards for Cyber Soldiers—as It Has Always Done," *War on the Rocks*, January 24, 2018, https://warontherocks.com/2018/01/pentagon-adjust-standards-cyber-soldiers-always-done/; and Ryan Pallas, "How to Boil a Frog: The Dangers of Downsizing in the U.S. Military," *War on the Rocks*, August 26, 2024, https://warontherocks.com/2024/08/how-to-boil-a-frog-the-dangers-of-downsizing-in-the-u-s-military/.

66. Milley's statement can be found in US Department of the Army, "The U.S. Army Holistic Health and Fitness Operating Concept," October 1, 2020, 3, https://www.army.mil/e2/downloads/rv7/acft/h2f_operating_concept.pdf.

67. See Nicholas Slayton, "The Marine Corps' Fitness Test from the 1950s Wasn't That Hard," *Task & Purpose*, October 30, 2022, https://taskandpurpose.com/military-life/marine-corps-1956-physical-fitness-test/.

68. The specific components of this approach at the primary military training level are spelled out in Jennifer Wolf, "Holistic Fitness During Primary Military Education Attendance," *Strength in Knowledge: The Warrant Officer Journal* 1, no. 3 (2023): 14–17. See also Drew Brooks, "The Army's Next Crisis: Americans Aren't Fit Enough to Fight," *Task & Purpose*, February 27, 2018, https://taskandpurpose.com/news/army-physical-fitness-crisis/; Meghann Myers, "The Army's New Combat Readiness Test Is Part of a Holistic Push for Soldier Health and Nutrition," *Defense News*, October 8, 2017, https://www.defensenews.com/news/your-army/2017/10/08/the-armys-new-combat-readiness-test-is-part-of-a-holistic-push-for-soldier-health-and-nutrition/; and Kyle Rempfer, "Citing 13 Brigades-Worth of Non-Deployable Troops, Army Looks to Holistic Health Solutions," *Defense News*, September 30, 2020, https://www.defensenews.com/news/your-army/2020/09/30/citing-13-brigades-worth-of-non-deployable-troops-army-looks-to-holistic-health-solutions/.

69. See "The U.S. Army Holistic Health and Fitness Operating Concept." For a real world example of how the system works, see Mikala K. Bruno et al., "Taking Holistic Health and Fitness to the Fight: A Case Example of One of the First H2F Teams to Deploy on Rotation to the Republic of Korea," *Military Review: The Professional Journal of the U.S. Army* Online Exclusive (April 2025): 1–11.

70. See Alex Hass, "Vitamin D Supplementation Program," *Marine Corps Gazette* 108, no. 8 (2024): 59–61.

71. On the efforts of nutritionists within the US military services, see Gina Harkins, "The Military Is Overhauling Troops' Chow as Obesity Rates Soar," Military.com, August 19, 2018, https://www.military.com/daily-news/2018/08/19/military-overhauling-troops-chow-obesity-rates-soar.html. Jupe's statement is also quoted in this report.

72. Stephen L. Rhoades, "Are We Too Fat to Fight? Changing Chow Hall Offerings Can Help," *Stars and Stripes*, November 21, 2023, https://www.stripes.com/opinion/2023-11-21/military-chow-hall-healthy-troops-12126933.html.

73. Tara Copp, "Goodbye, Tape Test? Sweeping DoD Fitness Review Underway," *Military Times*, February 7, 2018, https://www.militarytimes.com/news/your-military /2018/02/05/goodbye-tape-test-sweeping-dod-fitness-review-underway/.

74. Chief of Naval Operations, "NAVADMIN 205/10—ESTABLISHMENT OF THE CYBER WARFARE ENGINEER DESIGNATOR," Navy Specific Administrative Message (NAVADMIN), June 10, 2010, https://navadmin-viewer.github.io/?type =NAVADMIN&year=2010&number=252.

75. Kathleen Curthoys, "Army Offers Direct Commissions to Boost Cyber Force," *Army Times*, December 6, 2017, https://www.armytimes.com/news/your-army /2017/12/06/army-offers-direct-commissions-to-boost-cyber-force/.

76. Lauren Woods, "Airmen Selected to Participate in New Cyberspace Direct Appointment Program," Air Force Reserve Command, February 14, 2018, https:// www.afrc.af.mil/News/Article-Display/Article/1442503/airmen-selected-to -participate-in-new-cyberspace-direct-appointment-program/.

77. Jeff Schogol, "Every Marine a Rifleman No More?," *Marine Corps Times*, May 7, 2017, https://www.marinecorpstimes.com/news/your-marine-corps/2017/05/07 /every-marine-a-rifleman-no-more/.

78. David H. Berger, "Talent Management 2030" (Commandant, US Marine Corps, November 2021), https://www.hqmc.marines.mil/Portals/142/Users/183/35/4535/Ta lent%20Management%202030_November%202021.pdf?ver=E88HXGUdUQoiB -edNPKOaA%3d%3d. For additional context, see Jeff Schogol, "Marine Corps Plan Calls for Some Future Marines to Skip Boot Camp," *Task & Purpose*, November 23, 2021, https://taskandpurpose.com/news/marine-corps-plan-cyber-skip-boot -camp/. On the full spectrum of Berger's ideas for Marine recruiting, see David H. Berger, "Recruiting Requires Bold Changes," *U.S. Naval Institute Proceedings* 148, no. 11 (2022), https://www.usni.org/magazines/proceedings/2022/november /recruiting-requires-bold-changes.

79. See, for instance, Brian Sattler, "Recruit Medical Standards Are Out of Touch," *U.S. Naval Institute Proceedings* 150, no. 3 (2024), https://www.usni.org/magazines /proceedings/2024/march/recruit-medical-standards-are-out-touch.

80. This statement by Sylvester is quoted in Jeff Schogol, "The Army's Yearslong Fight Over Its Controversial New Fitness Test Isn't Over Yet," *Task & Purpose*, June 28, 2023, https://taskandpurpose.com/news/army-combat-fitness-test-problems -congress/.

81. See Steve Beynon, "The Army Needs Hundreds of Officers to Leave Combat Arms," Military.com, December 2, 2024, https://www.military.com/daily-news /2024/12/02/army-needs-hundreds-of-officers-leave-combat-arms.html.

82. Emma Moore, "The ACFT and the Problems with the Military's Cult of Physical Fitness," Military.com, December 16, 2019, https://www.military.com/daily-news /2019/12/16/acft-and-problems-militarys-cult-physical-fitness.html.

83. Jacquelyn Schneider, "Blue Hair in the Gray Zone," *War on the Rocks*, January 10, 2018, https://warontherocks.com/2018/01/blue-hair-gray-zone/. For a similar argument in favor of basing physical fitness standards on job specialty realities, see Jahara Matisek, "Change Physical Standards to Recognize Most Jobs Don't Require 'Combat Fitness,'" *Task & Purpose*, November 17, 2016, https://taskandpurpose

.com/news/change-physical-standards-recognize-jobs-dont-require-combat -fitness/.

84. John Cordle, "Physical Fitness Standards Should Be Tied to the Mission," *U.S. Naval Institute Proceedings* 151, no. 7 (2025), https://www.usni.org/magazines /proceedings/2025/july/physical-fitness-standards-should-be-tied-mission.

85. John G. Sotos, "Physical Fitness Programs Don't Fit Today's Fight," *U.S. Naval Institute Proceedings* 145, no. 6 (2019), https://www.usni.org/magazines/proceedings /2019/june/physical-fitness-programs-dont-fit-todays-fight.

86. Jahara Matisek, "Physical Fitness Is Not the Key to Winning America's Future Wars," Modern War Institute (at West Point), February 7, 2017, https://mwi.usma .edu/physical-fitness-not-key-winning-americas-future-wars/. Those wishing to learn more about the roots of American war failures should read Harlan Ullman, *Anatomy of Failure: Why America Loses Every War It Starts* (Naval Institute Press, 2017). The Vietnam War offers perhaps the most famous illustration of this point. For a case study from this perspective on that conflict, see Martin Clemis, "The Enduring Lessons of Vietnam: Implications for US Strategy and Policy," *The US Army War College Quarterly: Parameters* 55, no. 2 (2025): 117–33, https://doi.org/10 .55540/0031-1723.3347.

87. For example, see Nick LoRusso, "Maintain the Standards: Why Physical Fitness Will Always Matter in War," Modern War Institute (at West Point), February 14, 2017, https://mwi.usma.edu/maintain-standards-physical-fitness-will-always -matter-war/. For a specific rebuttal to Schneider's contentions earlier in this paragraph, see Mark Cancian, "Blue-Haired Soldiers? Just Say No," *War on the Rocks*, January 18, 2018, https://warontherocks.com/2018/01/blue-haired-soldiers -just-say-no/.

88. Rainey quote from Anonymous, "Baptism by Fire," *The Economist* 448, no. 9354 (2023): S12.

89. "H.R. 2670: National Defense Authorization Act for Fiscal Year 2024," § 577 (2023), https://www.govinfo.gov/content/pkg/BILLS 118hr2670enr/uslm/BILLS -118hr2670enr.xml.

90. The US Marine Corps has found, for example, that higher levels of cardiovascular fitness among officers correlates with higher levels of academic performance. See Luke S. Gilman, "Endurance and Executive Function: Implications for Military Education from a Study of Marine Officer Fitness and Cognition," *International Perspectives on Military Education* 2 (2025): 127–46.

91. For a comment on the military relevance of this connection, see Misty Posey, "Physical Fitness Fuels Cognitive Power," *U.S. Naval Institute Proceedings* 146, no. 11 (2020), https://www.usni.org/magazines/proceedings/2020/november/physical-fitness-fuels -cognitive-power.

92. Lixia Zhao et al., "Trends in Severe Obesity Among Children Aged 2 to 4 Years in WIC: 2010 to 2020," *Pediatrics*, 2023, e2023062461, https://doi.org/10.1542/peds .2023-062461.

93. Samuel D. Emmerich et al., "Obesity and Severe Obesity Prevalence in Adults: United States, August 2021–August 2023," NCHS Data Briefs (National Center for Health Statistics, 2024), https://doi.org/10.15620/cdc/159281. The number is

projected to approach 50 percent in the coming years. See Zachary J. Ward et al., "Projected U.S. State-Level Prevalence of Adult Obesity and Severe Obesity," *New England Journal of Medicine* 381, no. 25 (2019): 2440–50, https://doi.org/10.1056/NEJMsa1909301.

Chapter 2

1. See A. J. Higgins, "PL04 from Ancient Greece to Modern Athens: 3000 Years of Doping in Competition Horses," *Journal of Veterinary Pharmacology and Therapeutics* 29, no. s1 (2006): 4–8, https://doi.org/10.1111/j.1365-2885.2006.00770_4.x.

2. See Paul Christesen, "Sport and Society in Sparta," in *A Companion to Sport and Spectacle in Greek and Roman Antiquity*, ed. Paul Christesen and Donald G. Kyle (John Wiley & Sons, Ltd., 2013), 148, https://doi.org/10.1002/9781118609965.ch17.

3. See Michael A. Rinella, *Pharmakon: Plato, Drug Culture, and Identity in Ancient Athens* (Lexington Books, 2010), xvii.

4. See Vivian Nutton, *Ancient Medicine*, Series of Antiquity (Routledge, 2004).

5. See Thierry Appelboom, Christine Rouffin, and Eric Fierens, "Sport and Medicine in Ancient Greece," *The American Journal of Sports Medicine* 16, no. 6 (November 1, 1988): 594–96, https://doi.org/10.1177/036354658801600607.

6. Michele Verroken, "Drug Use and Abuse in Sport," *Best Practice & Research Clinical Endocrinology & Metabolism* 14, no. 1 (2000): 1, https://doi.org/10.1053/beem.2000.0050.

7. See, for example, Kate Kelland, "Ancient Dopers Got Their Kicks from Raw Testicles," *Reuters*, August 1, 2012, https://www.reuters.com/article/us-oly-doping-history-day5-idUSBRE8700YC20120801.

8. See Fabian Kanz, "Dying in the Arena: The Osseous Evidence from Ephesian Gladiators," in *Roman Amphitheatres and Spectacula, a 21st-Century Perspective: Papers from an International Conference Held at Chester, 16th–18th February, 2009*, 218, https://dialnet.unirioja.es/servlet/articulo?codigo=4164418.

9. See Amy Sue Biondich and Jeremy David Joslin, "Coca: The History and Medical Significance of an Ancient Andean Tradition," *Emergency Medicine International* (April 2016): 2–3, https://doi.org/10.1155/2016/4048764.

10. See Grant Gearhart, "The Knight and War: Alternative Displays of Masculinity in El Passo Honroso de Suero de Quiñones, El Victorial, and the Historia de los Hechos Del Marqués de Cádiz" (PhD diss., University of North Carolina at Chapel Hill, 2015), 27–28.

11. See Peter Andreas, *Killer High: A History of War in Six Drugs* (Oxford University Press, 2020), 180, http://ebookcentral.proquest.com/lib/utxa/detail.action?docID=6006832. On the premodern use of drugs in war, see chapter 1 of Lukasz Kamienski, *Shooting Up: A Short History of Drugs and War* (Oxford University Press, 2016).

12. Bradley Hatfield, "2019 Annual Meeting to Focus on Optimization of Human Performance," *National Academy of Kinesiology Newsletter* 40, no. 2 (2019): 4. See also Mark S. Dyreson, "Optimizing Human Performance—A Brief History of Macro and Micro Perspectives," *Kinesiology Review* 9, no. 1 (2020): 4–12.

13. Kamienski, *Shooting Up*, 91.

14. Paul Dimeo, *A History of Drug Use in Sport: 1876–1976: Beyond Good and Evil* (Routledge, 2008), 3.

15. See David T. Courtwright, *Forces of Habit: Drugs and the Making of the Modern World* (Harvard University Press, 2009).

16. See Jeff Schogol, "How Coffee Helped the Union Army Win the Civil War," *Task & Purpose*, September 30, 2020, https://taskandpurpose.com/history/civil-war -union-army-coffee/.

17. This statement can be found in Jon Grinspan, "How Coffee Fueled the Civil War," Opinionator, *New York Times*, July 9, 2014, https://archive.nytimes.com /opinionator.blogs.nytimes.com/2014/07/09/how-coffee-fueled-the-civil-war/.

18. See C. Anthony Pfaff, "Soldier Enhancement Ethics and the Lessons of World War I," in *A Persistent Fire: The Strategic Ethical Impact of World War I on the Global Profession of Arms*, ed. Timothy S. Mallard and Nathan H. White (National Defense University Press, 2019), 297–315.

19. See Jeff Koehler, "In WWI Trenches, Instant Coffee Gave Troops a Much-Needed Boost," *NPR*, April 6, 2017, Food History & Culture, https://www.npr.org/sections /thesalt/2017/04/06/522071853/in-wwi-trenches-instant-coffee-gave-troops-a -much-needed-boost.

20. See Andreas, *Killer High*, 180.

21. Ranke quoted in Norman Ohler, *Blitzed: Drugs in Nazi Germany*, trans. Shaun Whiteside (Houghton Mifflin, 2017), 46.

22. Von Brauchitsch quoted in Ohler, *Blitzed: Drugs in Nazi Germany*, 63. See p. 67 of the same text for the 35 million tablet order size.

23. See James Pugh, "'Not . . . Like a Rum-Ration': Amphetamine Sulphate, the Royal Navy, and the Evolution of Policy and Medical Research During the Second World War," *War in History* 24, no. 4 (November 1, 2017): 498–519, https://doi.org/10.1177 /0968344516643348; James Pugh, "The Royal Air Force, Bomber Command and the Use of Benzedrine Sulphate: An Examination of Policy and Practice During the Second World War," *Journal of Contemporary History* 53, no. 4 (October 1, 2018): 740–61, https://doi.org/10.1177/0022009416652717.

24. See Nicolas Rasmussen, "Medical Science and the Military: The Allies' Use of Amphetamine During World War II," *Journal of Interdisciplinary History* 42, no. 2 (October 2011): 205–33, https://doi.org/10.1162/JINH_a_00212.

25. Pendergrast's statement is quoted in Sarah Sicard, "Coffee Was So Important to the Army That GI Joe Gave It His Name," *Task & Purpose*, April 7, 2017, https:// taskandpurpose.com/news/coffee-important-army-gi-joe-gave-name/.

26. Kamienski, *Shooting Up*, 146.

27. See Lukasz Kamienski, "The Drugs That Built a Super Soldier," *The Atlantic*, April 8, 2016, https://www.theatlantic.com/health/archive/2016/04/the-drugs-that -built-a-super-soldier/477183/.

28. Lee N. Robins, "The Vietnam Drug User Returns," Special Action Monograph Series A, Number 2 (Special Action Office for Drug Abuse Prevention, 1973), 26, https://prhome.defense.gov/Portals/52/Documents/RFM/Readiness/DDRP/docs /35%20Final%20Report.%20The%20Vietnam%20drug%20user%20returns.pdf.

29. Geisler quote from Rhonda Cornum et al., "Stimulant Use in Extended Flight Operations," *Airpower Journal* (Maxwell AFB, United States) 11, no. 1 (1997): 55.

30. David L. Emonson and Rodger D. Vanderbeek, "The Use of Amphetamines in U.S. Air Force Tactical Operations During Desert Shield and Storm," *Aviation, Space, and Environmental Medicine* 66, no. 3 (1995): 260–63.

31. Cornum et al., "Stimulant Use in Extended Flight Operations," 57.

32. See David N. Kenagy et al., "Dextroamphetamine Use During B-2 Combat Missions," *Aviation, Space, and Environmental Medicine* 75, no. 5 (2004): 381–86.

33. See Jon Bonné, "'Go Pills': A War on Drugs?," NBC News, January 13, 2003, https://www.nbcnews.com/id/wbna3071789.

34. See Richard Martin, "It's Wake-Up Time," *Wired*, November 1, 2003, https://www.wired.com/2003/11/sleep/.

35. See Elliot Borin, "The U.S. Military Needs Its Speed," *Wired*, February 10, 2003, https://www.wired.com/2003/02/the-u-s-military-needs-its-speed/.

36. Schmidt quote from Bryan Smith, "Harry Schmidt's War," *Chicago Magazine*, June 28, 2007, https://www.chicagomag.com/Chicago-Magazine/April-2005/Harry-Schmidts-War/.

37. John Hoberman, *Dopers in Uniform: The Hidden World of Police on Steroids* (University of Texas Press, 2017), 216.

38. See Richard Sisk, "Robert Bales Among 8 Former Troops, Contractors Petitioning Trump for Pardons," Military.com, December 4, 2020, https://www.military.com/daily-news/2020/12/04/robert-bales-among-8-former-troops-contractors-petitioning-trump-pardons.html.

39. See Dave Philipps, "Death in Navy SEAL Training Exposes a Culture of Brutality, Cheating and Drugs," *New York Times*, August 30, 2022, https://www.nytimes.com/2022/08/30/us/navy-seal-training-death.html.

40. See Dave Philipps, "Navy Will Start Testing SEALs for Illicit Drug Use," *New York Times*, September 29, 2023, https://www.nytimes.com/2023/09/29/us/navy-seal-drug-testing.html.

41. See the following report sponsored by the US National Science Foundation: Mihail C. Roco and William Sims Bainbridge, *Converging Technologies for Improving Human Performance: Nanotechnology, Biotechnology, Information Technology and Cognitive Science* (National Science Foundation, 2002), https://obamawhitehouse.archives.gov/sites/default/files/microsites/ostp/bioecon-%28%23%20023SUPP%29%20NSF-NBIC.pdf.

42. John Caldwell et al., *The Efficacy of Modafinil for Sustaining Alertness and Simulator Flight Performance in F-117 Pilots During 37 Hours of Continuous Wakefulness* (US Air Force Research Laboratory, 2004).

43. See Michael J. Minzenberg and Cameron S. Carter, "Modafinil: A Review of Neurochemical Actions and Effects on Cognition," *Neuropsychopharmacology* 33, no. 7 (2008): 1477–1502, https://doi.org/10.1038/sj.npp.1301534.

44. This quote from Wolfowitz's memorandum is quoted in Anthony P. Tvaryanas et al., "Managing the Human Weapon System: A Vision for an Air Force Human-Performance Doctrine," *Air & Space Power Journal* (Maxwell AFB, United States) 23, no. 2 (2009): 34.

45. This definition from the Office of Net Assessment is quoted in Patricia A. Deuster et al., "Human Performance Optimization: An Evolving Charge to the Department

of Defense," *Military Medicine* 172, no. 11 (November 2007): 1133, https://doi.org/10.7205/milmed.172.11.1133.

46. See Damien McElroy, "Mumbai Attacks: Terrorists Took Cocaine to Stay Awake During Assault," *The Telegraph*, December 2, 2008, https://www.telegraph.co.uk/news/worldnews/asia/india/3540964/Mumbai-attacks-Terrorists-took-cocaine-to-stay-awake-during-assault.html.

47. Peter Holley, "The Tiny Pill Fueling Syria's War and Turning Fighters into Superhuman Soldiers," *Washington Post*, November 19, 2015. In fairness, Holley's depiction was based largely on a documentary produced by BBC Arabic: *The Drug Fueling Conflict in Syria* (BBC Arabic, 2015), https://www.youtube.com/watch?v=7ke13JNlpBQ.

48. German Lopez, "Captagon, ISIS's Favorite Amphetamine, Explained," *Vox*, November 20, 2015, https://www.vox.com/world/2015/11/20/9769264/captagon-isis-drug.

49. Joseph El Khoury, "The Use of Stimulants in the Ranks of Islamic State: Myth or Reality of the Syrian Conflict," *Studies in Conflict & Terrorism* 43, no. 8 (August 2, 2020): 679–87, https://doi.org/10.1080/1057610X.2018.1495291.

50. See Josh Meyer and Kim Hjelmgaard, "Were the Hamas Attacks on Israel So Brutal Because the Killers Were High on the Drug Captagon?," *USA Today*, November 2, 2023, https://www.usatoday.com/story/news/world/2023/11/02/hamas-captagon-drug-use-idf-claim/71288873007/; and Jay Solomon, "Some Hamas Killers Were High on Amphetamine, Officials Say," Semafor, November 1, 2023, https://www.semafor.com/article/10/31/2023/hamas-killers-were-high-on-amphetamine-officials-say.

51. Jim Mattis, *Summary of the 2018 National Defense Strategy* (US Department of Defense, 2018), 3.

52. Alex MacCalman et al., "The Hyper-Enabled Operator," *Small Wars Journal*, June 6, 2019, https://smallwarsjournal.com/jrnl/art/hyper-enabled-operator.

53. For a representative example of such discussions, see Chad A. Alford and Sean Chang, "Net Benefit of Performance-Enhancing Drugs Within U.S. Army Special Forces" (master's thesis, Naval Postgraduate School, 2020).

54. Chitty quote from David Larter, "'Performance Enhancing Drugs' Considered for Special Operations Soldiers," *Defense News*, May 16, 2017, https://www.defensenews.com/digital-show-dailies/sofic/2017/05/16/performance-enhancing-drugs-considered-for-special-operations-soldiers/.

55. Chitty quoted in Matthew Sussis, "Special Operations Command Begins Research on Performance-Enhancing Drugs—National Security Zone," *MEDILL National Security Zone* (blog), February 9, 2018, https://nationalsecurityzone.medill.northwestern.edu/blog/2018/conflicts-terrorism/special-operations-command-begins-research-on-performance-enhancing-drugs/.

56. Paul Scharre and Lauren Fish, *Human Performance Enhancement*, Super Soldiers (Center for a New American Security, 2018), 11.

57. See, for example, Scharre and Fish, *Human Performance Enhancement*; Clayton J. Aune, Building the Hyper-Capable Operator: Should the Military Enhance Its Special Operations Warriors?, research paper (Naval War College, n.d.), accessed August 22, 2023, https://apps.dtic.mil/sti/citations/AD1120842.

58. See Jai Galliott and Mianna Lotz, *Super Soldiers: The Ethical, Legal and Social Implications* (Routledge, 2016); Dave Shunk, "Ethics and the Enhanced Soldier of the Near Future," *Military Review* 95, no. 1 (February 2015): 91–98; and Patrick Lin et al., *Enhanced Warfighters: Risk, Ethics, and Policy* (Greenwell Foundation, January 1, 2013), https://ethics.calpoly.edu/Greenwall_report.pdf.

59. See Vania Nikolova, "American Runners Have Never Been Slower (Mega Study)," RunRepeat, September 21, 2021, https://runrepeat.com/american-runners-have -never-been-slower-mega-study.

60. See, for example, Alan M. Nevill and Gregory Whyte, "Are There Limits to Running World Records?," *Medicine & Science in Sports & Exercise* 37, no. 10 (2005): 1785–88, https://doi.org/10.1249/01.mss.0000181676.62054.79; A. M. Nevill et al., "Are There Limits to Swimming World Records?," *International Journal of Sports Medicine*, May 29, 2007, 1012–17, https://doi.org/10.1055/s-2007-965088; and Geoffroy Berthelot et al., "Has Athletic Performance Reached Its Peak?," *Sports Medicine* 45, no. 9 (September 1, 2015): 1263–71, https://doi.org/10.1007/s40279-015 -0347-2.

61. See Nigel Balmer et al., "Evolution and Revolution: Gauging the Impact of Technological and Technical Innovation on Olympic Performance," *Journal of Sports Sciences* 30, no. 11 (July 1, 2012): 1075–83, https://doi.org/10.1080/02640414.2011 .587018.

62. On development work on powered exoskeletons, see Sydney J. Freedberg Jr., "SOCOM Tests Sarcos Exoskeleton (No, It Isn't 'Iron Man')," *Breaking Defense*, March 18, 2019, https://breakingdefense.sites.breakingmedia.com/2019/03/socom -tests-sarcos-exoskeleton-no-it-isnt-iron-man/. On a partnership between the Army and Vanderbilt University on the testing of an unpowered exoskeleton, see US Army Combat Capabilities Development Command (DEVCOM), Public Affairs Office, "Army, Academia Collaborate on Exoskeleton to Reduce Soldier Injuries," US Army website, August 17, 2022, https://www.army.mil/article/259429 /army_academia_collaborate_on_exoskeleton_to_reduce_soldier_injuries.

63. See Carol Kuntz, "Genomes: The Era of Purposeful Manipulation Begins," A Report of the CSIS Strategic Technologies Program (Center for Strategic and International Studies [CSIS], 2022), https://www.csis.org/analysis/genomes-era -purposeful-manipulation-begins.

64. *Final Report: National Security Commission on Artificial Intelligence* (National Security Commission on Artificial Intelligence's [NSCAI], 2021), 52, https://www .nscai.gov/wp-content/uploads/2021/03/Full-Report-Digital-1.pdf.

65. Jason Mazanov, "Anti-Doping in Sport and Human Enhancing Technologies in Army," *Land Power Forum, Australian Army Research Centre* (blog), October 13, 2017, https://researchcentre.army.gov.au/library/land-power-forum/anti-doping -sport-and-human-enhancing-technologies-army.

66. John P. Jumper, "Global Strike Task Force: A Transforming Concept, Forged by Experience," *Aerospace Power Journal* 15, no. 1 (2001): 28.

67. Travis Pike, "Energy Drinks—The Unsung Hero of the Global War on Terror," Sandboxx, June 7, 2023, https://www.sandboxx.us/blog/energy-drinks-the-unsung -hero-of-the-global-war-on-terror/. On the use of caffeine by the US military for

performance enhancement, see Scharre and Fish, *Human Performance Enhancement*, 8.

68. This issue was discussed by a pair of former US military aviators at the following site: *How Do Fighter Pilots Prepare for Long Sorties/Ocean Crossing?*, 2023, https://www.youtube.com/watch?v=342Qie3FPjo.

69. *Human Performance*, Technical Report JSR-07-625 (Jason Program Office, The MITRE Corporation, 2008), 15, https://apps.dtic.mil/sti/citations/ADA480029.

70. Paul Scharre et al., *Super Soldiers: Summary of Findings and Recommendations*, Super Soldiers (Center for a New American Security, November 2018), 6.

71. See Erik Patton et al., "Fremen Stillsuits and Heated Bubble Gyms: Preparing the Army to Fight and Win on a Rapidly Warming Battlefield," Modern War Institute (at West Point), June 19, 2024, https://mwi.westpoint.edu/fremen-stillsuits-and-heated-bubble-gyms-preparing-the-army-to-fight-and-win-on-a-rapidly-warming-battlefield/; and Erik M. Patton et al., "Preparing for Hot Conflicts: Army Training and Operations in a Warming World," *Military Review: The Professional Journal of the U.S. Army* 105, no. 1 (February 2025): 132–44.

72. See Rick Blank and Tyler Patterson, "Relearning the Timeless Lessons of Land Operations in Asia," Modern War Institute (at West Point), February 28, 2025, https://mwi.westpoint.edu/relearning-the-timeless-lessons-of-land-operations-in-asia/.

73. Kuntz, "Genomes," 4–5.

74. Kuntz, "Genomes," 11.

75. Mozer is quoted in Alex Wilkins, "US Space Force Scientist Says 'Human Augmentation' Is Necessary in Next Decade," *Metro*, April 29, 2021, https://metro.co.uk/2021/04/29/us-space-force-scientist-says-military-human-augmentation-necessary-14493027/.

76. Between the completion of the manuscript upon which this book is based and its publication, President Donald Trump issued an executive order adding the US Department of War as a secondary title for the US Department of Defense: see https://www.war.gov/News/News-Stories/Article/Article/4295826/trump-renames-dod-to-department-of-war/.

77. See US Army, *Field Manual FM 7-22: Holistic Health and Fitness* (Headquarters, Department of the Army, 2020), https://armypubs.army.mil/epubs/DR_pubs/DR_a/ARN30964-FM_7-22-001-WEB-4.pdf; and Consensus Conference Panel, "Recommended Amount of Sleep for a Healthy Adult: A Joint Consensus Statement of the American Academy of Sleep Medicine and Sleep Research Society," *Sleep* 38, no. 6 (2015): 843–44, https://doi.org/10.5665/sleep.4716.

78. Cameron H. Good et al., "Sleep in the United States Military," *Neuropsychopharmacology* 45, no. 1 (January 2020): 176–91, https://doi.org/10.1038/s41386-019-0431-7.

79. See chapter 2 of Christopher Coker, *Future War* (John Wiley & Sons, 2015).

80. See Tyson Grier et al., "Physical and Behavioral Characteristics of Soldiers Acquiring Recommended Amounts of Sleep per Night," *Sleep Health* 9, no. 5 (October 2023): 626–33, https://doi.org/10.1016/j.sleh.2023.03.003.

81. Vincent Mysliwiec et al., "Sleep Disorders in US Military Personnel: A High Rate of Comorbid Insomnia and Obstructive Sleep Apnea," *Chest* 144, no. 2 (August 2013): 549–57, https://doi.org/10.1378/chest.13-0088.

82. See Riley Ceder, "18th Airborne Orders Soldiers on Staff Duty to Get 4 Hours of Sleep," *Army Times*, May 21, 2024, https://www.armytimes.com/news/your-army/2024/05/21/18th-airborne-orders-soldiers-on-staff-duty-to-get-4-hours-of-sleep/.

83. Dave Nixon, "A Lack of Sleep Is Breaking the US Military," *Military Times*, December 12, 2023, https://www.militarytimes.com/opinion/2023/12/12/a-lack-of-sleep-is-breaking-the-us-military/.

84. Daniel J. Taylor et al., "Internet and In-Person Cognitive Behavioral Therapy for Insomnia in Military Personnel: A Randomized Clinical Trial," *Sleep* 40, no. 6 (June 2017): 1–12, https://doi.org/10.1093/sleep/zsx075.

85. See Wendy M. Troxel et al., *Sleep in the Military: Promoting Healthy Sleep Among U.S. Servicemembers* (RAND Corporation, 2015), 19, https://www.ncbi.nlm.nih.gov/pmc/articles/PMC5158299/.

86. See John P. Cordle, "Fatigue Is the Navy's Black Lung Disease," *U.S. Naval Institute Proceedings* 146, no. 1 (January 2020), https://www.usni.org/magazines/proceedings/2020/january/fatigue-navys-black-lung-disease.

87. *Comprehensive Review of Recent Surface Force Incidents* (US Fleet Forces Command, US Navy, October 26, 2017), https://s3.documentcloud.org/documents/4172114/Comprehensive-Review-of-Recent-Surface-Force.pdf. See also Erin Patterson, "Ship Collisions: Address the Underlying Causes, Including Culture," *U.S. Naval Institute Proceedings* 143, no. 8 (2017), https://www.usni.org/magazines/proceedings/2017/august/ship-collisions-address-underlying-causes-including-culture.

88. See Megan Eckstein, "Fleet Finding New Sleep-Sensitive Watch Schedules Boosts Crew Performance, Efficiency," *US Naval Institute (USNI) News*, July 15, 2019, https://news.usni.org/2019/07/15/fleet-finding-new-sleep-sensitive-watch-schedules-boosts-crew-performance-efficiency. The 2017 standards were refined three years later in a more comprehensive plan for crew readiness and endurance. See Gidget Fuentes, "Latest Surface Navy Sleep Policy Aims for Better-Rested, More Alert, Healthier Crews," *US Naval Institute (USNI) News*, January 28, 2021, https://news.usni.org/2021/01/28/latest-surface-navy-sleep-policy-aims-for-better-rested-more-alert-healthier-crews.

89. This point is made in Matt Clark, "The Army Has a Physical Fitness Problem, Part 2: Toward a More Combat-Ready Force," Modern War Institute (at West Point), January 31, 2020, https://mwi.usma.edu/army-physical-fitness-problem-part-2-toward-combat-ready-force/. See also David Nixon and Porter Riley, "Sleep and Performance: Why the Army Must Change Its Sleepless Culture," *Military Review: The Professional Journal of the U.S. Army* 103, no. 6 (December 2023): 147–57.

90. US Department of Defense, "Study on Effects of Sleep Deprivation on Readiness of Members of the Armed Forces," *US Naval Institute (USNI) News*, March 2021, 1, https://news.usni.org/2021/03/03/pentagon-report-on-sleep-deprivation-and-military-readiness.

91. US Department of Defense, "Study on Effects of Sleep Deprivation."

92. See, for example, Tung Wai Auyeung et al., "Sleep Duration and Disturbances Were Associated with Testosterone Level, Muscle Mass, and Muscle Strength—A Cross-Sectional Study in 1274 Older Men," *Journal of the American Medical Directors Association* 16, no. 7 (July 1, 2015): 630.e1–630.e6, https://doi.org/10.1016

/j.jamda.2015.04.006; and Y. Harrison and J. A. Horne, "One Night of Sleep Loss Impairs Innovative Thinking and Flexible Decision Making," *Organizational Behavior and Human Decision Processes* 78, no. 2 (May 1, 1999): 128–45, https://doi .org/10.1006/obhd.1999.2827.

93. See Carl D. Smith et al., "Sleep Restriction and Cognitive Load Affect Performance on a Simulated Marksmanship Task," *Journal of Sleep Research* 28, no. 3 (2019): e12637, https://doi.org/10.1111/jsr.12637.

94. John Cordle and Robert Sweetman, "Human Factors Meets New Technology in 2025," Center for International Maritime Security, August 30, 2021, https://cimsec .org/human-factors-meets-new-technology-in-2025/.

95. See, for instance, Christopher Woolsey, "Mandate Uninterrupted Rest Hours for a Safer Navy," *U.S. Naval Institute Proceedings* 149, no. 11 (November 2023), https:// www.usni.org/magazines/proceedings/2023/november/mandate-uninterrupted -rest-hours-safer-navy.

96. Elsa B. Kania, "Minds at War: China's Pursuit of Military Advantage Through Cognitive Science and Biotechnology," *PRISM* 8, no. 3 (2019): 83.

97. A 2014 high-level report from the Australian Army contained the following language, for example: "By 2025 we may face adversaries with scientifically enhanced cognitive capacity. The land force will need to develop a better understanding of enhancing human capabilities with, for example, improved human-machine interfaces or better fusing of technology with biology. Targeted research in these areas will be necessary to explore opportunities and threats associated with enhancing human decision-making and performance in the future operating environment." Australian Army, *Future Land Warfare Report 2014* (Modernisation and Strategic Planning Division—Australian Army Headquarters, April 2014), 13, https:// researchcentre.army.gov.au/library/other/future-land-warfare-report-2014.

98. See Patrick Tucker, "Soldiers Can Now Steer Robot Dogs with Brain Signals," Defense One, March 22, 2023, https://www.defenseone.com/technology/2023/03 /soldiers-can-now-steer-robot-dogs-brain-signals/384338/.

99. See Nancy S. Jecker and Andrew Ko, "Brain-Computer Interfaces Could Allow Soldiers to Control Weapons with Their Thoughts and Turn Off Their Fear— but the Ethics of Neurotechnology Lags Behind the Science," *The Conversation*, December 2, 2022, http://theconversation.com/brain-computer-interfaces -could-allow-soldiers-to-control-weapons-with-their-thoughts-and-turn-off -their-fear-but-the-ethics-of-neurotechnology-lags-behind-the-science-194017; Mark Wess, "Brain-Machine Interface—the New 'BMI,'" *U.S. Naval Institute Proceedings* 149, no. 5 (May 2023), https://www.usni.org/magazines/proceedings/2023 /may/brain-machine-interface-new-bmi; and Anika Binnendijk et al., "Brain-Computer Interfaces: U.S. Military Applications and Implications, An Initial Assessment" (RAND Corporation, 2020), https://www.rand.org/pubs/research _reports/RR2996.html.

100. See Tucker, "Soldiers Can Now Steer Robot Dogs with Brain Signals."

101. See Jecker and Ko, "Brain-Computer Interfaces Could Allow Soldiers to Control Weapons with Their Thoughts and Turn Off Their Fear."

102. For such an argument, see Héloïse Goodley, *Pharmacological Performance Enhancement and the Military: Exploring an Ethical and Legal Framework for*

"Supersoldiers," research paper (Chatham House, The Royal Institute of International Affairs, 2020), 10–12, https://policycommons.net/artifacts/1423207/pharma cological-performance-enhancement-and-the-military/2037472/.

103. For an interesting historical overview on the place of fear in battle, see Gregory A. Daddis, "Understanding Fear's Effect on Unit Effectiveness," *Military Review: The Professional Journal of the U.S. Army* 84, no. 5 (2004): 22–27.

104. Such an example is outlined in Irik Johnson et al., "How Well Is an Officer Handling the Pressure of Battle? Wearables May Be Able to Tell," US Naval Institute, accessed October 10, 2023, https://www.usni.org/magazines/proceedings /sponsored/how-well-officer-handling-pressure-battle-wearables-may-be-able.

105. Quote from Patrick Tucker, "Special Operations Command Made a Mind-Reading Kit For Elite Troops," Defense One, December 11, 2019, https://www.defenseone.com /technology/2019/12/specops-lab-made-mind-reading-kit-elite-troops/161830/.

106. The Ukrainian soldiers are quoted in David Hambling, "Are Drugs Making Russian Soldiers Act Like Zombies?," *Forbes*, February 2, 2023, https://www.forbes .com/sites/davidhambling/2023/02/02/are-drugs-making-russian-soldiers-act -like-zombies/.

107. See Sarah El Sirgany et al., "Captured Russian Soldiers Tell of Low Morale, Disarray and Horrors of Trench Warfare," CNN, July 6, 2023, https://www.cnn.com /2023/07/06/europe/captured-russian-soldiers-ukraine-intl-cmd/index.html.

108. For more on the point that the American military needs greater progress in these areas, see Daniel Herlihy, "Cognitive Performance Enhancement for Multi-Domain Operations," *The US Army War College Quarterly: Parameters* 52, no. 4 (2022): 75–95, https://doi.org/10.55540/0031-1723.3188.

Chapter 3

1. For fascinating studies on the long history of war-gaming, see Martin Van Creveld, *Wargames: From Gladiators to Gigabytes* (Cambridge University Press, 2013); and Matthew B. Caffrey Jr., *On Wargaming: How Wargames Have Shaped History and How They May Shape the Future*, Naval War College Newport Papers 43 (US Naval War College Press, 2019). Those interested in the myriad connections between gaming and armed conflict should consult Ed Halter, *From Sun Tzu to XBox: War and Video Games* (PublicAffairs, 2006). For contemporary perspectives on the utility of war-gaming, see the following special issue: *Journal of Advanced Military Studies* 12, no. 2 (Fall 2021), https:// www.usmcu.edu/Outreach/Marine-Corps-University-Press/MCU-Journal /JAMS-Vol-12-No-2/. The history of US military war-gaming can be explored at the online collection "Wargaming and Crisis Simulation Initiative Collection" (Hoover Institution, Stanford University), accessed November 17, 2024, https:// wargaming.hoover.org/.

2. See Andrew P. Janco, "Training in the Amusements of Mars: Peter the Great, War Games and the Science of War, 1673–1699," *Russian History* 30, no. 1/2 (2003): 35–112.

3. On German war-gaming during this period, consult Paul Schuurman, "A Game of Contexts: Prussian-German Professional Wargames and the Leadership Concept of Mission Tactics 1870–1880," *War in History* 28, no. 3 (2021): 504–24, https://doi

.org/10.1177/0968344519855104. Interestingly, as recently as 2013 the US Army's Command and General Staff College has used the Prussian game in the classroom. See Richard A. McConnell and Mark T. Gerges, "Seeing the Elephant: Improving Leader Visualization Skills Through Simple War Games," *Military Review: The Professional Journal of the U.S. Army,* online exclusive (October 2018): 1–9, https://www.armyupress.army.mil/Journals/Military-Review/Online-Exclusive/2018-OLE/Oct/Seeing-the-Elephant/.

4. Colomb delivered a lecture on the topic: P. H. Colomb, "Broadside Fire, and a Naval War Game.," *Royal United Services Institution. Journal* 23, no. 101 (1879): 507–35, https://doi.org/10.1080/03071847909417112. A short overview of naval war-gaming is provided in John Prados, "Waging War with Cardboard Navies," *Naval History* 20, no. 4 (2006), https://www.usni.org/magazines/naval-history-magazine/2006/august/waging-war-cardboard-navies.

5. See Francis J. McHugh, "Eighty Years of War Gaming," *Naval War College Review* 21, no. 7 (1969): 88.

6. On war-gaming at the RAND Corporation, consult John R. Emery, "Moral Choices Without Moral Language: 1950s Political-Military Wargaming at the RAND Corporation," *Texas National Security Review* 4, no. 4 (2021): 12–31, https://doi.org/10.26153/tsw/17528. Robert's contributions to war-gaming at Avalon Hill are covered in Rick Britton, "How Gettysburg Inspired Modern War Gaming," HistoryNet, October 20, 2022, https://www.historynet.com/how-gettysburg-inspired-modern-war-gaming/.

7. See chapter 4 of Benjamin M. Jensen et al., *Information in War: Military Innovation, Battle Networks, and the Future of Artificial Intelligence* (Georgetown University Press, 2022).

8. See Roger Smith, "The Long History of Gaming in Military Training," *Simulation & Gaming* 41, no. 1 (February 1, 2010): 8–12, https://doi.org/10.1177/1046878109334330.

9. For a contemporaneous overview of the potential benefits of war-gaming for military historians, military practitioners, and game hobbyists alike, see Stephen P. Glick and L. Ian Charters, "War, Games, and Military History," *Journal of Contemporary History* 18, no. 4 (1983): 567–82. On the role of military computers during the Cold War years, consult Bryan Leese, "The Cold War Computer Arms Race," *Journal of Advanced Military Studies* 14, no. 2 (Fall 2023): 102–20, https://doi.org/10.21140/mcuj.20231402006.

10. This was preceded by a cultural shift on technological matters that took place in the American military during the 1980s. See chapter 3 of Dima Adamsky, *The Culture of Military Innovation: The Impact of Cultural Factors on the Revolution in Military Affairs in Russia, the US, and Israel* (Stanford University Press, 2010); and Leese, "The Cold War Computer Arms Race." In terms of military digital gaming during the '80s, after the 1981 release of *Battlezone* by the Atari corporation, a group of retired US Army leaders connected with a think tank approached the company to see whether it might be possible to modify the arcade game into a training system for the Bradley Fighting Vehicle, which had just entered service. See Tony Temple, "Bradley Trainer: Atari's Top Secret Military Project," *The Arcade Blogger,* October 28, 2016, https://arcadeblogger.com/2016/10/28/bradley-trainer-ataris-top-secret-military-project/.

11. See Eric H. Arnett, "Welcome to Hyperwar," *Bulletin of the Atomic Scientists* 48, no. 7 (September 1992): 14–21, https://doi.org/10.1080/00963402.1992.11460097; and Michael E. O'Hanlon, *Technological Change and the Future of Warfare* (Brookings Institution Press, 2000).

12. For a detailed chronicle of the war, consult Rick Atkinson, *Crusade: The Untold Story of the Persian Gulf War* (Houghton Mifflin, 1993). A shorter though excellent account can be found in chapter 2 of Jonathan M. House, *A Military History of the New World Disorder, 1989–2022* (University of Oklahoma Press, 2025).

13. The following text by an American general serves as a representative example of this thinking: William E. Odom, *America's Military Revolution: Strategy and Structure After the Cold War* (American University Press, 1993).

14. Andrew W. Marshall, "Some Thoughts on Military Revolutions—Second Version" (ONA Memorandum for the Record, August 23, 1993), https://stacks.stanford.edu /file/druid:yx275qm3713/yx275qm3713.pdf. Under Marshall, Andrew Krepinevich published the seminal study on the revolution in military affairs then taking place. The Center for Strategic and Budgetary Assessments has issued a reprint of that study: Andrew F. Krepinevich, "The Military-Technical Revolution: A Preliminary Assessment" (Office of Net Assessment, 1992), https://csbaonline .org/research/publications/the-military-technical-revolution-a-preliminary -assessment/publication/1.

15. See Evelyn Richards, "Lowdown on High Tech Weapons in Gulf War," *Washington Post*, January 27, 1991, https://www.washingtonpost.com/archive/politics/1991 /01/27/lowdown-on-high-tech-weapons-in-gulf-war/54219640-02e7-4270-b328 -bfa05bf90974/.

16. On the history of war-gaming and the Marine Corps, consult Sebastian J. Bae and Ian T. Brown, "Promise Unfulfilled: A Brief History of Educational Wargaming in the Marine Corps," *Journal of Advanced Military Studies* 12, no. 2 (2021): 45–80.

17. Charles Krulak, "Marine Corps Order 1500.55," April 12, 1997, 1, https://www .marines.mil/Portals/1/Publications/MCO%201500.55.pdf.

18. See Bae and Brown, "Promise Unfulfilled," 63; Rob Riddell, "Doom Goes to War," *Wired*, April 1, 1997, https://www.wired.com/1997/04/ff-doom/; and Michael J. Jernigan, "Marine Doom," *Marine Corps Gazette* 81, no. 8 (August 1997): 19–20.

19. The Army finally closed the game down in May 2022. See Andy Chalk, "America's Army Is Finally Closing for Good," *PC Gamer*, February 8, 2022, https://www .pcgamer.com/americas-army-is-finally-closing-for-good/. On the ethical dimensions of the game, see Marcus Schulzke, "Rethinking Military Gaming: America's Army and Its Critics," *Games and Culture* 8, no. 2 (March 1, 2013): 59–76, https:// doi.org/10.1177/1555412013478686. For a penetrating study of military employment of digital gaming technologies, consult Corey Mead, *War Play: Video Games and the Future of Armed Conflict* (Houghton Mifflin Harcourt, 2013).

20. Max Boot, "The New American Way of War," *Foreign Affairs* 82, no. 4 (2003): 41–58, https://doi.org/10.2307/20033648.

21. Krogh quotes from Anna Mulrine, "How Can Army Keep Soldiers Fighting Fit After Afghanistan? Avatars," *Christian Science Monitor*, May 3, 2012, https://www .csmonitor.com/USA/Military/2012/0503/How-can-Army-keep-soldiers-fighting -fit-after-Afghanistan-Avatars.

22. Robert Work, "Deputy Secretary of Defense Speech, CNAS Defense Forum," US Department of Defense, December 14, 2015, https://www.defense.gov/News/Speeches/Speech/Article/634214/cnas-defense-forum/. Astute observers had been aware of the growing importance of robotics in war fighting. To give one example, see P. W. Singer, *Wired for War: The Robotics Revolution and Conflict in the 21st Century* (Penguin, 2009).

23. Matthew Rosenberg and John Markoff, "The Pentagon's 'Terminator Conundrum': Robots That Could Kill on Their Own," *The New York Times*, October 25, 2016, https://www.nytimes.com/2016/10/26/us/pentagon-artificial-intelligence-terminator.html.

24. For an excellent review essay on the book-length scholarship on military unmanned aerial vehicles, see John R. Emery and Hadley Biggs, "Human, All Too Human: Drones, Ethics, and the Psychology of Military Technologies," *Political Psychology* 43, no. 3 (2022): 605–13, https://doi.org/10.1111/pops.12809. The development of drone technology predated 9/11. On the evolution of development efforts from World War I through the 1990s, see Katharine Hall Kindervater, "The Emergence of Lethal Surveillance: Watching and Killing in the History of Drone Technology," *Security Dialogue* 47, no. 3 (2016): 223–38.

25. A useful guide to drone warfare written during this period is Sarah E. Kreps, *Drones: What Everyone Needs to Know* (Oxford University Press, 2016).

26. See David Zucchino, "Drone Pilots Have a Front-Row Seat on War, from Half a World Away," *Los Angeles Times*, February 21, 2010, https://www.latimes.com/world/la-xpm-2010-feb-21-la-fg-drone-crews21-2010feb21-story.html.

27. Jeremiah Gertler, *U.S. Unmanned Aerial Systems*, CRS Report for Congress (Congressional Research Service, January 3, 2012), https://sgp.fas.org/crs/natsec/R42136.pdf.

28. US Air Force, *United States Air Force Unmanned Aircraft Systems Flight Plan 2009–2047* (Headquarters, US Air Force, 2009), 82.

29. The documentary film, directed by Tonje Hessen Schei, is *Drone* (Flimmer Film, Radiator Film, Volt Film, 2014), 1h18m.

30. See, for example, Jacqueline M. Wheatcroft et al., "Unmanned Aerial Systems (UAS) Operators' Accuracy and Confidence of Decisions: Professional Pilots or Video Game Players?," *Cogent Psychology* 4, no. 1 (2017), https://doi.org/10.1080/23311908.2017.1327628.

31. *Drone.*

32. These downsides are a key theme in the following work written by a commanding officer of an American drone unit: Wayne Phelps, *On Killing Remotely: The Psychology of Killing with Drones* (Little, Brown and Co., 2021). For additional insights rendered by a rigorous analysis of the issue, consult Chaitra M. Hardison et al., *Stress and Dissatisfaction in the Air Force's Remotely Piloted Aircraft Community: Focus Group Findings*, Research Reports (RAND Corporation, April 27, 2017), https://www.rand.org/pubs/research_reports/RR1756.html.

33. See Alan W. Dowd, "Drone Wars: Risks and Warnings," *The US Army War College Quarterly: Parameters* 42/43, no. 4/1 (2012): 7–16.

34. Bagwell quoted in Patrick Wintour, "RAF Urged to Recruit Video Game Players to Operate Reaper Drones," *The Guardian*, December 9, 2016, https://www

.theguardian.com/world/2016/dec/09/royal-air-force-recruit-video-game-players
-operate-reaper-drones.

35. Kwast quote from Oriana Pawlyk, "Here's How the Air Force Plans to Recruit
Teenage Gamers," Military.com, May 25, 2018, https://www.military.com/defense
tech/2018/05/25/heres-how-air-force-plans-recruit-teenage-gamers.html.

36. Chip Pons, "Flying Training Reimagined as First PTN Class Graduates," Air Force
website, August 6, 2018, https://www.af.mil/News/Article-Display/Article/1594573
/flying-training-reimagined-as-first-ptn-class-graduates/.

37. See Diana Correll, "Navy's New 'Project Avenger' Flight Training Program Aims
to Produce Stronger Aviators," *Navy Times*, May 26, 2021, https://www.navytimes
.com/news/your-navy/2021/05/25/navys-new-project-avenger-flight-training
-program-aims-to-produce-stronger-aviators/.

38. Daniels's statement is quoted in Patrick Tucker, "The Future of Flight Training Is a
Gamer Headset That Watches You Play," *Defense One*, December 9, 2020, https://
www.defenseone.com/technology/2020/12/future-flight-training-gamer-headset
-watches-you-play/170630/.

39. See Joseph Trevithick, "A-10 Warthog Pilots Are Using the Digital Combat Simulator Video Game to Train in VR," *The War Zone (on The Drive)*, May 14, 2021,
https://www.thedrive.com/the-war-zone/40620/a-10-warthog-pilots-are-using
-the-digital-combat-simulator-video-game-to-train-in-vr.

40. Glowa quoted from Jacob Stephens, "355th TRS Implements New VR Simulators,"
Defense Visual Information Distribution Service, October 28, 2020, https://www
.dvidshub.net/news/386384/355th-trs-implements-new-vr-simulators.

41. Manning quote from Stephens, "355th TRS Implements New VR Simulators."

42. A recent article in the US Navy's leading professional journal recommended Acey
Deucey as a learning mechanism, for instance. See Robert C. Rubel, "Acey Deucey
Can Teach Military Strategy," *U.S. Naval Institute Proceedings* 149, no. 3 (2023),
https://www.usni.org/magazines/proceedings/2023/march/acey-deucey-can
-teach-military-strategy.

43. See Joseph "Jay" Arnold, "Build Teams with Board Games," *U.S. Naval Institute
Proceedings* 144, no. 1 (2018), https://www.usni.org/magazines/proceedings/2018
/january/build-teams-board-games.

44. See Joseph "Jay" Arnold, "Littoral Commander: Indo-Pacific," *U.S. Naval Institute
Proceedings* 149, no. 7 (2023), https://www.usni.org/magazines/proceedings/2023
/july/littoral-commander-indo-pacific.

45. Ian Strebel and Matt McKenzie, "Embrace the Nerd: Dungeons & Dragons and
Military Intelligence," *War on the Rocks*, August 18, 2023, https://warontherocks
.com/2023/08/embrace-the-nerd-dungeons-dragons-and-military-intelligence/.

46. Scott Kuhn, "Soldiers Maintain Readiness Playing Video Games," Defense Visual
Information Distribution Service (DVIDS), April 29, 2020, https://www.dvidshub
.net/news/368773/soldiers-maintain-readiness-playing-video-games.

47. Kuhn quoted in Ryan Pickrell, "US Army Tankers Are Playing Video Games
Online to Train for Tank Warfare During the Coronavirus Pandemic," *Business
Insider*, May 1, 2020, https://www.businessinsider.com/coronavirus-us-army-tank
-crews-play-video-games-to-train-2020-4. For a brief overview of how this and
similar games can be used for such training, see Logan Nye, "5 Video Games

Worth Using as 'Hip-Pocket Training,'" Military.com, July 14, 2025, https://www
.military.com/off-duty/games/5-video-games-worth-using-hip-pocket-training
.html.

48. Quote from Jamie Hunter, "Fighter Pilots Warn of Newly Trained Pilots' Lack of
Actual Flying Experience," *The War Zone (on The Drive)*, July 22, 2020, https://
www.thedrive.com/the-war-zone/35033/fighter-pilots-warn-of-lack-of-new
-aviators-flying-experience-as-force-remodels-training.

49. See Rachel Cohen, "Can the Air Force Train New Pilots Without Planes?," *Air
Force Times*, November 16, 2022, https://www.airforcetimes.com/news/your-air
-force/2022/11/16/can-the-air-force-train-new-pilots-without-planes/.

50. "Gaming Technology: The Future of Training," Lockheed Martin, November 28,
2022, https://www.lockheedmartin.com/en-us/news/features/2022/gaming-tech
nology-the-future-of-training.html.

51. See Sean Carberry, "Air Force Wants to Sift Through Simulation Data to Boost
Training," *National Defense*, June 2023, 13.

52. For a good starting point on the evolving future battlefield, see Christian Brose,
The Kill Chain: Defending America in the Future of High-Tech Warfare (Hachette
Books, 2020).

53. For the debate on video games within the US military, see Verma Pranshu, "Inside
the Pentagon's Long Debate: Do Gamers Make Good Troops?," *Washington
Post*, June 10, 2022, https://www.washingtonpost.com/video-games/2022/06/10
/military-gaming/.

54. Robert C. Rubel, "Research & Debate—The Medium Is the Message: Weaving
Wargaming More Tightly into the Fabric of the Navy," *Naval War College Review*
72, no. 4 (2019): 151.

55. Mark Van Horn, "Big Data War Games: Necessary for Winning Future Wars,"
Military Review: The Professional Journal of the U.S. Army, September 2016, 2,
https://www.armyupress.army.mil/Journals/Military-Review/Online-Exclusive
/2016-Online-Exclusive-Articles/Big-Data War-Games/.

56. Brian O'Rourke, "Wargaming at MCU: A Small Step for Marines, a Giant Leap for
the Marine Corps," *U.S. Naval Institute* 148, no. 11 (November 2022), https://www
.usni.org/magazines/proceedings/2022/november/wargaming-mcu.

57. See Todd South, "New War-Gaming Center Will Plug into Marine Units with a Realistic Experience," *Marine Corps Times*, May 12, 2022, https://www.marinecorpstimes
.com/news/modern-day-marine/2022/05/12/new-war-gaming-center-will-plug-into
-marine-units-with-a-more-realistic-experience/.

58. See Nicholas Slayton, "The Air Force Is Letting Troops Play This Wargame on Its
Secure Networks," *Task & Purpose*, October 27, 2024, https://taskandpurpose.com
/culture/air-force-command-video-game/. For more context, see Daniel Michaels,
"A Million People Play This Video Wargame. So Does the Pentagon," *Wall Street
Journal*, October 27, 2024, https://www.wsj.com/politics/national-security/a
-million-people-play-this-video-wargame-so-does-the-pentagon-e6388f50?st
=ZkysKC.

59. As an example, see Stacie Pettyjohn et al., "Dangerous Straits: Wargaming a
Future Conflict over Taiwan" (Center for a New American Security, June 2022),
https://www.cnas.org/publications/reports/dangerous-straits-wargaming-a

-future-conflict-over-taiwans; and Mark F. Cancian et al., *The First Battle of the Next War: Wargaming a Chinese Invasion of Taiwan*, A Report of the CSIS International Security Program (Washington, DC: Center for Strategic and International Studies [CSIS], January 2023), https://www.csis.org/analysis/first-battle-next-war-wargaming-chinese-invasion-taiwan.

60. US Joint Chiefs of Staff, "Developing Today's Joint Officers for Tomorrow's Ways of War: The Joint Chiefs of Staff Vision and Guidance for Professional Military Education and Talent Management," May 1, 2020, 6, https://www.jcs.mil/Portals/36/Documents/Doctrine/education/jcs_pme_tm_vision.pdf?ver=2020-05-15-102429-817. A former secretary of defense and a former vice chairman of the Joint Chiefs of Staff outlined the ideas behind this and other policy changes on wargaming in Robert Work and Paul Selva, "Revitalizing Wargaming Is Necessary to Be Prepared for Future Wars," *War on the Rocks*, December 8, 2015, https://warontherocks.com/2015/12/revitalizing-wargaming-is-necessary-to-be-prepared-for-future-wars/.

61. For a fascinating essay on this matter, see Hyokwon Jung, "A Glimpse into the Future Battlefield with AI-Embedded Wargames," *U.S. Naval Institute Proceedings* 150, no. 6 (2024), https://www.usni.org/magazines/proceedings/2024/june/glimpse-future-battlefield-ai-embedded-wargames.

62. For this argument, see Sebastian J. Bae, "Put Educational Wargaming in the Hands of the Warfighter," *War on the Rocks*, July 13, 2023, https://warontherocks.com/2023/07/put-educational-wargaming-in-the-hands-of-the-warfighter/.

63. For an essay that focuses on these limits, see Marcel Plichta, "You Think It's a Game? What Video Games Can—and Can't—Teach About Strategy and History," Modern War Institute (at West Point), October 8, 2021, https://mwi.westpoint.edu/you-think-its-a-game-what-video-games-can-and-cant-teach-about-strategy-and-history/.

64. Gowipy's quote is from "NAWCTSD's Mobile NEST Expands Navy Training Capabilities," Naval Air Systems Command, April 18, 2025, https://www.navair.navy.mil/news/NAWCTSDs-Mobile-NEST-Expands-Navy-Training-Capabilities/Fri-04182025-0959.

65. Christiansen's quote is also from "NAWCTSD's Mobile NEST Expands Navy Training Capabilities."

66. Claudia Sanchez-Bustamante, "Why Today's 'Gen Z' Is at Risk for Boot Camp Injuries," Defense Visual Information Distribution Service (DVIDS), February 22, 2022, https://web.archive.org/web/20220222221823/https://www.dvidshub.net/news/414340/why-todays-gen-z-risk-boot-camp-injuries.

67. Perez quote taken from the following Navy press release: Bobby Cummings, "Video Games Can Enhance Warrior Cognitive Performance," US Navy website, February 17, 2022, https://www.navy.mil/Press-Office/News-Stories/Article/2938975/video-games-can-enhance-warrior-cognitive-performance/.

68. See Sebastian J. Bae, ed., *Forging Wargamers: A Framework for Wargaming Education* (Marine Corps University Press, 2022).

69. Patrick Molenda, "Winning the Next War—Virtually," *U.S. Naval Institute Proceedings* 149, no. 7 (2023), https://www.usni.org/magazines/proceedings/2023/july/winning-next-war-virtually.

70. Nick Moran and Arnel P. David, "Why Gamers Will Win the Next War," Modern War Institute (at West Point), June 30, 2022, https://mwi.usma.edu/why-gamers-will-win-the-next-war/.

71. Moran and David, "Why Gamers Will Win the Next War."

72. This is not to say that US military leaders haven't devoted considerable thought to this work of fiction. To give just one example of their interest, it remained on the US Marine Corps commandant's annual suggested professional reading list for many years from 1988 onward. See "Call to Action: Ender's Game," *Marine Corps Association* (blog), June 10, 2019, https://mca-marines.org/blog/2019/06/10/call-to-action-enders-game/. The commandant's Professional Reading Program (including the suggested lists for past years) can be found at "Research Guides: CMC Professional Reading Program 2020: Home," accessed September 2, 2022, https://grc-usmcu.libguides.com/usmc-reading-list-2020/home. For a detailed study on talent management in the context of the skills embodied by Ender, see Susan F. Bryant and Andrew Harrison, *Finding Ender: Exploring the Intersections of Creativity, Innovation, and Talent Management in the U.S. Armed Forces*, No. 31, Strategic Perspectives, edited by Thomas F. Lynch III (Center for Strategic Research, Institute for National Strategic Studies, National Defense University, 2019).

73. Moran and David, "Why Gamers Will Win the Next War."

74. Daher quoted in CNN, "Hezbollah Video Game: War with Israel," August 16, 2007, http://edition.cnn.com/2007/WORLD/meast/08/16/hezbollah.game.reut/.

75. Ruth Sherlock, "Hezbollah Designed a Video Game to Appeal to the U.S.," *NPR*, July 16, 2018, https://www.npr.org/2018/07/16/629588453/hezbollah-designed-a-video-game-to-appeal-to-the-u-s.

76. See CNN International, "Web Video Game Aim: 'Kill' Bush Characters," September 18, 2006, http://edition.cnn.com/2006/WORLD/meast/09/18/bush.game/.

77. See Adam Clark Estes, "Syrian Rebels Now Have a Tank Powered by a Playstation Controller," The Atlantic, December 10, 2012, https://www.theatlantic.com/international/archive/2012/12/syrian-rebels-now-have-tank-powered-playstation-controller/320687/.

78. See Nicolas Schillinger, "Playing Soldiers: The War Game in Late Qing and Republican China," *Journal of Chinese Military History* 9, no. 1 (2020): 38–64, https://doi.org/10.1163/22127453-BJA10003.

79. See Zhang Zihan, "Lock and Load," *Global Times*, October 7, 2012, https://www.globaltimes.cn/content/736758.shtml; see also CNN, "Chinese Online Game Lets Players Fight Japan over Disputed Islands," August 5, 2013, https://www.cnn.com/2013/08/05/world/asia/china-online-game-fight-japan-diaoyu-senkaku.

80. See Elsa Kania and Ian Burns McCaslin, *Learning Warfare from the Laboratory: China's Progression in Wargaming and Opposing Force Training*, Military Learning and The Future of War Project (Institute for the Study of War, September 2021), https://www.understandingwar.org/report/learning-warfare-laboratory-china%E2%80%99s-progression-wargaming-and-opposing-force-training; Ryan D. Martinson, "The PLA Navy's Blue Team Center Games for War," *U.S. Naval Institute* 150, no. 5 (May 2024), https://www.usni.org/magazines/proceedings/2024/may/pla-navys-blue-team-center-games-war.

81. An insightful commentary from this perspective is offered in Adam Elkus, "War Is a Video Game, and We're Losing," *War on the Rocks*, March 20, 2014, https://warontherocks.com/2014/03/war-is-a-video-game-and-were-losing/.

82. Brian Kerg, "Maneuver Chess: Integrating Think-and-Fight Drills into Military Training," *U.S. Naval Institute Proceedings* 150, no. 11 (November 2024), https://www.usni.org/magazines/proceedings/2024/november/maneuver-chess -integrating-think-and-fight-drills-military.

Chapter 4

1. See Linda Grant De Pauw, "Women in Combat: The Revolutionary War Experience," *Armed Forces and Society* 7, no. 2 (1981): 209–26; and Holly A. Mayer, "Bearing Arms, Bearing Burdens: Women Warriors, Camp Followers and Home-Front Heroines of the American Revolution," in *Gender, War and Politics: Transatlantic Perspectives, 1775–1830*, ed. Karen Hagemann et al., War, Culture and Society, 1750–1850 (Palgrave Macmillan UK, 2010), 169–87, https://doi.org/10.1057 /9780230283046_9.

2. See DeAnne Blanton and Lauren M. Cook, *They Fought Like Demons: Women Soldiers in the American Civil War* (Louisiana State University Press, 2002).

3. On these contributions, see Susan Zeiger, *In Uncle Sam's Service: Women Workers with the American Expeditionary Force, 1917–1919* (University of Pennsylvania Press, 2004); Evelyn Monahan and Rosemary Neidel-Greenlee, *A Few Good Women: America's Military Women from World War I to the Wars in Iraq and Afghanistan* (Knopf Doubleday Publishing Group, 2010); and Lynn Dumenil, *The Second Line of Defense: American Women and World War I* (University of North Carolina Press, 2017).

4. See Elizabeth Cobbs, *The Hello Girls: America's First Women Soldiers* (Harvard University Press, 2017).

5. On these contributions, begin with the excellent Lena S. Andrews, *Valiant Women: The Extraordinary American Servicewomen Who Helped Win World War II* (HarperCollins, 2023). See also Doris Weatherford, *American Women and World War II* (Facts on File, 1990); and Margaret Vining, "Women Join the Armed Forces: The Transformation of Women's Military Work in World War II and After (1939–1947)," in *A Companion to Women's Military History*, ed. Barton Hacker and Margaret Vining, History of Warfare 74 (Brill, 2012), 233–89, https://doi.org/10.1163/9789004206823 _008.

6. On the evolving place of women in America's armed forces during this period, see Tanya L. Roth, *Her Cold War: Women in the U.S. Military, 1945–1980* (University of North Carolina Press, 2021).

7. An excellent overview of the history of this transition can be found in William A. Taylor, "From WACs to Rangers: Women in the U.S. Military Since World War II," *Marine Corps University Journal*, Special Issue: Gender Integration (2018): 78–101.

8. For a sophisticated analysis of the cultural construction of the American military body as a masculine form, see Christina S. Jarvis, *The Male Body at War: American Masculinity During World War II* (Northern Illinois University Press, 2004). See also Aaron B. O'Connell, *Underdogs: The Making of the Modern Marine Corps* (Harvard University Press, 2012), particularly pp. 212–222; and Mark Ryland Folse,

The Globe and Anchor Men: U.S. Marines and American Manhood in the Great War Era, Modern War Studies (University Press of Kansas, 2024). For a work on the image of women in the armed forces that both reinforced and questioned the masculine military norm, see Sarah E. Patterson, "'Beauty Isn't Prerequisite for Girl Marines': Images of Female Marines During World War II," *Marine Corps History* 8, no. 1 (2022): 5–20, https://doi.org/10.35318/mch.2022080101. A fantastic recent essay on the two sides of the debate on women in combat is provided in M. Alejandra Parra-Orlandoni, "Women in the Ranks, but Not in the Clear," War on the Rocks, May 29, 2025, https://warontherocks.com/2025/05/women-in-the -ranks-but-not-in-the-clear/.

9. See Melissa T. Brown, *Enlisting Masculinity: The Construction of Gender in US Military Recruiting Advertising During the All-Volunteer Force*, Oxford Studies in Gender and International Relations (Oxford University Press, 2012).

10. For such an argument, see Leisa D. Meyer, *Creating GI Jane: Sexuality and Power in the Women's Army Corps During World War II* (Columbia University Press, 1996).

11. Among the branches, the US Marine Corps has been particularly active in making this argument. See the following summary of a 2015 Marine Corps study on the subject: US Marine Corps, "Marine Corps Force Integration Plan—Summary," September 2015, https://wiisglobal.org/wp-content/uploads/2013/05/Research-Sum mary-Final_9Sept15.pdf. For a broader assessment of the matter, consult Rebecca Jensen, "Opening Marine Infantry to Women: A Civil-Military Crisis?," *Marine Corps University Journal*, Special Issue: Gender Integration and the Military (2018): 132–55.

12. The higher musculoskeletal rates for women in the military are outlined in Bradley C. Nindl et al., "Operational Physical Performance and Fitness in Military Women: Physiological, Musculoskeletal Injury, and Optimized Physical Training Considerations for Successfully Integrating Women into Combat-Centric Military Occupations," *Military Medicine* 181, no. suppl_1 (January 2016): 50–62, https://doi.org/10.7205/MILMED-D-15-00382. On the argument that women perform less well in markmanship evaluations, see Dan Lamothe, "Marine Experiment Finds Women Get Injured More Frequently, Shoot Less Accurately Than Men," *Washington Post*, September 10, 2015, https://www.washingtonpost.com /news/checkpoint/wp/2015/09/10/marine-experiment-finds-women-get-injured -more-frequently-shoot-less-accurately-than-men/.

13. For one example of such critiques, see Heather MacDonald, "Women Don't Belong in Combat Units," Opinion, *Wall Street Journal*, January 13, 2019, https://www.wsj .com/articles/women-dont-belong-in-combat-units-11547411638.

14. See Daniel Mon-López et al., "Recent Changes in Women's Olympic Shooting and Effects in Performance," *PLOS ONE* 14, no. 5 (May 13, 2019): e0216390, https://doi .org/10.1371/journal.pone.0216390; and Daniel Mon-López et al., "Pistol and Rifle Performance: Gender and Relative Age Effect Analysis," *International Journal of Environmental Research and Public Health* 17, no. 4 (February 2020): 1365, https:// doi.org/10.3390/ijerph17041365.

15. Such is the testimony of former US Marine Lieutenant Colonel Kate Germano, an officer who greatly improved the performance of the women under her command

in the Marine Corps's Fourth Recruit Training Battalion. See Kate Germano and Kelly Kennedy, *Fight Like a Girl: The Truth Behind How Female Marines Are Trained* (Prometheus Books, 2018).

16. For a commentary on such historical examples, see Mary Raum, "Warrior Women: 3,000 Years in the Fight," *Joint Force Quarterly* 93 (Quarter 2019): 47–54.

17. See Øyvind Sandbakk et al., "Sex Differences in World-Record Performance: The Influence of Sport Discipline and Competition Duration," *International Journal of Sports Physiology and Performance* 13, no. 1 (2018): 2–8, https://doi.org/10.1123/ijspp .2017-0196; Petr Schlegel and Adam Křehký, "Performance Sex Differences in CrossFit®," *Sports* 10, no. 11 (2022): 165, https://doi.org/10.3390/sports10110165; and Kevin G. Keenan et al., "Girls in the Boat: Sex Differences in Rowing Performance and Participation," *PLOS ONE* 13, no. 1 (January 19, 2018): e0191504, https://doi.org /10.1371/journal.pone.0191504.

18. For a clarification of the various personnel roles within special operations forces (and the many contributions of women in them), see Nicole Alexander and Lyla Kohistany, "Dispelling the Myth of Women in Special Operations," Center for a New American Security, March 19, 2019, https://www.cnas.org/publications /commentary/dispelling-the-myth-of-women-in-special-operations. The unique skills that women bring to the table in Special Operations Forces are discussed in Frank Gasca et al., "Unique Capabilities of Women in Special Operations Forces," *Special Operations Journal* 1, no. 2 (2015): 105–11, https://doi.org/10.1080/23296151 .2015.1070613.

19. See David Walton, "Are Green Berets Turning Pink? Integrating Women into Special Forces," War on the Rocks, September 20, 2022, https://warontherocks.com /2022/09/are-green-berets-turning-pink-integrating-women-into-special-forces/.

20. See US Army Public Affairs, "Army Establishes New Fitness Test of Record to Strengthen Readiness and Lethality," US Army website, April 20, 2025, https:// www.army.mil/article/284799/army_establishes_new_fitness_test_of_record_to _strengthen_readiness_and_lethality.

21. See Walter T. Delpero et al., "Aviation-Relevant Epidemiology of Color Vision Deficiency," *Aviation, Space, and Environmental Medicine* 76, no. 2 (2005): 128–29.

22. See M. P. Murphy and G. A. Gates, "Hearing Loss: Does Gender Play a Role?," *Medscape Women's Health* 2, no. 10 (1997): 2; Karl E. Friedl, "Biomedical Research on Health and Performance of Military Women: Accomplishments of the Defense Women's Health Research Program (DWHRP)," *Journal of Women's Health* 14, no. 9 (November 2005): 793, https://doi.org/10.1089/jwh.2005.14.764; Gijsbert Stoet et al., "Are Women Better Than Men at Multi-Tasking?," *BMC Psychology* 1, no. 1 (2013): 1–18, https://doi.org/10.1186/2050-7283-1-18.

23. See Mark Thompson, "Army Women: Better Chopper Pilots Than the Guys?," *Time*, February 17, 2014, https://time.com/8404/army-women-helicopter-pilots/. Other data suggests that female pilots may, on the other hand, be associated with more dangerous accidents than their male counterparts; see Scott Burgess et al., "Characteristics of Helicopter Accidents Involving Male and Female Pilots," *International Journal of Aviation, Aeronautics, and Aerospace* 5, no. 2 (2018), https://doi .org/10.15394/ijaaa.2018.1216.

24. See Aaron M. White, "Gender Differences in the Epidemiology of Alcohol Use and Related Harms in the United States," *Alcohol Research: Current Reviews* 40, no. 2 (2020): 01, https://doi.org/10.35946/arcr.v40.2.01; and Allison Milner et al., "Shifts in Gender Equality and Suicide: A Panel Study of Changes over Time in 87 Countries," *Journal of Affective Disorders* 276 (2020): 495–500, https://doi.org/10.1016/j.jad.2020.07.105.

25. See National Institute of Diabetes and Digestive and Kidney Diseases, "Overweight & Obesity Statistics," accessed March 16, 2024, https://www.niddk.nih.gov/health-information/health-statistics/overweight-obesity.

26. For an insightful article on moderating the disadvantages that women might hold (as well as how to best leverage their advantages), see Katherine Kuzminski, "The Pentagon's New Opportunity to Boost Readiness Among Female Troops," *Military Times*, November 22, 2023, https://www.militarytimes.com/opinion/2023/11/22/the-pentagons-new-opportunity-to-boost-readiness-among-female-troops/.

27. For a fascinating case study on this issue within US airborne forces during the Second World War, see R. F. M. Williams, "'Our Problem Children': Masculinity and Its Discontents in American Parachute Units in World War II," *Journal of Military History* 87, no. 3 (2023): 675–702. The complexities of masculinity within the American military are thoughtfully handled in Aaron Belkin, *Bring Me Men: Military Masculinity and the Benign Facade of American Empire, 1898–2001* (Oxford University Press, 2012).

28. For an overview of the positive characteristics that women bring to direct combat units, see Ellen L. Haring, "What Women Bring to the Fight," *The US Army War College Quarterly: Parameters* 43, no. 2 (2013): 27–32.

29. See Anita Woolley and Thomas Malone, "What Makes a Team Smarter? More Women," *Harvard Business Review* 89, no. 6 (June 2011): 32–33.

30. See Kris Gonzalez, "Soldiers Bridge Cultural Gap: Female Interaction Critical to Deploying Units' Success," US Army website, March 11, 2010, https://www.army.mil/article/35644/soldiers_bridge_cultural_gap_female_interaction_critical_to_deploying_units_success; and Ginger E. Beals, "Women Marines in Counterinsurgency Operations: Lioness and Female Engagement Teams" (master's thesis, US Marine Corps Command and Staff College, Marine Corps University, 2010), https://apps.dtic.mil/sti/pdfs/ADA604399.pdf.

31. See Michele Norris, "Roles for Women in U.S. Army Expand," *NPR*, October 1, 2007, https://www.npr.org/2007/10/01/14869648/roles-for-women-in-u-s-army-expand.

32. See Maya Yang, "How Women Played Crucial Roles in Iraq—and Changed US Military Forever," *The Guardian*, March 21, 2023, https://www.theguardian.com/world/2023/mar/21/iraq-war-us-military-women.

33. See Gayle Tzemach Lemmon, "The Women of the Army Rangers' Cultural Support Teams," *At War, Notes from the Front Lines—New York Times* (blog), September 14, 2015, https://archive.nytimes.com/atwar.blogs.nytimes.com/2015/09/14/army-rangers-cultural-support-teams/. For a book-length narrative of the subject by the same author, see Gayle Tzemach Lemmon, *Ashley's War: The Untold Story of a Team of Women Soldiers on the Special Ops Battlefield* (HarperCollins, 2015).

Another insightful text—one based on oral histories and interviews—is James E. Wise Jr. and Scott Baron, *Women at War: Iraq, Afghanistan, and Other Conflicts* (Naval Institute Press, 2006).

34. Manning quote from Yang, "How Women Played Crucial Roles in Iraq." The involvement in this way of women in the American military has not been without critique. See the following for an analysis that called such women's involvement a part of a larger US unjust war: Jennifer Greenburg, *At War with Women: Military Humanitarianism and Imperial Feminism in an Era of Permanent War* (Cornell University Press, 2023).

35. See David Barno and Nora Bensahel, "Addressing the U.S. Military Recruiting Crisis," War on the Rocks, March 10, 2023, https://warontherocks.com/2023/03/addressing-the-u-s-military-recruiting-crisis/.

36. See Kelly Laco, "Republicans Move to Strike Dem Proposal to Include Women in the Military Draft" (Fox News, October 12, 2022), https://www.foxnews.com/politics/republicans-move-kill-dem-proposal-include-women-military-draft. For analyses on why women should be included in the draft, see Sydney Frankenberg and Hallie Lucas, "Draft Us Too, America," *U.S. Naval Institute Proceedings* 147, no. 6 (June 2021), https://www.usni.org/magazines/proceedings/2021/june/draft-us-too-america; Jordan Foley, "The Supreme Court Should End Male-Only Selective Service Registration," *U.S. Naval Institute Proceedings* 147, no. 5 (May 2021), https://www.usni.org/magazines/proceedings/2021/may/supreme-court-should-end-male-only-selective-service-registration; and Kelly M. Dickerson, "Protecting Posterity: Women and the Military Selective Service Act," *Military Review: The Professional Journal of the U.S. Army* 102, no. 5 (2022): 52–60.

37. See Benjamin Ramsey et al., "Retaining Female Leaders: A Key Readiness Issue," *Joint Force Quarterly* 104 (Quarter 2022): 81–88; Jatin G. Khona et al., "Retention of Women in the Naval Officer Corps," *U.S. Naval Institute* 148, no. 3 (2022), https://www.usni.org/magazines/proceedings/2022/march/retention-women-naval-officer-corps; and Natalia Best, "Integrating Women into the Long Blue Line," *U.S. Naval Institute Proceedings* 148, no. 7 (2022), https://www.usni.org/magazines/proceedings/2022/july/integrating-women-long-blue-line.

38. See US Department of Defense, *Department of Defense Annual Report on Sexual Assault in the Military: Fiscal Year 2022* (US Department of Defense, 2023), https://www.sapr.mil/sites/default/files/public/docs/reports/AR/FY22/DOD_Annual_Report_on_Sexual_Assault_in_the_Military_FY2022.pdf. On the problem of gender bias within the military in particular, consult Elizabeth M. Trobaugh, "Women, Regardless: Understanding Gender Bias in U.S. Military Integration," *Joint Force Quarterly* 88 (Quarter 2018): 46–53.

39. See Jim McClure, "Flight of the Phoenix," *U.S. Naval Institute Proceedings* 148, no. 9 (September 2022), https://www.usni.org/magazines/proceedings/2022/september/flight-phoenix; Felicia Sonmez, "Adm. Linda Fagan Becomes First Woman to Lead U.S. Coast Guard," *Washington Post*, June 1, 2022, https://www.washingtonpost.com/politics/2022/06/01/adm-linda-fagan-becomes-first-woman-lead-us-coast-guard/; and Niha Masih, "Biden Nominates Adm. Lisa Franchetti as First Woman to Lead U.S. Navy," *Washington Post*, July 22, 2023,

https://www.washingtonpost.com/national-security/2023/07/22/lisa-franchetti
-navy-biden-nomination/.

40. Burke's statement is quoted in Ben Werner, "Military Branches Are Doing More to Recruit Women into Active Duty," *US Naval Institute (USNI) News*, April 13, 2018, https://news.usni.org/2018/04/13/service-branches-want-more-women.

41. On the integration of women within the American military services, see Brenda Oppermann, "Women and Gender in the U.S. Military: A Slow Process of Integration," in *Women and Gender Perspectives in the Military: An International Comparison*, ed. Robert Egnell and Mayesha Alam (Georgetown University Press, 2019), 113–40.

42. On opposition to women in the nation's combat arms branches (specifically within the US Marine Corps), see Dan Lamothe, "How Big Is Opposition to Women in Combat Units Among Marines? This Report Explains," *Washington Post*, March 10, 2016, https://www.washingtonpost.com/news/checkpoint/wp/2016/03 /10/survey-details-depth-of-opposition-in-marine-corps-to-allowing-women-in -combat-jobs/. On Griest's accomplishments, see Michelle Tan, "Meet the Army's First Female Infantry Officer," *Army Times*, April 27, 2016, https://www.armytimes .com/news/your-army/2016/04/27/meet-the-army-s-first-female-infantry-officer/.

43. See Keith R. Boydston, "Fort Benning Graduates First Women Armor Officers," US Army website, December 2, 2016, https://www.army.mil/article/179006/fort _benning_graduates_first_women_armor_officers.

44. See Merrit Kennedy, "First Female Marine Completes Grueling Infantry Officer Course," *NPR*, September 25, 2017, https://www.npr.org/sections/thetwo-way /2017/09/25/553494258/first-female-marine-completes-grueling-infantry-officer -course.

45. See Gayle Tzemach Lemmon, "No Turning Back?: First Woman Makes Army's Elite 75th Ranger Regiment, Big Step for Women in Combat," *Defense One*, January 19, 2017, https://www.defenseone.com/ideas/2017/01/no-turning-back -first-woman-makes-armys-elite-75th-ranger-regiment-big-step-women-combat /134691/.

46. See Associated Press, "First Woman Completes Navy Special Warfare Training," NBC News, July 15, 2021, https://www.nbcnews.com/politics/politics-news/first -woman-completes-navy-special-warfare-training-n1274125.

47. See Joe Legros, "Sky Soldier Makes History as First Active-Duty Female Army Sniper," US Army website, December 5, 2023, https://www.army.mil/article/272115 /sky_soldier_makes_history_as_first_active_duty_female_army_sniper.

48. US Secretary of Defense Ash Carter to Acting Under Secretary of Defense for Personnel and Readiness, Chiefs of the Military Services, and Commander, US Special Operations Command, "Implementation Guidance for the Full Integration of Women in the Armed Forces," Memorandum for Secretaries of the Military Departments, December 3, 2015, https://dod.defense.gov/Portals/1/Documents /pubs/OSD014303-15.pdf.

49. Scales's observation on the "culture of the rank and file" is made in Robert H. Scales, *Scales on War: The Future of America's Military at Risk* (Naval Institute Press, 2016), 210. Moreover, the entirety of chapter 21 of this text deserves reading on the issue of women in direct combat units.

50. On this point, see Anthony King, "On Combat Effectiveness in the Infantry Platoon: Beyond the Primary Group Thesis," *Security Studies* 25, no. 4 (2016): 699–728, https://doi.org/10.1080/09636412.2016.1220205.

51. On this point, see Heather Venable, "How the Few Become the Valued: Improving Recruitment and Retention of Female Marines," *U.S. Naval Institute Proceedings* 146, no. 10 (2020), https://www.usni.org/magazines/proceedings/2020/october/how-few-become-valued-improving-recruitment-and-retention-female.

52. See Mary E. Gross et al., *Flight Suit Sizes for Women*, Final Report (US Air Force Research Laboratory, 2020).

53. This point is outlined in Mona Mamiche, "Adapting Military Equipment to Account for Gender Differences," NATO Association of Canada, October 10, 2019, https://natoassociation.ca/adapting-military-equipment-to-account-for-gender-differences/.

54. See Ema O'Connor and Vera Bergengruen, "Military Doctors Told Them It Was Just 'Female Problems' Weeks Later, They Were in the Hospital.," BuzzFeed News, March 8, 2019, https://www.buzzfeednews.com/article/emaoconnor/woman-military-doctors-female-problems-health-care.

55. On the need for appropriate gear for women, see M. Josephine Hessert, "Needed: Body Armor Built for Women," *U.S. Naval Institute Proceedings* 149, no. 6 (2023), https://www.usni.org/magazines/proceedings/2023/june/needed-body-armor-built-women. On the potential utility of optimized physical training for women and combat, see Nindl et al., "Operational Physical Performance and Fitness in Military Women."

56. On the trend of excluding women from human performance research, see Nash Anderson et al., "Under-Representation of Women Is Alive and Well in Sport and Exercise Medicine: What It Looks like and What We Can Do About It," *BMJ Open Sport & Exercise Medicine* 9, no. 2 (2023): e001606, https://doi.org/10.1136/bmjsem-2023-001606. On these attempts, see Kimberly Hefling, "Army Tests New Body Armor for Women," NBC News, April 22, 2011, https://www.nbcnews.com/id/wbna42721218; Ernesto Londoño, "A Decade into War, Body Armor Gets Curves," *Washington Post*, September 20, 2012, https://www.washingtonpost.com/world/national-security/a-decade-into-war-body-armor-gets-curves/2012/09/20/ffaa06c4-0353-11e2-91e7-2962c74e7738_story.html; Janelle Nanos, "Armor All: New Body Armor Issued for Women in the Military," *Boston Magazine*, September 26, 2013, https://www.bostonmagazine.com/news/2013/09/26/new-body-armor-women-military/; Kyle Jahner, "Start-Up May Revolutionize Fit of Body Armor, Other Gear," *Army Times*, December 10, 2014, https://www.armytimes.com/news/your-army/2014/12/10/start-up-may-revolutionize-fit-of-body-armor-other-gear/; and Travis J. Tritten, "Form-Fitted Body Armor Rolling Out as Combat Roles Expand for Women," *Stars and Stripes*, March 4, 2016, https://www.stripes.com/news/form-fitted-body-armor-rolling-out-as-combat-roles-expand-for-women-1.397554. A summary of the entirety of this history up until October 2020 is provided in Haley Britzky, "The Military Says It's Finally Designing Body Armor for Women for the Umpteenth Time This Decade," *Task & Purpose*, October 16, 2020, https://taskandpurpose.com/news/women-combat-gear-problem/.

57. Defense Advisory Committee on Women in the Services, *Executive Summary* (Defense Advisory Committee on Women in the Services, 2018), 3, https:// dacowits.defense.gov/Portals/48/Documents/Reports/2018/Annual%20Report /DACOWITS%20ES%202018.pdf.

58. See Richard Sisk, "Dunford Pushes Services to Move Faster on Body Armor Fitted for Women," Military.com, May 14, 2018, https://www.military.com/kitup /daily-news/2018/05/14/dunford-pushes-services-move-faster-body-armor-fitted -women.html.

59. See Jesse Rifkin, "Body Armor for Females Modernization Act Would Help Military Women's Uniforms and Protective . . . ," GovTrack Insider, January 13, 2020, https://govtrackinsider.com/body-armor-for-females-modernization-act-would -help-military-womens-uniforms-and-protective-23d66ea3853f.

60. See Daniel Boffey, "'Fighting Two Enemies': Ukraine's Female Soldiers Decry Harassment," *The Guardian*, August 4, 2023, https://www.theguardian.com/world /2023/aug/04/fighting-two-enemies-ukraine-female-soldiers-decry-harassment.

61. Reznikov's Twitter statement can be found at: "Oleksii Reznikov on Twitter," Twitter, August 5, 2023, https://twitter.com/oleksiireznikov/status/1687859613289349126. See also Nathan Rennolds, "Ukraine Unveils Military Uniforms for Women After Complaints That Ill-Fitting Men's Clothes Hindered Them on the Frontline," *Business Insider*, August 6, 2023, https://www.businessinsider.com/ukraine-russia -army-approves-female-military-uniforms-first-time-2023-8.

62. Goldfein quote from Bailee A. Darbasie, "New Direction for Female-Specific Flight Equipment," Air Force website, April 1, 2019, https://www.af.mil/News/Article -Display/Article/1801680/new-direction-for-female-specific-flight-equipment/.

63. See Todd South, "Here's the New Gear Being Readied by the Army's One-Stop-Shop for All Things Soldier," *Army Times*, October 11, 2021, https://www.armytimes .com/news/2021/10/11/heres-the-new-gear-being-readied-by-the-armys-one-stop -shop-for-all-things-soldier/.

64. See Sarah Sicard, "The Army Is Working on a Tactical Bra," *Military Times*, August 3, 2022, https://www.militarytimes.com/off-duty/military-culture/2022 /08/03/the-army-is-working-on-a-tactical-bra/; and Patricia Marx, "Is the Army's New Tactical Bra Ready for Deployment?," *The New Yorker*, June 19, 2023, https:// www.newyorker.com/magazine/2023/06/26/is-the-armys-new-tactical-bra-ready -for-deployment.

65. See Hope Hodge Seck, "Navy Developing Smart Bra, Undershirt for Fighter Pilots," Navy Times, December 27, 2022, https://www.navytimes.com/news/your -navy/2022/12/27/navy-developing-smart-bra-undershirt-for-fighter-pilots/.

66. See chapter 6, "A Cultural History of the Sports Bra," in Jaime Schultz, *Qualifying Times: Points of Change in U.S. Women's Sport* (University of Illinois Press, 2014), 149–66.

67. US Army Special Operations Command, "Breaking Barriers: Women in Army Special Operations" (US Army Special Operations Command, Department of the Army, 2021), 27, https://www.soc.mil/wia/women-in-arsof-report-2023.pdf.

68. These matters are discussed in some depth in Andrea M. Peters et al., "Rethinking Female Urinary Devices for the US Army," *The US Army War College Quarterly:*

Parameters 52, no. 1 (2022): 41–56, https://doi.org/10.55540/0031-1723.3128. On the issue for women in Special Operations Forces, see US Army Special Operations Command, "Breaking Barriers," 27.

69. Quotes from Rebecca Sirimarco-Lang, "SJAFB Tests New In-Flight Bladder Relief System," Air Combat Command, March 20, 2023, https://www.acc.af.mil/News /Article-Display/Article/3334916/sjafb-tests-new-in-flight-bladder-relief-system /. See also Thomas Novelly, "Female Air Force Pilots Would Be Able to Safely Pee In-Flight During Long Missions with New Tech Being Tested," Military.com, March 20, 2023, https://www.military.com/daily-news/2023/03/20/female -air-force-pilots-would-be-able-safely-pee-flight-during-long-missions-new-tech -being-tested.html; and Patty Nieberg, "Air Force Pilots Get a New Way to Pee at 30,000 Feet," *Task & Purpose*, April 17, 2025, https://taskandpurpose.com/news /air-force-pilots-pee-while-flying/.

70. See Rachel N. Weber, "Accommodating Difference: Gender and Cockpit Design in Military and Civilian Aviation," *Transportation Research Record* 1480 (1995): 51–56; and Rachel N. Weber, "Manufacturing Gender in Commercial and Military Cockpit Design," *Science, Technology, & Human Values* 22, no. 2 (April 1, 1997): 235–53, https://doi.org/10.1177/016224399702200204.

71. See Valerie Insinna, "To Get More Female Pilots, the Air Force Is Changing the Way It Designs Weapons," *Air Force Times*, August 19, 2020, https://www .airforcetimes.com/news/your-air-force/2020/08/19/to-get-more-female-pilots -the-air-force-is-changing-the-way-it-designs-weapons/. Roper's statement is also quoted in this article.

72. See John M. Donnelly, "F-35 Ejection Seats Could Endanger Many Pilots," *Roll Call*, October 16, 2015, https://www.rollcall.com/2015/10/16/exclusive-f-35-ejection -seats-could-endanger-many-pilots/.

73. Knapp quotes from Heather Mongilio, "USNI News Video: Naval Aviation Breaks Out the Tape Measure to Size Up a New Generation of Pilots," *US Naval Institute (USNI) News*, May 17, 2023, https://news.usni.org/2023/05/17/usni-news-video-naval -aviation-breaks-out-the-tape-measure-to-size-up-a-new-generation-of-pilots.

74. See the following reports on the issue by the Center for a New American Security think tank: Lauren Fish and Paul Scharre, *The Soldier's Heavy Load*, Super Soldiers (Center for a New American Security, 2018), https://www.cnas.org/publications /reports/the-soldiers-heavy-load-1; and Lauren Fish and Paul Scharre, *Soldier Protection Today*, Super Soldiers (Center for a New American Security, 2018), https://www.cnas.org/publications/reports/super-soldiers-1.

75. See James King, "The Overweight Infantryman," Modern War Institute, January 10, 2017, https://mwi.usma.edu/the-overweight-infantryman/.

76. See Ellen Haring, "Marines' Requirements for Infantry Officers Are Unrealistic, Army Colonel Says," *Marine Corps Times*, October 15, 2016, https://www .marinecorpstimes.com/opinion/2016/10/15/marines-requirements-for-infantry -officers-are-unrealistic-army-colonel-says/.

77. Fish and Scharre, *The Soldier's Heavy Load*, 2.

78. Linell Letendre, "Women Warriors: Why the Robotics Revolution Changes the Combat Equation," *PRISM: The Journal of Complex Operations* 6, no. 1 (2016): 91–103.

79. For one such argument, see Leanne K. Simpson, "Eight Myths About Women on the Military Frontline—and Why We Shouldn't Believe Them," *The Conversation*, April 1, 2016, http://theconversation.com/eight-myths-about-women-on-the-military-frontline-and-why-we-shouldnt-believe-them-55594. On the 1960s roots of military exoskeletons, see "News Notes: Future Soldier-Human Tank," *Armor* 70, no. 3 (1961): 54–55.

80. Olson quote from Hope Hodge Seck, "Marine Corps Shelves Futuristic Robo-Mule Due to Noise Concerns," Military.com, December 22, 2015, https://www .military.com/daily-news/2015/12/22/marine-corps-shelves-futuristic-robo-mule -due-to-noise-concerns.html.

81. This quote is from the following article in the US naval services' landmark professional journal: Walker J. Mills and Christopher Booth, "Marines Need a Few Good Mules," *U.S. Naval Institute Proceedings* 148, no. 4 (2022), https://www.usni .org/magazines/proceedings/2022/april/marines-need-few-good-mules.

82. See ITV News, *Robots Join Ukraine's Soldiers in the Battle to Rescue the Dead and Shield the Living*, 2023, https://www.youtube.com/watch?v=ye_OlAatH8I.

83. See Todd South, "Marines Eye Tactical Resupply Drone Prototypes," *Marine Corps Times*, May 10, 2021, https://www.marinecorpstimes.com/news/your-marine -corps/2021/05/10/marines-eye-tactical-resupply-drone-prototypes/; and Warren Duffie Jr., "Special Delivery: New Autonomous Flight Technology Means Rapid Resupply for Marines," Office of Naval Research, December 13, 2017, https://www .nre.navy.mil/media-center/news-releases/special-delivery-new-autonomous -flight-technology-means-rapid-resupply.

84. See Todd South, "New in 2019: The Army, Marines Will Put Four Robotic Gear Mules to the Test in Rugged, Austere Conditions," *Army Times*, January 2, 2019, https://www.armytimes.com/news/your-army/2019/01/02/new-in-2019-the -army-marines-will-put-four-robotic-gear-mules-to-the-test-in-rugged-austere -conditions/.

85. See Jared Keller, "Soldiers Are Finally Getting a 'Robotic Mule' to Haul All Their Heavy Sh*t for Them," *Task & Purpose*, November 6, 2022, https://taskandpurpose .com/tech-tactics/army-s-met-robotic-mule-fielding/.

86. See Jared Keller, "This Video of a Drone Airdropping an Armed Robot Dog Is the Stuff of Dystopian Nightmares," *Task & Purpose*, October 27, 2022, https:// taskandpurpose.com/tech-tactics/robot-dog-machine-gun-drone-airdrop -kestrel-defense-video/.

87. See Laura Heckmann, "Upgraded Kevlar to Lighten Loads for Troops," *National Defense*, 107, no. 835 (2023): 10.

88. See Hope Seck, "MARSOC Raiders to Deploy in Smaller, Tech-Loaded Teams as Conflict Gets More Complex," Sandboxx, July 17, 2023, https://www.sandboxx.us /blog/marsoc-raiders-to-deploy-in-smaller-tech-loaded-teams-as-conflict-gets -more-complex/.

89. See "MDM 22—Tomahawk Robotics RAID Plate," *Soldier Systems: An Industry Daily*, May 11, 2022, https://soldiersystems.net/2022/05/11/mdm-22-tomahawk -robotics-raid-plate/.

90. On this point, see Meghann Myers, "Experts, Data Point to Women as Best Military Recruiting Pool," *Military Times*, January 26, 2023, https://www.militarytimes

.com/news/your-military/2023/01/26/experts-data-point-to-women-as-best
-military-recruiting-pool/. On the fact that this has long been known, see the
following 1980 article in the US Army's professional journal: Judith H. Stiehm,
"Women and the Combat Exemption," *The US Army War College Quarterly:
Parameters* 10, no. 1 (1980), https://doi.org/10.55540/0031-1723.1229.

Chapter 5

1. Book-length treatments of the conflict in Ukraine include the following: Anna
   Arutunyan, *Hybrid Warriors: Proxies, Freelancers and Moscow's Struggle for
   Ukraine* (Hurst, 2022); Owen Matthews, *Overreach: The Inside Story of Putin's War
   Against Ukraine* (Mudlark, 2022); Alexander Etkind, *Russia Against Modernity*
   (Polity Press, 2023); Samuel Ramani, *Putin's War on Ukraine: Russia's Campaign
   for Global Counter-Revolution* (Oxford University Press, 2023); Hal Brands, ed.,
   War in Ukraine: Conflict, Strategy, and the Return of a Fractured World (Johns
   Hopkins University Press, 2024); and Mick Ryan, *The War for Ukraine: Strategy
   and Adaptation Under Fire* (Naval Institute Press, 2024).

2. On the difficulty of ascertaining the long-term lessons of the war in Ukraine,
   see Joseph Caddell, "Rewind and Reconnoiter: Joseph Caddell on the Evolving
   Historiography of National Security Events," *War on the Rocks*, November 16,
   2023, https://warontherocks.com/2023/11/rewind-and-reconnoiter-joseph-caddell
   -on-the-evolving-historiography-of-national-security-events/; David Barno
   and Nora Bensahel, "Learning from Real Wars: Gaza and Ukraine," *War on the
   Rocks*, December 6, 2023, https://warontherocks.com/2023/12/learning-from-real
   -wars-gaza-and-ukraine/; and Sandor Fabian, "The Illusion of Conventional War:
   Europe Is Learning the Wrong Lessons from the Conflict in Ukraine," Modern
   War Institute (at West Point), April 23, 2024, https://mwi.westpoint.edu/the
   -illusion-of-conventional-war-europe-is-learning-the-wrong-lessons-from-the
   -conflict-in-ukraine/. For attempts to identify the potential lessons of the conflict
   for US and allied policymakers, consult David Barno and Nora Bensahel, "The
   Other Big Lessons That the U.S. Army Should Learn from Ukraine," *War on the
   Rocks*, June 27, 2022, https://warontherocks.com/2022/06/the-other-big-lessons
   -that-the-u-s-army-should-learn-from-ukraine/; Howard Altman, "Top Ukraine
   War Lessons from USAF's Commander in Europe," *The War Zone (at The Drive)*,
   August 8, 2023, https://www.thedrive.com/the-war-zone/top-ukraine-war-lessons
   -from-usafs-commander-in-europe; Frank Hoffman, "American Defense Pri-
   orities After Ukraine," *War on the Rocks*, January 2, 2023, https://warontherocks
   .com/2023/01/american-defense-priorities-after-ukraine/; Matthew Van Wagenen
   and Arnel P. David, "Lessons from Ukraine Many Don't Want to Hear," Real-
   ClearDefense, March 11, 2023, https://www.realcleardefense.com/articles/2023
   /03/11/lessons_from_ukraine_many_dont_want_to_hear_886750.html; Yagil
   Henkin, "The 'Big Three' Revisited: Initial Lessons from 200 Days of War in
   Ukraine," *Expeditions with Marine Corps University Press* (November 2022),
   https://doi.org/10.36304/ExpwMCUP.2022.13; Herbert Bowsher, "Air Denial Lessons from Ukraine," *U.S. Naval Institute Proceedings* 149, no. 9 (September 2023),
   https://www.usni.org/magazines/proceedings/2023/september/air-denial-lessons
   -ukraine; Harry Halem, "Ukraine's Lessons for Future Combat: Unmanned

Aerial Systems and Deep Strike," *The US Army War College Quarterly: Parameters* 53, no. 4 (2023): 19–32, https://doi.org/10.55540/0031-1723.3252. Especially useful is this recent book-length treatment: John A. Nagl and Katie Crombe, *A Call to Action: Lessons from Ukraine for the Future Force* (US Army War College Press, 2024), https://press.armywarcollege.edu/monographs/968.

3. For the example of the American army in this regard, see Brian McAllister Linn, *The Echo of Battle: The Army's Way of War* (Harvard University Press, 2007); and Brian McAllister Linn, *Real Soldiering: The US Army in the Aftermath of War, 1815–1980*, Modern War Studies (University Press of Kansas, 2023).

4. Milley quoted in David Ignatius, "How the Algorithm Tipped the Balance in Ukraine," Opinion, *Washington Post*, December 19, 2022, https://www.wash-ingtonpost.com/opinions/2022/12/19/palantir-algorithm-data-ukraine-war/. For Milley's broader thoughts on the future of warfare, begin with Mark A. Milley, "Strategic Inflection Point: The Most Historically Significant and Fundamental Change in the Character of War Is Happening Now—While the Future Is Clouded in Mist and Uncertainty," *Joint Force Quarterly* 110 (Quarter 2023): 6–15.

5. Reznikov quote from The Atlantic Council, "A Conversation with Minister of Defense of Ukraine Oleksii Reznikov," July 19, 2022, https://www.atlanticcouncil.org/event/a-conversation-with-oleksii-reznikov/. See also Dan Parsons, "Ukraine Situation Report: Defense Chief Wants Advanced Weapons Testing Against Russians," *The War Zone (at The Drive)*, July 19, 2022, https://www.thedrive.com/the-war-zone/ukraine-situation-report-defense-chief-wants-advanced-weapons-testing-against-russians.

6. For a useful overview of why the first few days of the war turned out poorly for Russia, see Zack Beauchamp, "Why the First Few Days of War in Ukraine Went Badly for Russia," *Vox*, February 28, 2022, https://www.vox.com/22954833/russia-ukraine-invasion-strategy-putin-kyiv. A penetrating analysis of perhaps the most important battle during this early period in the war can be found in Liam Collins et al., "The Battle of Hostomel Airport: A Key Moment in Russia's Defeat in Kyiv," *War on the Rocks*, August 10, 2023, https://warontherocks.com/2023/08/the-battle-of-hostomel-airport-a-key-moment-in-russias-defeat-in-kyiv/.

7. See Lucian Kim, "How U.S. Military Aid Has Helped Ukraine Since 2014," *NPR*, December 18, 2019, https://www.npr.org/2019/12/18/788874844/how-u-s-military-aid-has-helped-ukraine-since-2014; and Daniel Michaels, "The Secret of Ukraine's Military Success: Years of NATO Training," *Wall Street Journal*, April 13, 2022, https://www.wsj.com/articles/ukraine-military-success-years-of-nato-training-11649861339.

8. On cultural resourcefulness within the Ukrainian military, see Alan W. Dowd, "Ukraine Reminds Us Necessity Is Still the Mother of Invention," RealClearDefense, October 7, 2023, https://www.realcleardefense.com/articles/2023/10/07/ukraine_reminds_us_necessity_is_still_the_mother_of_invention_984556.html. On the early development of a culture of learning going back to 2014 in the Ukrainian military, see Tom Dyson and Yuriy Pashchuk, "Organisational Learning During the Donbas War: The Development of Ukrainian Armed Forces Lessons-Learned Processes," *Defence Studies* 22, no. 2 (2022): 141–67, https://doi.org/10.1080/14702436.2022.2037427. On the Ukrainian "will to fight," consult Benjamin A. Okonofua, Nicole Laster-Loucks, and Andrew Johnson, "'Will to

Fight': Twenty-First-Century Insights from the Russo-Ukrainian War," *Military Review: The Professional Journal of the U.S. Army* 104, no. 3 (2024): 34–39; and, for its historical roots, Gregg W. Etter Sr. and David H. McElreath, "Why the Ukrainians Fight: The Holodomor (1932–33)," *Journal of Advanced Military Studies* 16, no. 1 (2025): 96–108, https://doi.org/10.21140/mcuj.20251601005.

9. For an overview on this matter, see Seth G. Jones et al., *Ukrainian Innovation in a War of Attrition*, CSIS Briefs (Washington, D.C.: Center for Strategic and International Studies (CSIS), 2023, https://www.csis.org/analysis/ukrainian-innovation-war-attrition.

10. For an excellent overview of these matters, consult Brian A. Hester et al., "Techcraft on Display in Ukraine," *War on the Rocks*, May 16, 2024, https://warontherocks.com/2024/05/techcraft-on-display-in-ukraine/.

11. See Michael T. Hackett and John A. Nagl, "A Long, Hard Year: Russia-Ukraine War Lessons Learned 2023," *The US Army War College Quarterly: Parameters* 54, no. 3 (2024): 43–45, https://doi.org/10.55540/0031-1723.3302.

12. Schmidt quote from Andrey Poznyakov and Frances Lopez, "Ukraine War: How Have Western Weapons Performed in Combat?," *Euronews*, July 13, 2023, https://www.euronews.com/2023/07/13/war-in-ukraine-how-have-western-weapons-performed-in-combat.

13. On this point, see Steven Feldstein, "Disentangling the Digital Battlefield: How the Internet Has Changed War," *War on the Rocks*, December 7, 2022, https://warontherocks.com/2022/12/disentangling-the-digital-battlefield-how-the-internet-has-changed-war/.

14. Shyam Sankar, "Ukraine's Software Warrior Brigade," *Wall Street Journal*, March 9, 2023.

15. Ryan Broderick, *The Most Online War of All Time Until the Next One*, February 24, 2022, https://www.thecontentmines.com/p/the-most-online-war-of-all-time-until-a13.

16. For one example, see Kyle Chayka, "Ukraine Becomes the World's 'First TikTok War,'" *The New Yorker*, March 3, 2022, https://www.newyorker.com/culture/infinite-scroll/watching-the-worlds-first-tiktok-war. For a counterargument regarding this label, see Kaitlyn Tiffany, "The Myth of the 'First TikTok War,'" *The Atlantic*, March 10, 2022, https://www.theatlantic.com/technology/archive/2022/03/tiktok-war-ukraine-russia/627017/. For an analysis of what the war in Ukraine can tell us about the future of information warfare, see Theodore W. Kleisner, "Tactical TikTok for Great Power Competition: Applying the Lessons of Ukraine's IO Campaign to Future Large-Scale Conventional Operations," *Military Review: The Professional Journal of the U.S. Army* 102, no. 4 (2022): 23–43. For a prescient overview of the place of social media in modern conflict, consult Peter Warren Singer and Emerson T. Brooking, *Likewar: The Weaponization of Social Media* (Houghton Mifflin Harcourt, 2018). Schmidt's remark can be found at *Eric Schmidt: The War in Ukraine Is the First Broadband War* (MSNBC, 2023), https://www.youtube.com/watch?v=G3ksdKxDsdk.

17. Fedorov quoted in Ignatius, "How the Algorithm Tipped the Balance in Ukraine."

18. See, for example, David Johnson, "The Tank Is Dead: Long Live the Javelin, the Switchblade, the . . . ?," *War on the Rocks*, April 18, 2022, https://warontherocks.com/2022/04/the-tank-is-dead-long-live-the-javelin-the-switchblade-the/.

To be fair, this argument predates the Russian invasion into Ukraine. For a representative example, consult Arash Heydarian Pashakhanlou, "AI, Autonomy, and Airpower: The End of Pilots?," *Defence Studies* 19, no. 4 (2019): 337–52, https://doi.org/10.1080/14702436.2019.1676156. For rebuttal, see Curtis A. Buzzard et al., "The Tank Is Dead . . . Long Live the Tank: The Persistent Value of Armored Combined Arms Teams in the 21st Century," *Military Review: The Professional Journal of the U.S. Army* 103, no. 6 (2023): 22–30.

19. See Joseph Trevithick, "Army Cancels High-Speed Armed Reconnaissance Helicopter Program," *The War Zone (at The Drive)*, February 8, 2024, https://www.twz.com/air/army-cancels-hight-speed-armed-reconnaissance-helicopter-program.

20. For the place of drones in the war, see Adam Lowther and Mahbube K. Siddiki, "Combat Drones in Ukraine," *Air & Space Operations Review* 1, no. 4 (2022): 3–13; Kerry Chávez, "Learning on the Fly: Drones in the Russian-Ukrainian War," *Arms Control Today* (Washington, United States) 53, no. 1 (2023): 6–11; Franz-Stefan Gady, "How an Army of Drones Changed the Battlefield in Ukraine," *Foreign Policy*, December 6, 2023, https://foreignpolicy.com/2023/12/06/ukraine-russia-war-drones-stalemate-frontline-counteroffensive-strategy/; and Mark Jacobsen, "Ukraine's Drone Strikes Are a Window into the Future of Warfare," Atlantic Council, September 14, 2023, https://www.atlanticcouncil.org/blogs/new-atlanticist/ukraines-drone-strikes-are-a-window-into-the-future-of-warfare/.

21. Fedorov quoted in Greg Myre, "How Ukraine Created an 'Army of Drones' to Take on Russia," *NPR*, June 20, 2023, https://www.npr.org/2023/06/20/1183050117/how-ukraine-created-an-army-of-drones-to-take-on-russia.

22. Jack Watling and Nick Reynolds, *Tactical Developments During the Third Year of the Russo–Ukrainian War* (Royal United Services Institute for Defence and Security Studies, 2025), 10.

23. See Jared Malsin, "Ukraine's Sea Drones Alter Balance of Power in Black Sea," *Wall Street Journal*, August 11, 2023, https://www.wsj.com/articles/ukraines-sea-drones-alter-balance-of-power-in-black-sea-391cebee; and Alfred F. Lord and Thomas R. Kunish, "Lessons from the War at Sea," in *A Call to Action: Lessons from Ukraine for the Future Force*, by John A. Nagl and Katie Crombie (US Army War College Press, 2024), 247–59, https://press.armywarcollege.edu/monographs/968. For an analysis regarding how such systems will be employed in the future, see Robbin Laird, *The Coming of Maritime Autonomous Systems: Empowering and Enhancing the Kill Web Force* (Independently published, 2024).

24. See Kateryna Bondar, *Understanding the Military AI Ecosystem of Ukraine* (Center for Strategic and International Studies [CSIS], November 2024), https://www.csis.org/analysis/understanding-military-ai-ecosystem-ukraine; and Kateryna Bondar, *Ukraine's Future Vision and Current Capabilities for Waging AI-Enabled Autonomous Warfare* (Center for Strategic and International Studies [CSIS], March 2025), https://www.csis.org/analysis/ukraines-future-vision-and-current-capabilities-waging-ai-enabled-autonomous-warfare.

25. See John Hudson and Kostiantyn Khudov, "The War in Ukraine Is Spurring a Revolution in Drone Warfare Using AI," *Washington Post*, July 26, 2023, https://www.washingtonpost.com/world/2023/07/26/drones-ai-ukraine-war-innovation/.

26. See Steve Brown, "Ukraine Trained AI for Its 'Spiders Web' Airfield Drone Attacks at Aviation Museum," Kyiv Post, June 2, 2025, https://www.kyivpost.com/post/53784. The nearly 5,000-mile distance is referenced in Laura Gozzi, "How Ukraine Carried Out Daring 'Spider Web' Attack on Russian Bombers," BBC News, June 2, 2025, https://www.bbc.com/news/articles/cq69qnvj6nlo.

27. See Olena Roshchina, "Russians Destroy 60–70% of Uncrewed Surface Vessels, but 30% Remains Dangerous for Russian Navy," *Ukrainska Pravda*, August 24, 2023, https://www.pravda.com.ua/eng/news/2023/08/24/7416940/.

28. See Jack Watling and Nick Reynolds, *Meatgrinder: Russian Tactics in the Second Year of Its Invasion of Ukraine*, Special Report (Royal United Services Institute for Defence and Security Studies, 2023), 18, https://www.rusi.orghttps://www.rusi.org.

29. See Volodymyr Zelenskyy, "I Signed a Decree Initiating the Establishment of a Separate Branch of Forces—the Unmanned Systems Forces—Address by the President of Ukraine," Official website of the President of Ukraine, February 6, 2024, https://www.president.gov.ua/en/news/pidpisav-ukaz-yakij-rozpochinaye-stvorennya-okremogo-rodu-si-88817; and Dylan Malyasov, "Russia Follows Ukraine in Creating Drone Forces," Defence Blog, December 16, 2024, https://defence-blog.com/russia-follows-ukraine-in-creating-drone-forces/. A useful overview of the place of Russian unmanned systems in the war is provided in Jeffrey Edmonds and Samuel Bendett, *Russia's Use of Uncrewed Systems in Ukraine* (Center for Naval Analyses, 2023), https://www.cna.org/reports/2023/05/russias-use-of-drones-in-ukraine.

30. Demchenko can be seen making this statement at *See How Gamers Are Outwitting and Helping to Kill Russian Soldiers* (CNN, 2022), https://www.youtube.com/watch?v=b166ecyNBCw.

31. James Marson, "The Nerdy Gamers Who Became Ukraine's Deadliest Drone Pilots," *The Wall Street Journal*, November 5, 2024.

32. See Myre, "How Ukraine Created an 'Army of Drones.'"

33. See *Ukrainian Military Training: Forces Partner with Video Game Maker* (Al Jazeera English, August 6, 2022), https://www.youtube.com/watch?v=Ackgv_gmGnQ.

34. See Thomas Newdick, "Ukrainian MiG-29 Pilot's Front-Line Account of the Air War Against Russia," *The War Zone (at The Drive)*, April 1, 2022, https://www.thedrive.com/the-war-zone/45019/fighting-russia-in-the-sky-mig-29-pilots-in-depth-account-of-the-air-war-over-ukraine.

35. On this initiative, see Brian Bennett and Simon Shuster, "Exclusive: Ukraine's Secret Effort to Train for U.S. Jets," *Time*, August 19, 2022, https://time.com/6207115/ukraine-train-fighter-pilots-russia/.

36. See GlobalData Thematic Intelligence, "The Metaverse Is Preparing Ukrainian Fighter Pilots for Combat," *Airforce Technology* (blog), May 30, 2023, https://www.airforce-technology.com/comment/metaverse-ukrainian-fighter-pilots/.

37. A copy of the US Air Force study can be found at Michael Weiss and James Rushton, "U.S. Could Train Ukrainian Pilots to Fly F-16s in 4 Months," Yahoo News, May 18, 2023, https://news.yahoo.com/exclusive-us-could-train-ukrainian-pilots-to-fly-f-16s-in-4-months-184136820.html.

38. People's Project.com—Ukraine's Military and Civil Crowdfunding, "Sabre Remote Weapon Station," accessed May 8, 2023, https://www.peoplesproject.com /en/sabre-remote-weapon-station/.

39. Kelsey Vlamis, "Ukrainian Soldiers Are Using a Handheld Video Game Console to Operate Machine Gun Mounts, per Reports. The US Navy Has Also Utilized Xbox Controllers in the Past," *Business Insider*, May 1, 2023, https://www.businessinsider .com/ukraine-soldiers-video-game-consoles-steam-deck-machine-gun-turrets -2023-5.

40. Singer quote from Jared Keller, "The US Military Will Fight the Next Big War with Xbox-Style Video Game Controllers," *Task & Purpose*, March 22, 2023, https:// taskandpurpose.com/tech-tactics/us-military-video-game-controllers-war/.

41. Quote from "UkraineWorld on Twitter," Twitter, August 11, 2023, https://twitter .com/ukraine_world/status/1690014709208977408.

42. Serhiy's statements are provided in the following coverage of the event: Howard Altman, "Video Games Helped Ukrainian Bradley Gunner Win Duel with Russian T-90M Tank," *The War Zone (at The Drive)*, January 20, 2024, https://www .thedrive.com/the-war-zone/video-games-helped-ukrainian-bradley-gunner -win-duel-with-russian-t-90m-tank.

43. This goes back to Russia's intervention in 2014 in the Donbass region of eastern Ukraine. See Joseph Trevithick, "Ukrainian Officer Details Russian Electronic Warfare Tactics Including Radio 'Virus,'" *The War Zone (at The Drive)*, October 30, 2019, https://www.thedrive.com/the-war-zone/30741/ukrainian-officer -details-russian-electronic-warfare-tactics-including-radio-virus.

44. Kieran Devine, "Ukraine War: Mobile Networks Being Weaponised to Target Troops on Both Sides of Conflict," Sky News, January 4, 2023, https://news.sky .com/story/ukraine-war-mobile-networks-being-weaponised-to-target-troops -on-both-sides-of-conflict-12577595.

45. Roman Horbyk, "'The War Phone': Mobile Communication on the Frontline in Eastern Ukraine," *Digital War* 3, no. 1 (2022): 17, https://doi.org/10.1057/s42984-022 -00049-2.

46. See "Education That Makes Developers in Ukraine the Best in the World," N-iX, September 9, 2019, https://www.n-ix.com/tech-education-developers-in-ukraine/.

47. Eliot A. Cohen and Wesley Clark, "Transcript: Reflections on the Ukraine War," Transcript, Center for Strategic and International Studies, February 20, 2024, https://www.csis.org/analysis/reflections-ukraine-war.

48. The quote from the anonymous US veteran can be found in Jeff Schogol, "US Vet on the Front Line Says Ukrainians Need Better Infantry Gear to Beat the Russians," *Task & Purpose*, June 27, 2022, https://taskandpurpose.com/news /ukraine-american-military-veteran-infantry-gear/.

49. Such is the case regarding a possible future war between the United States and China, for example. See Iskander Rehman, *Planning for Protraction: A Historically Informed Approach to Great-Power War and Sino-US Competition* (Routledge, 2023).

50. See Ellie Cook, "How Russia and Ukraine's Losses Compare," *Newsweek*, February 23, 2024, https://www.newsweek.com/russia-ukraine-losses-casualties-tanks -death-toll-anniversary-1864726.

51. On what Western policymakers might learn from the matter, see Thomas Haydock and Jack Meeker, "Lessons in Reconstitution from the Russia-Ukraine War: Gaining Asymmetric Advantage Through Transformative Reconstitution," *Military Review: The Professional Journal of the U.S. Army* 105, no. 1 (2025): 26–41.

52. For the technological implications of this context, see Shashank Joshi, "Ypres with AI," *The Economist*, July 8, 2023 (Special Report: Warfare after Ukraine), S3–S5.

53. See Thibault Spirlet, "Hundreds of Russian Soldiers Are Going AWOL and Refusing to Fight as Morale Plummets, UK Intelligence Says," *Business Insider*, August 30, 2023, https://www.businessinsider.com/hundreds-russian-soldiers-awol-refusing-fight-uk-intel-2023-8.

54. See Aila Slisco, "Russia's Military Has a Major Desertion Problem," *Newsweek*, December 5, 2023, https://www.newsweek.com/russias-military-has-major-desertion-problem-1849729.

55. On the training of Ukrainian troops, see Jim Garamone, "Training Key to Ukrainian Advantages in Defending Nation," US Department of Defense, September 6, 2022, https://www.defense.gov/News/News-Stories/Article/article/3149975/training-key-to-ukrainian-advantages-in-defending-nation/. An overview of the training of Russian troops is provided in Michael Connell et al., "Training in the Russian Armed Forces" (Center for Naval Analyses, 2023), https://www.cna.org/reports/2023/09/training-in-the-russian-armed-forces.

56. The challenges that Ukraine has faced in this process are outlined in Andrew Milburn, "Time Is Not on Kyiv's Side: Training, Weapons, and Attrition in Ukraine," Modern War Institute (at West Point), June 27, 2022, https://mwi.westpoint.edu/time-is-not-on-kyivs-side-training-weapons-and-attrition-in-ukraine/. Russia's problems, on the other hand, are the focus of Michael G. Anderson, "A People Problem: Learning from Russia's Failing Efforts to Reconstitute Its Depleted Units in Ukraine," Modern War Institute (at West Point), January 26, 2023, https://mwi.westpoint.edu/a-people-problem-learning-from-russias-failing-efforts-to-reconstitute-its-depleted-units-in-ukraine/.

57. James Robb, "Ukraine Serves as Lesson in the Value of Training," *National Defense*, 2022, 3.

58. Zelensky's Twitter statement can be found at "Володимир Зеленський on Twitter," Twitter, August 11, 2023, https://twitter.com/ZelenskyyUa/status/1689973671476023297.

59. See Eric Schmitt et al., "Ukraine's Forces and Firepower Are Misallocated, U.S. Officials Say," *New York Times*, August 22, 2023, https://www.nytimes.com/2023/08/22/us/politics/ukraine-counteroffensive-russia-war.html.

60. See Oleh Hukovskyy et al., "The Combat Path: Sustaining Mental Readiness in Ukrainian Soldiers," *The US Army War College Quarterly: Parameters* 54, no. 2 (2024): 21–36, https://doi.org/10.55540/0031-1723.3285.

61. See Tanisha M. Fazal, "Ukraine's Military Medicine Is a Critical Advantage," *Foreign Policy*, October 31, 2022, https://foreignpolicy.com/2022/10/31/ukraine-military-medicine-russia-war/. The overloading of Russia's medical system in the conflict is outlined in Gabriela Iveliz Rosa-Hernandez and Patrick Enoch, *We Need a Medic!: The Russian Military Medicine Experience in Ukraine* (Center for

Naval Analyses, November 2024), https://www.cna.org/reports/2024/11/we-need -a-medic.

62. See Quil Lawrence, "Ukrainian Soldiers Benefit from U.S. Prosthetics Expertise but Their War Is Different," *NPR*, February 7, 2023, https://www.npr.org/2023/02 /07/1153472827/ukrainian-soldiers-benefit-from-u-s-prosthetics-expertise-but -their-war-is-diffe.

63. See Alexander Motyl, "Ukraine's Military Advantage? How Quick It Treats Its Wounded," Opinion, *EUobserver* (Brussels), April 20, 2023, https://euobserver .com/opinion/156933.

64. UK Ministry of Defence, "Latest Defence Intelligence Update on the Situation in Ukraine," Twitter, July 10, 2023, https://twitter.com/DefenceHQ/status /1678279648847822848.

65. See Xander Landen, "Putin Mobilizing Medically, Mentally 'Unfit' Troops to Ukraine: General," *Newsweek*, September 22, 2022, https://www.newsweek.com /vladimir-putin-mobilizing-medically-mentally-unfit-troops-ukraine-general -jack-keane-1745492.

66. Keane quote from Landen, "Putin Mobilizing Medically, Mentally 'Unfit' Troops."

67. Dara Massicot, "The Russian Military's Looming Personnel Crises of Retention and Veteran Mental Health," *The RAND Blog, The RAND Corporation*, June 1, 2023, https://www.rand.org/blog/2023/06/the-russian-militarys-looming -personnel-crises-of-retention.html.

68. Stephen Biddle, "Back in the Trenches," *Foreign Affairs*, August 10, 2023, https:// www.foreignaffairs.com/ukraine/back-trenches. For more of Biddle's observations on the technological nature of modern warfare, consult Stephen Biddle, *Military Power: Explaining Victory and Defeat in Modern Battle* (Princeton University Press, 2004), https://doi.org/10.1515/9781400837823. A recent monograph by political scientist Ben Connable makes a compelling argument on the enduring nature of modern ground combat. See Ben Connable, *Ground Combat: Puncturing the Myths of Modern War* (Georgetown University Press, 2025).

69. On the point that war remains a fundamentally human endeavor, see Bonnie L. Rushing and Kyleanne Hunter, "The Human Weapon System in Gray Zone Competition," *Journal of Advanced Military Studies* 14, no. 1 (2023): 255–71, https://doi .org/10.21140/mcuj.20231401011. For a penetrating critique of the Russian (and earlier Soviet) military forces from this perspective, see Robert McKeown, "Assessing Military Capability: More Than Just Counting Guns," *U.S. Naval Institute Proceedings* 148, no. 12 (2022), https://www.usni.org/magazines/proceedings /2022/december/assessing-military-capability-more-just-counting-guns.

70. George S. Patton, *War as I Knew It* (Houghton Mifflin Harcourt, 1947), 335.

71. On this point, the author of this book is indebted to an anonymous outsider reviewer of the manuscript. See "Anonymous Review of Manuscript 'Bodies at War: Technological Change and the Evolving Physicality of Combat,'" 2025.

72. On this point, see Joshua Suthoff, "Reimagining Combat Power for Tomorrow's Battlefield: The Enhanced Brigade Combat Team," Modern War Institute (at West Point), April 18, 2025, https://mwi.westpoint.edu/reimagining-combat-power-for -tomorrows-battlefield-the-enhanced-brigade-combat-team/.

73. Eric Schmidt, "Innovation Power: Why Technology Will Define the Future of Geopolitics," *Foreign Affairs* 102, no. 2 (2023): 43.

74. Ignatius, "How the Algorithm Tipped the Balance in Ukraine."

Chapter 6

1. See Dave Philipps, "With Few Able and Fewer Willing, U.S. Military Can't Find Recruits," *New York Times*, July 14, 2022, https://www.nytimes.com/2022/07/14/us/us-military-recruiting-enlistment.html. For a useful overview of the national security implications of the matter, see the following report by an American think tank: Thomas W. Spoehr, *The Administration and Congress Must Act Now to Counter the Worsening Military Recruiting Crisis*, No. 5283, Issue Brief (Center for National Defense, The Heritage Foundation, July 28, 2022), http://report.heritage.org/ib5283.

2. See Courtney Kube and Molly Boigon, "Every Branch of the Military Is Struggling to Make Its 2022 Recruiting Goals, Officials Say," NBC News, June 27, 2022, https://www.nbcnews.com/news/military/every-branch-us-military-struggling-meet-2022-recruiting-goals-officia-rcna35078.

3. See Heather Mongilio, "Military Services Competing Over the Same Recruiting Pool of Less Than 500,000," *US Naval Institute (USNI) News*, February 14, 2023, https://news.usni.org/2023/02/14/military-services-competing-over-the-same-recruiting-pool-of-less-than-500000.

4. Tillis quoted in Roxana Tiron, "US Military Faces Biggest Recruiting Hurdles in 50 Year," Bloomberg Government, September 21, 2022, https://about.bgov.com/news/us-military-services-face-biggest-recruiting-hurdles-in-50-years/.

5. James Stavridis, "US Military's Recruiting Woes Are a National-Security Crisis," Bloomberg.com, July 4, 2023, https://www.bloomberg.com/opinion/articles/2023-07-04/us-military-recruiting-crisis-is-a-national-security-emergency.

6. See Heather Mongilio, "Navy Needs to Fill About 9,000 At-Sea Billets in More Than a Dozen Ratings, Says Personnel Command," *US Naval Institute (USNI) News*, October 25, 2022, https://news.usni.org/2022/10/25/navy-needs-to-fill-about-9000-at-sea-billets-in-more-than-a-dozen-ratings-says-personnel-command.

7. See John Grady, "Navy Sees 'Difficult Times' with Recruiting Goals for Nuclear, Cyber Sailors," *US Naval Institute (USNI) News*, February 15, 2018, https://news.usni.org/2018/02/15/navy-wants-congressional-permission-bring-new-cyber-sailors-higher-ranks.

8. See US Government Accountability Office, *Coast Guard: Recruitment and Retention Challenges Persist* (US Government Accountability Office, May 11, 2023), https://www.gao.gov/products/gao-23-106750; and Steve Beynon, "Army National Guard Can't Retain Enough Soldiers, Even as Active Duty Meets Goals," Military.com, June 22, 2023, https://www.military.com/daily-news/2023/06/22/army-national-guard-cant-retain-enough-soldiers-even-active-duty-meets-goals.html.

9. See Lolita C. Baldor, "U.S. Military Recruiting Rebounds After Several Tough Years, but Challenges Remain," *Los Angeles Times*, September 26, 2024, https://www.latimes.com/world-nation/story/2024-09-26/u-s-military-recruiting-rebounds-after-several-tough-years-but-challenges-remain.

10. See Lolita C. Baldor, "US Army Is Slashing Thousands of Posts in Major Revamp to Prepare for Future Wars," AP News, February 27, 2024, https://apnews.com/article/army-cuts-soldiers-recruiting-shortfall-9f2f41cbe512f6330ce6008709e3435b.

11. The problem is outlined in the following UK House of Commons report: House of Commons Committee of Public Accounts, *Skill Shortages in the Armed Forces.* Fifty-Ninth Report of Session 2017–19 (House of Commons of the United Kingdom, September 12, 2018).

12. Mackenzie quote from Neil Murphy, "We Need Soldiers, Not Robots: British Army's Plea for New Recruits," *The National*, July 1, 2022, https://www.thenationalnews.com/world/uk-news/2022/07/01/we-need-soldiers-not-robots-british-armys-plea-for-new-recruits/.

13. See "German Military Recruitment," Army Technology, March 18, 2022, https://www.army-technology.com/comment/german-military-recruitment/.

14. François Thériault et al., *Health and Lifestyle Information Survey of Canadian Armed Forces Personnel 2013/2014—Regular Force Report*, ed. Barbara Strauss and Jeff Whitehead (Department of National Defense, 2016), 150.

15. See Ashley Burke, "Canadian Military Reports Sagging Recruitment as NATO Ramps Up Deployment in Eastern Europe," CBC News, March 23, 2022, https://www.cbc.ca/news/politics/canadian-armed-forces-staff-shortfall-1.6395131; and Andrew Latham, "The Canadian Armed Forces Recruitment Crisis," RealClearDefense, December 4, 2024, https://www.realcleardefense.com/articles/2024/12/04/the_canadian_armed_forces_recruitment_crisis_1076308.html.

16. Lisa Visentin Wright Shane, "ADF Boost Hard to Achieve without Overhaul of Recruitment and Retention Policies," *The Sydney Morning Herald*, March 10, 2022, https://www.smh.com.au/politics/federal/adf-boost-won-t-be-achieved-without-overhaul-of-recruitment-and-retention-policies-20220310-p5a3bx.html.

17. Those looking for long-term solutions would do well to begin with Matthew Weiss, *We Don't Want YOU, Uncle Sam: Examining the Military Recruiting Crisis with Generation Z* (Night Vision Publishing, 2023). For thoughts on the subject on a shorter term, see the following essay by a recently retired US Marine Corps Commandant: David H. Berger, "Recruiting Requires Bold Changes," *U.S. Naval Institute Proceedings* 148, no. 11 (2022), https://www.usni.org/magazines/proceedings/2022/november/recruiting-requires-bold-changes. See also Christopher McMahon and Colin Bernard, "Storm Clouds on the Horizon—Challenges and Recommendations for Military Recruiting and Retention," *Naval War College Review* 72, no. 3 (2019): 84–100.

18. On this point, see Stephen Losey, "Air Force Meets Recruitment Goals, Eyes 20% Increase in 2025," *Defense News*, September 17, 2024, https://www.defensenews.com/news/your-air-force/2024/09/17/air-force-meets-recruitment-goals-eyes-20-increase-in-2025/; and Patty Nieberg, "Army Ups Recruiting Goal to 61,000 Soldiers in 2025, an 11% Jump," *Task & Purpose*, October 14, 2024, https://taskandpurpose.com/news/army-new-recruiting-goal-61000/.

19. The term "generational shift" on the issue of military recruitment is pulled from Thomas Spoehr and Katherine Kuzminski, "Bad Idea: Relying on the Same Old Solutions to Meet the Military Recruitment Challenge," Center for Strategic and

International Studies, March 10, 2023, https://defense360.csis.org/bad-idea-relying
-on-the-same-old-solutions-to-meet-the-military-recruitment-challenge/. On
the fact that recruiting problems have been the historical norm rather than the
exception for the American military, consult Brian McAllister Linn, "A Historical
Perspective on Today's Recruiting Crisis," *The US Army War College Quarterly:
Parameters* 53, no. 3 (2023): 7–20, https://doi.org/10.55540/0031-1723.3239. This is
also far from the first generational shift that has affected military recruiting. On
this point, see Corie Weathers, *Military Culture Shift: The Impact of War, Money,
and Generational Perspective on Morale, Retention, and Leadership* (Elva Resa
Publishing, 2023).

20. See Andrew Swick and Emma Moore, "DoD Must Consider Non-Traditional
Approaches to Military Service," *The Hill* (blog), February 28, 2018, https://thehill
.com/opinion/national-security/376005-dod-must-consider-non-traditional
-approaches-to-military-service/. On the potential of a reinstituted draft to meet
US military operational requirements, consult Justin Lynch, "A Great Power War
Could Require More Troops, and More Quickly, Than the United States Can
Generate—Even with a Draft," Modern War Institute (at West Point), October 1,
2020, https://mwi.westpoint.edu/a-great-power-war-could-require-more-troops
-and-more-quickly-than-the-united-states-can-generate-even-with-a-draft/; and
Katherine L. Kuzminski and Taren Dillon Sylvester, *Back to the Drafting Board:
U.S. Draft Mobilization Capability for Modern Operational Requirements* (Center
for a New American Security, 2024), https://www.cnas.org/publications/reports
/back-to-the-drafting-board.

21. Thomas W. Spoehr, "It's Time for a National Security Strategy for Military
Recruiting," *Stars and Stripes*, August 17, 2023, https://www.stripes.com/opinion
/2023-08-17/national-security-strategy-military-recruiting-11076498.html.

22. See Rob Swain, "Retention Must Be a Special Evolution," *U.S. Naval Institute
Proceedings* 148, no. 12 (2022), https://www.usni.org/magazines/proceedings/2022
/december/retention-must-be-special-evolution; and Ian Clark and Kyle Atkinson, "Gen-Z Will Fight: But First, They Need to Know Why," *U.S. Naval Institute*
149, no. 1 (2023), https://www.usni.org/magazines/proceedings/2023/january/gen
-z-will-fight-first-they-need-know-why. For more on service thinking on the
recruitment of today's young Americans, see Nieberg, "Army Ups Recruiting
Goal to 61,000 Soldiers." A particularly good set of proposals on the subject can be
found in Richard Bell et al., "Stop Talking to Yourself: Military Recruiting in the
Modern Age," *Joint Force Quarterly* 115, no. 3 (2024): 53–56.

23. On this point, see Spoehr and Kuzminski, "Bad Idea."

24. See Lolita C. Baldor, "Army Sees Safety, Not 'Wokeness,' as Top Recruiting
Obstacle," *Army Times*, February 12, 2023, https://www.armytimes.com/news
/your-army/2023/02/12/army-sees-safety-not-wokeness-as-top-recruiting
-obstacle/.

25. The United States has a history of this policy going back at least to 1950. See Brad
Hardy, "Citizen Candidates: Cold War Naturalization, Military Service, and the
Lodge Act of 1950," *Journal of Military History* 87, no. 1 (2023): 169–88. For an
argument to reinstate and expand the policy after it was halted during the Trump
administration, see Collin Fox, "Restore MAVNI for Legal Aliens to Enter the

Military," *U.S. Naval Institute Proceedings* 147, no. 8 (2021), https://www.usni.org /magazines/proceedings/2021/august/restore-mavni-legal-aliens-enter-military. This process, of course, was often far from straightforward. For three case studies that together illuminate this point, see Cristina-Ioana Dragomir, *Making the Immigrant Soldier: How Race, Ethnicity, Class, and Gender Intersect in the US Military* (University of Illinois Press, 2023).

26. See, for example, Jaime Moore-Carrillo, "Air Force, Space Force Raise Max Enlistment Age to 42," *Air Force Times*, October 30, 2023, https://www.airforcetimes .com/news/your-air-force/2023/10/30/air-force-space-force-raise-max-enlistment -age-to-42/.

27. See Ryan Haberman and Michael Pollard, "The Army Should Be Looking for a Few Older Soldiers," *Defense One*, April 7, 2023, https://www.defenseone.com /ideas/2023/04/army-should-be-looking-few-older-soldiers/384875/.

28. On this point, see Kuzminski and Sylvester, *Back to the Drafting Board.*

29. See Stephen A. Cheney and Stephen N. Xenakis, *Perspective—Obesity's Increasing Threat to Military Readiness: The Challenge to U.S. National Security* (American Security Project, 2022), https://www.jstor.org/stable/resrep46869; and Matthew Wallin, "Briefing Note—The Military Recruiting Crisis: Obesity's Impact on the Shortfall," American Security Project, March 2023, https://www.americansecurityproject.org /briefing-note-the-military-recruiting-crisis-obesitys-impact-on-the-shortfall/.

30. See Kirsten Weir, "The Extra Weight of COVID-19," https://www.apa.org, July 1, 2021, https://www.apa.org/monitor/2021/07/extra-weight-covid.

31. David Poyer, "The War with China, as I Saw It," *Air Force Times*, November 27, 2020, https://www.airforcetimes.com/opinion/commentary/2020/11/27 /the-war-with-china-as-i-saw-it/.

32. Brandon Leshchinskiy and Andrew Bowne, "Digital Transformation Is a Cultural Problem, Not a Technological One," *War on the Rocks*, May 17, 2022, https:// warontherocks.com/2022/05/digital-transformation-is-a-cultural-problem-not-a -technological-one/.

33. On the need for combatants with relevant tech skills, see Jon Reisher, "Talent and Tech: Fielding and Wielding New Systems Requires the Right People," Modern War Institute (at West Point), December 20, 2024, https://mwi .westpoint.edu/talent-and-tech-fielding-and-wielding-new-systems-requires -the-right-people/.

34. Chris Dougherty, *More than Half the Battle* (Center for a New American Security, 2021), 1, https://www.cnas.org/publications/reports/more-than-half-the-battle.

35. See Grady, "Navy Sees 'Difficult Times.'"

36. See Lauren C. Williams, "1/4 of DOD Cyber Jobs Are Vacant. Here's the Plan to Fill Them," *Defense One*, August 4, 2023, https://www.defenseone.com/defense -systems/2023/08/pentagon-lays-out-plan-boost-its-cyber-workforce/389123/.

37. Christine Wormuth et al., "Uncle Sam Wants You for a Military Job That Matters; We Need Data Scientists, Coders and Engineers as Much as We Need Pilots, Submariners and Infantry," Opinion, *Wall Street Journal*, October 24, 2022, https:// www.proquest.com/docview/2727758691/citation/527B56E49FCD4CA6PQ/1.

38. A copy of the February 8, 2018, memorandum that is quoted here can be found at Sydney J. Freedberg Jr., "Mattis Upguns Infantry: Task Force to Invest Over \$1B,"

Breaking Defense, February 21, 2018, https://breakingdefense.com/2018/02/mattis
-upguns-infantry-close-combat-lethality-task-force/.

39. The reenvisioning of the Marine Corps is expressed in the following guidance
document: US Marine Corps, Department of the Navy, "Force Design 2030,"
March 2020, https://www.hqmc.marines.mil/Portals/142/Docs/CMC38%20
Force%20Design%202030%20Report%20Phase%20I%20and%20II.pdf?ver=2020
-03-26-121328-460. The reorientation of the service has not been without its critics.
See Stephen W. Larose, "A View from the Trenches on the Debate Wracking the
Marine Corps," *War on the Rocks*, May 6, 2022, https://warontherocks.com/2022
/05/a-view-from-the-trenches-on-the-debate-wracking-the-marine-corps/.

40. See Todd South, "12-Man Rifle Squads, Including a Squad Systems Operator, Commandant Says," *Marine Corps Times*, May 4, 2018, https://www.marinecorpstimes
.com/news/your-marine-corps/2018/05/04/12-man-rifle-squads-including-a
-squad-systems-operator-commandant-says/; and Jay Price, "To Keep Up with
Modern Combat, Marines Add Drone Operators to Infantry Units," WUNC
91.5, North Carolina Public Radio, June 4, 2018, https://www.wunc.org/military
/2018-06-04/to-keep-up-with-modern-combat-marines-add-drone-operators-to
-infantry-units.

41. See Price, "To Keep Up with Modern Combat, Marines Add Drone Operators to
Infantry Units."

42. Jeong Soo Kim, "Beyond Force Design 2030: Preparing for Fourth-Generation
Combat," *U.S. Naval Institute Proceedings* 148, no. 11 (2022), https://www.usni.org
/magazines/proceedings/2022/november/beyond-force-design-2030-preparing
-fourth-generation-combat.

43. On this potential, see Zachary Schwartz, "Infantry Battalions as Sensor Webs for
the Fleet," *U.S. Naval Institute Proceedings* 150, no. 9 (2024), https://www.usni.org
/magazines/proceedings/2024/september/infantry-battalions-sensor-webs-fleet.

44. Heckl's statement is quoted in John Grady, "Marines Considering Autonomous
Systems for Almost Everything, General Says," *US Naval Institute (USNI) News*,
September 6, 2023, https://news.usni.org/?p=105549.

45. US Marine Corps Headquarters, "Force Design 2030 Annual Update," June 2023,
14, https://www.marines.mil/Portals/1/Docs/Force_Design_2030_Annual_Update
_June_2023.pdf.

46. Chirhart quote from Hope Hodge Seck, "'Robot Marines' in Every Formation:
Corps' Robotics Chief Casts Vision," *The War Zone (at The Drive)*, May 7, 2024,
https://www.twz.com/sea/robot-marines-in-every-formation-corps-robotics
-chief-casts-vision.

47. Scott Humr, "Bridging the Marine Corps's Digital Chasm," Modern War Institute (at West Point), November 20, 2024, https://mwi.westpoint.edu/bridging-the
-marine-corpss-digital-chasm/.

48. Such technologies are already undergoing experimentation at the small-unit
level. See Todd South, "31st MEU Tests Artificial Intelligence Sensing Gear to
Help Marines, Soldiers See Invisible Threats," *Marine Corps Times*, February 17,
2021, https://www.marinecorpstimes.com/news/your-marine-corps/2021/02/17
/31st-meu-tests-artificial-intelligence-sensing-gear-to-help-marines-soldiers-see
-invisible-threats/.

49. For more on this matter, see Heather Mongilio, "Marine Corps Personnel Change Was Key to New Force Design, Says CMC Berger," *US Naval Institute (USNI) News*, July 6, 2023, https://news.usni.org/2023/07/06/marine-corps-personnel -change-was-key-to-new-force-design-says-cmc-berger.

50. This point is emphasized in the following guidance document from the service: US Marine Corps Headquarters, "Training and Education 2030" (US Marine Corps, January 2023), https://www.marines.mil/Portals/1/Docs/Training%20and%20Edu cation%202030.pdf.

51. Regarding the Coast Guard on the issue, see Stew Magnuson, "Why the Coast Guard Lags When It Comes to Unmanned Systems," *National Defense*, July 7, 2023, https://www.nationaldefensemagazine.org/articles/2023/7/7/why-the-coast -guard-lags-when-it-comes-to-unmanned-systems.

52. Interestingly, the two services partnered in an overarching vision on unmanned systems. See US Department of the Navy, *Unmanned Campaign Framework* (US Department of the Navy, 2021).

53. See John Keller, "Boeing Starts Producing MQ-25 Stingray Unmanned Tanker Aircraft to Support Navy Aircraft Carrier Operations," Military Aerospace, November 8, 2022, https://www.militaryaerospace.com/sensors/article/14285399 /unmanned-tanker-aircraft-carrier.

54. For this development, see Brandi Vincent, "Hicks Unveils DOD's New 'Replicator' Initiative to Counter China via Autonomous Tech," *DefenseScoop*, August 28, 2023, https://defensescoop.com/2023/08/28/hicks-unveils-dods-new-replicator -initiative-to-counter-china-via-autonomous-tech/.

55. See Patty Nieberg, "The Army Is Planning for a New Robotics Technician MOS," *Task & Purpose*, October 31, 2024, https://taskandpurpose.com/news/robot-tech -warrant-officer/.

56. See Patrick Tucker, "The Pentagon Will Host a 'Top Gun' School for Ukraine-Style Attack Drones," *Defense One*, July 18, 2025, https://www.defenseone.com /technology/2025/07/pentagon-will-host-top-gun-school-ukraine-style-attack -drones/406818/.

57. See Sam Howell, "The United States' Quantum Talent Shortage Is a National Security Vulnerability," *Foreign Policy*, July 31, 2023, https://foreignpolicy.com/2023/07 /31/us-quantum-technology-china-competition-security/; *Final Report: National Security Commission on Artificial Intelligence* (National Security Commission on Artificial Intelligence's [NSCAI], 2021), 121, https://www.nscai.gov/wp-content /uploads/2021/03/Full-Report-Digital-1.pdf; Iain Cruickshank, "An AI-Ready Military Workforce," *Joint Force Quarterly* 110 (Quarter 2023): 46–53; and Brent Spillner, "Artificial Intelligence Is a Human-Centric Endeavor," *U.S. Naval Institute Proceedings* 149, no. 10 (2023), https://www.usni.org/magazines/proceedings /2023/october/artificial-intelligence-human-centric-endeavor. On the battlefield impact that artificial intelligence (as well as the high-end computing power that will enable it) will have, see Paul Scharre, *Four Battlegrounds: Power in the Age of Artificial Intelligence* (W.W. Norton, 2023).

58. See Leon L. Robert Jr. and Carl J. Wojtaszek, "Closing the Gap: Officer Advanced Education STEM+M (Management)," *The US Army War College Quarterly: Parameters* 54, no. 2 (2024): 111–27, https://doi.org/10.55540/0031-1723.3290; and

Scott A. Humr, "Take the Conn! Steering a Course for Technical Talent in Modern Naval Warfare," Center for International Maritime Security, September 3, 2024, https://cimsec.org/take-the-conn-steering-a-course-for-technical-talent-in-modern-naval-warfare/.

59. An example of such a critique is Lauren C. Williams, "To Solve National Security Problems, the US May Have to Rethink Higher Education," *Defense One*, November 2, 2023, https://www.defenseone.com/threats/2023/11/solve-national-security-problems-us-may-have-rethink-higher-education/391710/.

60. See James E. Bevins, "Incentivizing Innovation: Promoting Technical Competency to Win Future Wars," *Air & Space Operations Review* 1, no. 3 (2022): 22–37.

61. See Brian J. Fry, "Taking the Brakes Off Uniformed Scientists and Engineers," *Air & Space Operations Review* 1, no. 1 (2022): 18–33.

62. Megan O'Keefe, "Grow, Borrow, Recruit, and Reorganize: How the Military Can Get the Personnel It Needs for Digital War," Modern War Institute (at West Point), June 9, 2022, https://mwi.westpoint.edu/grow-borrow-recruit-and-reorganize-how-the-military-can-get-the-personnel-it-needs-for-digital-war/.

63. See Nand Mulchandani and Lieutenant General (Ret.) John N. T. "Jack" Shanahan, *Software-Defined Warfare: Architecting the DOD's Transition to the Digital Age*, a Report of the CSIS Strategic Technologies Program (Center for Strategic and International Studies, 2022), https://www.csis.org/analysis/software-defined-warfare-architecting-dods-transition-digital-age.

64. On this point, see Rebecca Slayton, "What Is a Cyber Warrior? The Emergence of U.S. Military Cyber Expertise, 1967–2018," *Texas National Security Review* 4, no. 1 (2020): 61–96.

65. See, again, O'Keefe, "Grow, Borrow, Recruit, and Reorganize." On the challenges of relying on such reserve forces, see Andrew Lewis Chadwick, *Part-Time Soldiers: Reserve Readiness Challenges in Modern Military History*, Studies in Civil-Military Relations (University Press of Kansas, 2023).

66. See Gregg Curley, "The Provision of Cyber Manpower: Creating a Virtual Reserve," *Marine Corps University Journal* 9, no. 1 (2018): 191–217; and Jeffrey J. Fair, "America's Cyber Auxiliary: Building Capacity and Future Operators," *The Cyber Defense Review* 7, no. 2 (2022): 57–66.

67. US Department of the Navy, *Department of the Navy Cyber Strategy* (US Department of the Navy, 2023), 3, https://media.defense.gov/2023/Nov/21/2003345095/-1/-1/0/DEPARTMENT%20OF%20THE%20NAVY%20CYBER%20STRATEGY.PDF.

68. See Todd South, "Army Boot Camp Will Soon Include Counter-Drone Training," C4ISRNet, November 15, 2023, https://www.c4isrnet.com/news/your-army/2023/11/15/army-boot-camp-will-soon-include-counter-drone-training/; and Jeff Schogol, "Army Revamps Basic Training to Simulate Battlefield Stalked by Drone Swarms," *Task & Purpose*, October 17, 2024, https://taskandpurpose.com/news/army-training-drones-conceal/.

69. See Mark Pomerleau, "Army Trying to Expose Entire Force to Electromagnetic Warfare During Training," *DefenseScoop*, August 17, 2023, https://defensescoop.com/2023/08/17/army-trying-to-expose-entire-force-to-electromagnetic-warfare-during-training/.

70. Abdul Subhani, "Is the Army Ready to Think Like a Tech Company? Why the Service Needs to Value Coding the Same Way as Shooting," Modern War Institute (at West Point), August 17, 2023, https://mwi.westpoint.edu/is-the-army-ready-to -think-like-a-tech-company-why-the-service-needs-to-value-coding-the-same -way-as-shooting/.

71. Jan De Backer, "The Physical Fitness of Soldiers: A Critical Factor in Modern Warfare," *Surfers and Chess Players: Everything About Healthcare and Technology* (blog), December 13, 2023, https://www.linkedin.com/pulse/physical-fitness -soldiers-critical-factor-modern-qs9nf/.

72. US Department of Defense, Office of the Under Secretary of Defense for Personnel and Readiness, "DOD Instruction 1308.03: DOD Physical Fitness/Body Composition Program" March 10, 2022, https://www.esd.whs.mil/portals/54/documents /dd/issuances/dodi/130803p.pdf.

73. On the creation and early operation of the new branch, see Forrest L. Marion, *Standing Up Space Force: The Road to the Nation's Sixth Armed Service* (US Naval Institute Press, 2023), https://www.usni.org/press/books/standing-space-force.

74. See Greg Hadley, "Everything You Need to Know About the Space Force's Fitness Tracker PT Study," *Air & Space Forces Magazine*, May 25, 2023, https://www .airandspaceforces.com/space-forces-fitness-tracker-pt-study/.

75. David Roza, "The Space Force Wants to Try Ditching Physical Fitness Tests for a Year," *Task & Purpose*, March 18, 2022, https://taskandpurpose.com/news/space -force-fitness-test/. For an argument from another service in favor of adopting wearable devices for physical fitness training, see Juan Garcia, "Bring the Navy's Physical Fitness Assessment into the 21st Century," *U.S. Naval Institute Proceedings* 145, no. 7 (2019), https://www.usni.org/magazines/proceedings/2019/july /bring-navys-physical-fitness-assessment-21st-century.

76. These issues are discussed in Adam T. Biggs and Dale W. Russell, "The Challenges of Wearable Technology," *U.S. Naval Institute Proceedings* 149, no. 20 (2023), https://www.usni.org/magazines/proceedings/2023/october/challenges-wearable -technology.

77. For a frightening look at one dimension of this security challenge, see Ronald J. Deibert, "The Autocrat in Your iPhone: How Mercenary Spyware Threatens Democracy," *Foreign Affairs* 102, no. 1 (2023): 72–88.

78. See Mariya Knight et al., "Russian Commander Killed While Jogging May Have Been Tracked on Strava App," CNN, July 11, 2023, https://www.cnn.com/2023/07 /11/europe/russian-submarine-commander-killed-krasnador-intl/index.html.

79. See Richard Pérez-Peña and Matthew Rosenberg, "Strava Fitness App Can Reveal Military Sites, Analysts Say," *New York Times*, January 29, 2018, https://www .nytimes.com/2018/01/29/world/middleeast/strava-heat-map.html.

80. Patrick M. Shanahan, "Use of Geolocation-Capable Devices, Applications, and Services," Memorandum, August 3, 2018, https://media.defense.gov/2018/Aug/06 /2001951064/-1/-1/1/GEOLOCATION-DEVICES-APPLICATIONS-SERVICES .PDF.

81. See John Vandiver, "Army Field-Tests AI System That Shields Wireless Network Use from Foes," *Stars and Stripes*, November 22, 2023, https://www.stripes.com /branches/army/2023-11-22/ai-army-electronic-warfare-12133758.html.

82. See Andrew Guthart, "Twitter Trackers Jeopardize Military Aircraft," *U.S. Naval Institute Proceedings* 148, no. 9 (2022), https://www.usni.org/magazines /proceedings/2022/september/twitter-trackers-jeopardize-military-aircraft-0. On the national security risks of open broadcasts from military aircraft transponders, see Jan Tegler, "Air Hazars: Open Source Flight Tracking Called Threat to Military Aircraft," *National Defense*, February 2023, 22–25. Proposals have been made to help secure this potential problem. See Joseph Trevithick, "Tracking U.S. Military Aircraft Online Could Become Much Harder," *The War Zone (at The Drive)*, July 19, 2023, https://www.thedrive.com/the-war-zone/tracking-u-s-military-aircraft -online-could-become-much-harder.

83. On the security risks associated with the personal social media activities of US military personnel, see Timothy McGeehan, "Web Storm Rising," *U.S. Naval Institute Proceedings* 148, no. 3 (2022), https://www.usni.org/magazines/proceedings /2022/march/web-storm-rising.

84. On this example, see Will McGee, "Social Media Puts EABO at Risk," *U.S. Naval Institute Proceedings* 148, no. 4 (2022), https://www.usni.org/magazines /proceedings/2022/april/social-media-puts-eabo-risk.

85. US Army Asymmetric Warfare Group, *Russian New Generation Warfare Handbook* (US Army Asymmetric Warfare Group, 2016), 46–47, https://info.publicintelligence .net/AWG-RussianNewWarfareHandbook.pdf. See also Joseph Trevithick, "New US Army Manual Shows It's Worried About Russia's Hybrid Warfare Tactics," *The War Zone (at The Drive)*, June 29, 2019, https://www.thedrive.com/the-war-zone /14647/new-us-army-manual-shows-its-worried-about-russias-hybrid-warfare -tactics.

86. This statement from an anonymous special operations specialist is quoted in Hope Seck, "MARSOC Raiders to Deploy in Smaller, Tech-Loaded Teams as Conflict Gets More Complex," *Sandboxx News*, July 17, 2023, https://www.sandboxx.us/ blog/marsoc-raiders-to-deploy-in-smaller-tech-loaded-teams-as-conflict-gets- more-complex/.

87. See Robert Labrenz, "Drill Emission Control as a Main Battery," *U.S. Naval Institute Proceedings* 149, no. 6 (2023), https://www.usni.org/magazines/proceedings /2023/june/drill-emission-control-main-battery; Michael Holdridge, "Mission Command Means Emissions Control," *U.S. Naval Institute Proceedings* 147, no. 5 (2021), https://www.usni.org/magazines/proceedings/2021/may/mission -command-means-emissions-control; and Patrick A. Goldman, "In the Digital Age, Make Ships Go Dark," *U.S. Naval Institute Proceedings* 148, no. 8 (2022), https://www.usni.org/magazines/proceedings/2022/august/digital-age-make -ships-go-dark.

88. For more on this point, see Sam J. Tangredi, "Fighting When the Network Dies," *U.S. Naval Institute Proceedings* 149, no. 3 (2023), https://www.usni.org/magazines /proceedings/2023/march/fighting-when-network-dies; and Nick Danby, "Carrier Strike Groups Should Be Ready to Go Dark in Conflict," *War on the Rocks*, August 29, 2023, https://warontherocks.com/2023/08/carrier-strike-groups-should -be-ready-to-go-dark-in-conflict/.

89. Berger quote from Christopher Woody, "Russia's War in Ukraine Shows Why Troops Need to Learn to Put Their Phones Away, Top US Marine General Says,"

Business Insider, December 31, 2022, https://www.businessinsider.com/russia-ukraine-war-shows-battlefield-phone-risk-top-marine-says-2022-12.

90. See Zachary Reuther, "The Coast Guard Must Adopt the Physical Fitness Test," *U.S. Naval Institute Proceedings* 149, no. 1 (2023), https://www.usni.org/magazines/proceedings/2023/january/coast-guard-must-adopt-physical-fitness-test.

91. Krogh quote from Anna Mulrine, "How Can Army Keep Soldiers Fighting Fit After Afghanistan? Avatars," *Christian Science Monitor*, May 3, 2012, https://www.csmonitor.com/USA/Military/2012/0503/How-can-Army-keep-soldiers-fighting-fit-after-Afghanistan-Avatars.

92. US Army Public Affairs, "Army Announces Creation of Future Soldier Preparatory Course," US Army website, July 26, 2022, https://www.army.mil/article/258758/army_announces_creation_of_future_soldier_preparatory_course.

93. Haley Britzky, "The Army's New Pre-Basic-Training Boot Camp Might Actually Build Better Soldiers," *Task & Purpose*, October 19, 2022, https://taskandpurpose.com/news/army-future-soldier-prep-course-recruitment/.

94. See Davis Winkie, "Army Expands 'Prep Course' for Low-Scoring Applicants After Pilot," *Army Times*, January 9, 2023, https://www.armytimes.com/news/your-army/2023/01/09/army-expands-prep-course-for-low-scoring-applicants-after-pilot/.

95. Center for a New American Security, "Virtual Fireside Chat: Recruitment, Retention, and Quality of Life in the Force with the Hon. Christine Wormuth, Secretary of the Army," with Katherine Kuzminski and Christine Wormuth, November 18, 2022, https://www.cnas.org/events/virtual-fireside-chat-recruitment-retention-and-quality-of-life-in-the-force-honorable-christine-wormuth-secretary-of-the-army.

96. See Doug G. Ware, "Army's Basic Training Prep Course to Become Permanent Part of Recruiting Strategy," *Stars and Stripes*, September 1, 2023, https://www.stripes.com/branches/army/2023-09-01/army-basic-training-prep-course-recruiting-11237624.html.

97. See Stew Smith, "Avoid These Training Activities When Preparing to Serve," Military.com, September 17, 2021, https://www.military.com/military-fitness/avoid-these-training-activities-when-preparing-serve.

98. This argument is made in David Barno and Nora Bensahel, "Addressing the U.S. Military Recruiting Crisis," *War on the Rocks*, March 10, 2023, https://waron-therocks.com/2023/03/addressing-the-u-s-military-recruiting-crisis/.

99. See Rebecca Kheel, "Navy Follows Army in Offering Prep Courses to Recruits Who Don't Meet Fitness, Academic Standards," Military.com, March 22, 2023, https://www.military.com/daily-news/2023/03/22/navy-follows-army-offering-prep-courses-recruits-who-dont-meet-fitness-academic-standards.html.

100. See Diana Correll, "Navy Prep Course Aims to Launch Hundreds of Recruits into the Fleet," *Navy Times*, July 3, 2023, https://www.navytimes.com/news/your-navy/2023/07/03/navy-prep-course-aims-to-launch-hundreds-of-recruits-into-the-fleet/.

101. See Hadiyah Brendel, "MIRROR Project Creates Body-Worn Sensor Technology Predictive of Musculoskeletal Injury," Defense Visual Information Distribution Service (DVIDS), February 23, 2023, https://www.dvidshub.net/news/439242/mirror

-project-creates-body-worn-sensor-technology-predictive-musculoskeletal
-injury.

102. See Meghann Myers, "Get Ready for Dietitians, Physical Therapists and More in Every Army Battalion," *Defense News*, October 9, 2018, https://www.defensenews .com/news/your-army/2018/10/09/get-ready-for-dietitians-physical-therapists -and-more-in-every-army-battalion/.

103. An example of such an undertaking is covered in Haley Britzky, "The Army Is Developing an Alternate Combat Fitness Test for Soldiers with Permanent Injuries," *Task & Purpose*, May 21, 2019, https://taskandpurpose.com/news/army -combat-fitness-test-alternative/.

104. Helton quote from Brendel, "MIRROR Project Creates Body-Worn Sensor Technology."

105. See Hope Seck, "Researchers Want to Prevent Injuries in Soldiers Before They Happen," *Sandboxx News*, April 5, 2023, https://www.sandboxx.us/blog/researchers -want-to-prevent-injuries-in-soldiers-before-they-happen/.

106. Katherine L. Kuzminski, "The Military Needs to Make Human-Performance Optimization Part of Daily Ops," Center for a New American Security, August 1, 2024, https://www.cnas.org/publications/commentary/the-military-needs-to-make -human-performance-optimization-part-of-daily-ops.

107. See Annette Boyle, "DHA Adds New Weapons in the Military's War on Obesity in Servicemembers," US Medicine, June 18, 2018, https://www.usmedicine.com /clinical-topics/obesity/dha-adds-new-weapons-in-the-militarys-war-on-obesity -in-servicemembers/.

108. Linn, "A Historical Perspective on Today's Recruiting Crisis," 8–9.

109. Pope quote from Gidget Fuentes, "Pilot Course Aims to Build Marines' Skills as Communicators for the Future Fight," *US Naval Institute (USNI) News*, February 28, 2023, https://news.usni.org/2023/02/28/pilot-course-aims-to-build-marines -skills-as-communicators-for-the-future-fight.

110. H.R. 2670: National Defense Authorization Act for Fiscal Year 2024, § 577 (2023), https://www.govinfo.gov/content/pkg/BILLS-118hr2670enr/uslm/BILLS-118 hr2670enr.xml.

111. See Todd South, "New Military Simulations for Shooting, Trench War, Drones Unveiled," *Marine Corps Times*, November 22, 2023, https://www .marinecorpstimes.com/flashpoints/2023/11/22/new-military-simulations-for -shooting-trench-war-drones-unveiled/.

112. Størdal's quote is from Sean Carberry, "Siloed Simulators: Improved Training at NATO a Bureaucracy, Not Technology Problem," *National Defense* 107, no. 835 (2023): 25.

113. See South, "New Military Simulations for Shooting."

114. See Hyokwon Jung, "A Glimpse into the Future Battlefield with AI-Embedded Wargames," *U.S. Naval Institute Proceedings* 150, no. 6 (2024), https://www.usni. org/magazines/proceedings/2024/june/glimpse-future-battlefield-ai-embedded- wargames.

115. Berger's explanation is quoted in Megan Eckstein, "US Navy, Marines Look for Training Systems with Accurate Adversaries, Ability to Track Individual Performance," *Defense News*, November 30, 2021, https://www.defensenews.com

/naval/2021/11/30/navy-marines-looking-for-training-systems-with-accurate-adversaries-ability-to-track-individuals-performance/.

116. Ritzema's quote as well as a description of the drone simulator can be found in Meghann Myers, "Army Scrambles to Improve Drone Training," *Defense One*, May 30, 2025, https://www.defenseone.com/technology/2025/05/army-scrambles-improve-drone-training/405706/.

117. See Colin Schultz, "A Military Contractor Just Went Ahead and Used an Xbox Controller for Their New Giant Laser Cannon," *Smithsonian Magazine*, September 9, 2014, https://www.smithsonianmag.com/smart-news/military-contractor-just-went-ahead-and-used-xbox-controller-their-new-giant-laser-cannon-180952647/.

118. See David Hambling, "Game Controllers Driving Drones, Nukes," *Wired*, July 19, 2008, https://www.wired.com/2008/07/wargames/; Nathan Hodge, "Future Warbot Powered by Xbox Controller," *Wired*, June 12, 2009, https://www.wired.com/2009/06/future-warbot-powered-by-xbox-controller/; Leore Dayan and Noah Smith, "A New Israeli Tank Features Xbox Controllers, AI Honed by 'StarCraft II' and 'Doom,'" *Washington Post*, July 28, 2020, https://www.washingtonpost.com/video-games/2020/07/28/new-israeli-tank-features-xbox-controllers-ai-honed-by-starcraft-ii-doom/; and Travis M. Andrews, "The Navy's Adding a New Piece of Equipment to Nuclear Submarines: Xbox Controllers," *Washington Post*, September 25, 2017, https://www.washingtonpost.com/news/morning-mix/wp/2017/09/25/the-navys-adding-a-new-piece-of-a-equipment-to-nuclear-submarines-xbox-controllers/.

119. McElvogue's quote can be found in *Breaking Defense*, "Better User Interface and Experience Create Faster Decision-Making in Multi-Domain Operations," October 28, 2024, http://breakingdefense.com/2024/10/better-user-interface-and-experience-create-faster-decision-making-in-multi-domain-operations/.

120. On this point, see David H. Lewis, "Build This, Not That," *U.S. Naval Institute Proceedings* 149, no. 10 (2023), https://www.usni.org/magazines/proceedings/2023/october/build-not.

121. See Vikram Mittal and Andrew Davidson, "Combining Wargaming with Modeling and Simulation to Project Future Military Technology Requirements," *IEEE Transactions on Engineering Management* 68, no. 4 (2021): 1195–1207, https://doi.org/10.1109/TEM.2020.3017459.

122. Amy Robinson, "Army Seeks Gamers' Input to Help Shape Future Force," US Department of Defense, August 23, 2017, https://www.defense.gov/News/News-Stories/Article/Article/1286948/army-seeks-gamers-input-to-help-shape-future-force/.

123. Dyess quote from Jon Harper, "Army Creates Video Game to Prototype New Weapons," *National Defense* 102, no. 769 (2017):, 36, 35. See also Adin Dobkin, "The Video Game That Could Shape the Future of War," *The Atlantic*, October 26, 2017, https://www.theatlantic.com/technology/archive/2017/10/operation-overmatch/544062/.

124. For more on the use of such technologies in actual war-fighting, consult Akshat Upadhyay, "Virtual and Augmented Reality and Warfare: Fighting War as a Computer Game?," in *Future Warfare and Critical Technologies: Evolving Tactics and Strategies*, ed. Rajeswari Pillai Rajagopalan and Sameer Patil (Observer Research Foundation and Global Policy Journal, 2024), 28–42.

125. Jared Keller, "The US Military Will Fight the Next Big War with Xbox-Style Video Game Controllers," *Task & Purpose*, March 22, 2023, https://taskandpurpose.com /tech-tactics/us-military-video-game-controllers-war/.

126. See UK Ministry of Defense and Harriett Baldwin, "Xbox Controllers, Hoverbikes and Robotic Trucks Trialled by British and American Armies," Gov.uk, November 14, 2017, https://www.gov.uk/government/news/xbox-controllers-hoverbikes -and-robotic-trucks-trialled-by-british-and-american-armies.

127. Marine Corps Headquarters, "Expeditionary Advanced Base Operations (EABO)," US Marine Corps Flagship, August 2, 2021, https://www.marines.mil /News/News-Display/Article/2708120/expeditionary-advanced-base-operations -eabo/.

128. See James Winnefeld, "NMESIS Now," *U.S. Naval Institute Proceedings* 147, no. 11 (2021), https://www.usni.org/magazines/proceedings/2021/november/nmesis-now.

129. See Keller, "The US Military Will Fight the Next Big War."

130. See Keller, "The US Military Will Fight the Next Big War."

131. On this point, see Ashley Roque, "Army Eyeing First Human Machine Integrated Formations in 2027, Common Controller for Robotics," *Breaking Defense*, April 14, 2025, https://breakingdefense.com/2025/04/army-eyeing-first-human-machine-inte grated-formations-in-2027-common-controller-for-robotics/.

132. See Arie Egozi, "IAI Wins Israeli Carmel Land Vehicle Competition; Company Wants American Partner," *Breaking Defense*, October 11, 2021, https:// breakingdefense.com/2021/10/iai-wins-israeli-carmel-land-vehicle-competition -company-wants-american-partner/.

133. Dayan and Smith, "A New Israeli Tank Features Xbox Controllers."

134. Bolthouse quoted in Thomas M. Hunt, "The Digital Games That Militaries Play," Stratfor.com, January 8, 2018, https://worldview.stratfor.com/article/digital-games -militaries-play.

135. US Space Force, "U.S. Space Force Vision for a Digital Service," May 2021, https:// media.defense.gov/2021/May/06/2002635623/-1/-1/1/USSF%20VISION%20FOR%20 A%20DIGITAL%20SERVICE%202021%20(2).pdf; Courtney Albon, "Space Force's Digital Push Focuses on 'Spaceverse,'" *Defense News*, October 5, 2022, https://www .defensenews.com/battlefield-tech/space/2022/10/05/space-forces-digital-push -focuses-on-spaceverse/. On the many possibilities of the metaverse concept for the military, see Jennifer McArdle and Caitlin Dohrman, "The Full Potential of a Military Metaverse," *War on the Rocks*, February 18, 2022, https://warontherocks .com/2022/02/the-full-potential-of-a-military-metaverse/.

136. This was suggested in a 2015 Carnegie Mellon University study funded by the Defense Department: Christopher Herr and Dennis M Allen, *Video Games as a Training Tool to Prepare the Next Generation of Cyber Warriors* (Software Engineering Institute, Carnegie Mellon University, 2015). For insights into how cyber war-gaming might be employed, see the various contributions in Frank L. Smith III, Nina A. Kollars, and Benjamin H. Schechter, eds., *Cyber Wargaming: Research and Education for Security in a Dangerous Digital World* (Georgetown University Press, 2024).

137. See Jeff Schogol, "Russia Jams US GPS-Guided Weapons Given to Ukraine, Leaked Info Shows," *Task & Purpose*, April 13, 2023, https://taskandpurpose.com /news/russia-ukraine-gps-weapons-jam/.

138. See Daniel J. Finkenstadt and Erik Helzer, "Gamified Learning Can Be Effective," *U.S. Naval Institute Proceedings* 149, no. 3 (2023), https://www.usni.org/magazines /proceedings/2023/march/gamified-learning-can-be-effective.

Conclusion

1. See GBD 2021 US Obesity Forecasting Collaborators, "National-Level and State-Level Prevalence of Overweight and Obesity Among Children, Adolescents, and Adults in the USA, 1990–2021, and Forecasts up to 2050," *The Lancet* 404, no. 10469 (2024): 2278–98, https://doi.org/10.1016/S0140-6736(24)01548-4.

2. See Jon Harper, "Gen. Milley Anticipates New 'Joint Futures' Organization Will Come to Fruition," *DefenseScoop*, June 30, 2023, https://defensescoop.com/2023 /06/30/gen-milley-anticipates-new-joint-futures-organization-will-come-to -fruition/.

3. See Thomas Newdick, "Navy's Aviation Boss Lays Out Big Vision for Drone-Packed Carriers Of The Future," *The War Zone (at The Drive)*, April 1, 2021, https:// www.twz.com/40007/navys-aviation-boss-lays-out-big-vision-for-drone-packed -carrier-air-wings-of-the-future.

4. On this point, see Riley Ceder, "MARSOC Is Fusing Traditionally Rugged Marines with Tech-Curious Ones," *Military Times*, April 30, 2025, https://www .militarytimes.com/land/2025/04/30/marsoc-is-fusing-traditionally-rugged -marines-with-tech-curious-ones/; and Bradley C. Tibbetts et al., *Heritage Meets Hardware: The Fusion of SOF Tradecraft with Modern Technology*, Futures Seminar Report, Irregular Advantage Initiative (US Army War College, 2025), https:// media.defense.gov/2025/Jun/02/2003730500/-1/-1/0/4_HERITAGEMEETSHARD WARE(USAWCFUTURES_SOCOMJ7FINALPROJECT).pdf.

5. Christopher Hurd, "Army Futures Command General Lays Out Continuous Transformation Plan," US Army website, April 2, 2024, https://www.army .mil/article/275040/army_futures_command_general_lays_out_continuous _transformation_plan.

6. The literature on these issues is voluminous. Interested readers might begin with Robert H. Latiff, *Future War: Preparing for the New Global Battlefield* (Alfred A. Knopf, 2017); and Robert H. Latiff, *Future Peace: Technology, Aggression, and the Rush to War* (University of Notre Dame Press, 2022).

7. For useful discussions on such matters, see C. D. Meyers, "GI, Robot: The Ethics of Using Robots in Combat," *Public Affairs Quarterly* 25, no. 1 (2011): 21–36; Joseph O. Chapa, *Is Remote Warfare Moral?: Weighing Issues of Life and Death from 7,000 Miles* (PublicAffairs, 2022); Michael J. Boyle, *The Drone Age: How Drone Technology Will Change War and Peace* (Oxford University Press, 2020); and Deane Baker, *Should We Ban Killer Robots?* (Polity Press, 2022).

8. For such a perspective, consult Michael P. Ferguson, "Ghost in the Machine: Coming to Terms with the Human Core of Unmanned War," *Texas National Security Review* 8, no. 2 (2025): 27–46. Helping balance against this is the fact that while drones increase the risk of combat, research shows that their use tends to limit the level of conflict intensity. See Erik Lin-Greenberg, "Wargame of Drones: Remotely Piloted Aircraft and Crisis Escalation," *Journal of Conflict Resolution* 66, no. 10 (2022): 1737–65, https://dspace.mit.edu/handle/1721.1/148674.

9. This argument is outlined in Nathan Gabriel Wood, "The Problem with Killer Robots," *Journal of Military Ethics* 19, no. 3 (2020): 220–40, https://doi.org/10.1080/15027570.2020.1849966.

10. Deputy Secretary of Defense Ashton B. Carter, "U.S. Department of Defense Directive 3000.09: Autonomy in Weapon Systems," US Department of Defense, November 21, 2012, https://www.hsdl.org/?view&did=726163.

11. Tyson Wetzel, "'Killer Robots' Are Coming. Is the US Ready for the Consequences?," Atlantic Council, June 17, 2022, https://www.atlanticcouncil.org/content-series/automating-the-fight/killer-robots-are-coming-is-the-us-ready-for-the-consequences/.

12. Although the article is dated, the following provides a useful summary of the ethical debates on AI-driven autonomous weapons systems: Amitia Etzioni and Oren Etzioni, "Pros and Cons of Autonomous Weapons Systems," *Military Review: The Professional Journal of the U.S. Army* 97, no. 3 (2017): 72–81. On Russia's and China's positions on these issues, see Anna Nadibaidze, "Great Power Identity in Russia's Position on Autonomous Weapons Systems," *Contemporary Security Policy* 43, no. 3 (2022): 407–35, https://doi.org/10.1080/13523260.2022.2075665; Gregory C. Allen, "One Key Challenge for Diplomacy on AI: China's Military Does Not Want to Talk," Center for Strategic and International Studies, May 20, 2022, https://www.csis.org/analysis/one-key-challenge-diplomacy-ai-chinas-military-does-not-want-talk; and Thomas Christian Bächle and Jascha Bareis, "'Autonomous Weapons' as a Geopolitical Signifier in a National Power Play: Analysing AI Imaginaries in Chinese and US Military Policies," *European Journal of Futures Research* 10, no. 1 (2022), https://doi.org/10.1186/s40309-022-00202-w.

13. See Defense Advanced Research Projects Agency (DARPA), "DARPA Exploring Ways to Assess Ethics for Autonomous Weapons," December 19, 2024, https://www.darpa.mil/news/2024/asimov-approaches.

14. On the films and other fictional accounts of such matters, consult Teresa Heffernan, "Autonomous Weapons in Fiction and the Fiction of Autonomous Weapons," in *The Realities of Autonomous Weapons*, ed. Thomas Christian Bächle and Jascha Bareis (Bristol University Press, 2025). The risks of artificial intelligence systems are most acute in connection to nuclear weapons systems. See Jacob Stokes et al., *Averting AI Armageddon* (Center for a New American Security, 2025), https://www.cnas.org/publications/reports/averting-ai-armageddon.

15. The event is described in Tim Robinson and Stephen Bridgewater, "Highlights from the RAeS Future Combat Air & Space Capabilities Summit," Royal Aeronautical Society, May 26, 2023, https://www.aerosociety.com/news/highlights-from-the-raes-future-combat-air-space-capabilities-summit/.

16. Hamilton's clarification is outlined in Joseph Trevithick, "Air Force Colonel Now Says Drone That Turned on Its Operator Was a 'Thought Experiment,'" *The War Zone (at The Drive)*, June 1, 2023, https://www.thedrive.com/the-war-zone/artificial-intelligence-enabled-drone-went-full-terminator-in-air-force-test.

17. For a critique regarding the probability of keeping humans in operational control of such systems, see Stephen Gerras and Andrew A. Hill, "Lucy in the Chocolate Factory: On the Inevitability of Killer Machines," *War Room*, September 12, 2024, https://warroom.armywarcollege.edu/articles/the-chocolate-factory/.

18. Dean's quote is from Jon Harper, "Army Taps General Dynamics, American Rheinmetall for Next Phases of Optionally Manned Fighting Vehicle Program," *DefenseScoop*, June 26, 2023, https://defensescoop.com/2023/06/26/army -taps-general-dynamics-american-rheinmetall-for-next-phases-of-optionally -manned-fighting-vehicle-program/.

19. On this point, see Daniel Vardiman, "Is Cutting-Edge Military Tech Really Cheaper Than Manpower?," Atlantic Council, September 20, 2022, https:// www.atlanticcouncil.org/content-series/automating-the-fight/is-cutting -edge-military-tech-really-cheaper-than-manpower/; and Andrew Herr, "Will Humans Matter in the Wars of 2030?," *Joint Force Quarterly* 77 (2nd Quarter 2015): 76–83.

20. See Ashley Roque, "More Soldiers, Control Vehicles May Be Required to Maneuver Army's Early Robotic Combat Vehicles," *Breaking Defense*, October 31, 2024, https://breakingdefense.com/2024/10/more-soldiers-control-vehicles-may-be -required-to-maneuver-armys-early-remote-combat-vehicles/.

21. On the matter, see Stephen Walsh, "Maximize the Two-Seat Super Hornet for the Peer Fight," *U.S. Naval Institute Proceedings* 145, no. 9 (2019), https://www.usni .org/magazines/proceedings/2019/september/maximize-two-seat-super-hornet -peer-fight; and Graham Scarbro, "Naval Flight Officers' Unmanned Future," *U.S. Naval Institute Proceedings* 147, no. 9 (2021), https://www.usni.org/magazines /proceedings/2021/september/naval-flight-officers-unmanned-future.

22. In respect to the continuing trend toward single-seat tactical aircraft, it is instructive that the US Marine Corps is eliminating the weapons systems officer position entirely. See Tyler Rogoway, "Marines Winding Down Weapon Systems Officer Position, F/A-18Ds to Fly with Pilot Only," The Drive, July 19, 2019, https://www .thedrive.com/the-war-zone/29078/marines-winding-down-weapon-system -officers-position-f-a-18ds-to-fly-with-pilot-only. For insights into how to optimize the two-seat tactical aircraft that remain in the Navy for the management of this complexity, consult Walsh, "Maximize the Two-Seat Super Hornet."

23. On this point, see Patrick Tucker, "'Swarm Pilots' Will Need New Tactics—and Entirely New Training Methods: Air Force Special-Ops Chief," *Defense One*, May 10, 2024, https://www.defenseone.com/policy/2024/05/swarm-pilots-will -need-new-tacticsand-entirely-new-training-methods-air-force-special-ops-chief /396477/. Bauernfeind's quote is also found in this piece.

24. On the impact of autonomous weapons systems, see Paul Scharre, *Army of None: Autonomous Weapons and the Future of War* (W.W. Norton, 2018). For the ongoing importance of human actors within this context, see Avi Goldfarb and Jon R. Lindsay, "Prediction and Judgment: Why Artificial Intelligence Increases the Importance of Humans in War," *International Security* 46, no. 3 (2021): 7–50, https://doi.org/10.1162/isec_a_00425.

25. On this reality, see Kelsey Kernstine, "US Army Recruiting: Could Female Soldiers Close the Gap?," *NewsNation*, April 17, 2023, https://www.newsnationnow .com/us-news/military/us-army-recruiting-could-female-soldiers-close-the-gap/. Though its focus goes beyond US combatant forces, also see Christine Martinez, "Women, Peace, and Security: An Underutilized Tool in Countering the People's Republic of C," *Journal of Indo-Pacific Affairs* 6, no. 7 (2023): 110–16.

26. Hegseth's comments in this regard can be found at: *Pete Hegseth—Secretary of Defense Nominee | SRS #143*, YouTube video, Shawn Ryan Show, 2024, https://www.youtube.com/watch?v=DoN5ovwB8s4.

27. See Carrie Keller-Lynn, "Israel Is Putting More Women on the Front Line to Help Fix Its Manpower Problem," *Wall Street Journal*, June 8, 2025, https://www.wsj.com/world/middle-east/israel-is-putting-more-women-on-the-front-line-to-help-fix-its-manpower-problem-a962fe79.

28. Act to Amend Title 10, United States Code, to Simplify Laws Relating to Members of the Army, Navy, Air Force, and Marine Corps, and for Other Purposes., Pub. L. No. Public Law 90–235, 753 (1968). https://www.congress.gov/90/statute/STATUTE-81/STATUTE-81-Pg753.pdf.

29. National Defense Authorization Act for Fiscal Year 2006, Pub. L. No. Public Law 109–163, 3136 (2006). https://www.congress.gov/109/plaws/publ163/PLAW-109publ163.pdf.

30. See Thomas Claburn, "US Mulls Drafting Gray-Haired Hackers During Times of Crisis," *The Register*, March 21, 2018, https://www.theregister.com/2018/03/21/uncle_sam_mulls_drafting_grayhaired_hackers_during_times_of_crisis/.

31. For the argument that personnel levels should become smaller in the US military, see Shawn M. Cook et al., "Developing Alternative Manning Strategies to Maintain the Combat Effectiveness of the Joint Force," *Joint Force Quarterly* 117, no. 2 (2025): 58–63.

32. This argument is made using the much earlier Korean War as an example in Andrew J. Forney, "Davy Crockett and the Boy Scouts: The Korean War and Mismanaging Protracted Conflict," *Texas National Security Review* 8, no. 1 (2024): 45–48.

33. See Matthew Fandre, "Medical Changes Needed for Large-Scale Combat Operations: Observations from Mission Command Training Program Warfighter Exercises," *Military Review: The Professional Journal of the U.S. Army* 100, no. 3 (2020): 37; and Jeff Schogol and Patty Nieberg, "War with China 'Would Result in Large-Scale Casualties,' Army General Says," *Task & Purpose*, July 24, 2025, https://taskandpurpose.com/news/us-china-war-casualties/.

34. The logistical challenges of the Pacific are outlined in Carmelia Scott-Skillern, "Army Logistics in the Pacific," New America, September 13, 2023, http://newamerica.org/future-security/reports/army-logistics-in-the-pacific/.

35. See Matt White, "No 'Golden Hour'? How Army Medicine Is Changing for the Next War," *Task & Purpose*, June 20, 2023, https://taskandpurpose.com/news/golden-hour-army-medical-training-ukraine/; and George A. Barbee, "The Strategic Survivability Triad: The Future of Military Medicine in Support of Combat Power," *Joint Force Quarterly* 107 (October 2022): 102–15.

36. See Heather Mongilio, "Navy Medicine Is Preparing for the Future of Expeditionary Combat," *US Naval Institute (USNI) News*, February 22, 2023, https://news.usni.org/2023/02/22/navy-medicine-is-preparing-for-the-future-of-expeditionary-combat. For more on US military planning on medical evacuations in the Indo-Pacific, consult Mahdi Al-Husseini et al., "Bridging Sky and Sea: Joint Strategies for Medical Evacuation in the Indo-Pacific," *The US Army War College Quarterly: Parameters* 55, no. 1 (2025): 117–32, https://doi.org/10.55540/0031-1723.3334.

37. See Patty Nieberg, "Blood Brothers: In the Pacific, US Medics May Soon Use Blood from Allies," *Task & Purpose*, November 6, 2023, https://taskandpurpose.com /history/blood-agreements-china-allies/.

38. See Olivia Peduzzi, "The Promising Virtual Mental Health Pilot," *U.S. Naval Institute Proceedings* 147, no. 2 (2021), https://www.usni.org/magazines/proceedings /2021/february/promising-virtual-mental-health-pilot; and Matthew Thayer Hall, "Drones Can Speed Medical Care, Search and Rescue," *U.S. Naval Institute Proceedings* 147, no. 2 (2021), https://www.usni.org/magazines/proceedings/2021 /february/drones-can-speed-medical-care-search-and-rescue.

39. See Benjamin P. Donham, "It's Not Just About the Algorithm: Development of a Joint Medical Artificial Intelligence," *Joint Force Quarterly* 111 (2023): 50–57.

40. See Tom Mehlhorn and Allen Garner, "Electric Plasmas Can Extend the Golden Hour," *U.S. Naval Institute* 149, no. 4 (2023), https://www.usni.org/magazines /proceedings/2023/april/electric-plasmas-can-extend-golden-hour.

41. See Justin Lynch, "A Great Power War Could Require More Troops, and More Quickly, Than the United States Can Generate—Even with a Draft," Modern War Institute, October 1, 2020, https://mwi.westpoint.edu/a-great-power-war-could -require-more-troops-and-more-quickly-than-the-united-states-can-generate -even-with-a-draft/.

42. See Hunter Keeley, "Envisioning a Hellscape: Ukrainian Lessons for a Taiwan Drone Strategy," *U.S. Naval Institute Proceedings* 151, no. 4 (2025): 54–59.

43. On this point, see Tyler Hacker, "How the US Army Can Close Its Dangerous— and Growing—Small Drone Gap," Modern War Institute, March 6, 2024, https:// mwi.westpoint.edu/how-the-us-army-can-close-its-dangerous-and-growing -small-drone-gap/; Trevor Phillips-Levine and Walker D. Mills, "Outgunned in the Drone Fight: The U.S. Military Is Failing to Adopt the Next Machine Gun," *War on the Rocks*, March 6, 2024, https://warontherocks.com/2024/03/outgunned -in-the-drone-fight-the-u-s-military-is-failing-to-adopt-the-next-machine-gun/; Stacie L. Pettyjohn, "America's Eroding Airpower: Washington Must Upgrade Its Fleet of Planes, Drones, and Missiles," *Foreign Affairs*, March 10, 2025, https://www .foreignaffairs.com/united-states/americas-eroding-airpower; and Kerry Chávez and Ori Swed, "Emulating Underdogs: Tactical Drones in the Russia Ukraine War," *Contemporary Security Policy* 44, no. 4 (2023): 592–605, https://doi.org/10 .1080/13523260.2023.2257964. The need for such an emphasis on low-cost innova- tion extends beyond drones. See Jacquelyn Schneider, "Does Technology Win Wars?," *Foreign Affairs*, March 3, 2023, https://www.foreignaffairs.com/ukraine /does-technology-win-wars.

44. On this point, see Tucker, "'Swarm Pilots' Will Need New Tactics." An excellent overview of what it takes to optimize the use of unmanned systems on the battle- field is provided in the following report: Justin Bronk and Jack Watling, *Mass Pre- cision Strike: Designing UAV Complexes for Land Forces*, occasional paper (Royal United Services Institute for Defence and Security Studies, 2024).

45. On these points, see Antonio Salinas et al., "The Meaning of Drone-Enabled Infantry Striking Beyond Line of Sight," *War on the Rocks*, June 23, 2025, https:// warontherocks.com/2025/06/the-meaning-of-drone-enabled-infantry-striking -beyond-line-of-sight/.

46. On this point, see Mark Jbeily, "Trust, but Verify: Building Aviator Relationships with AI," *U.S. Naval Institute Proceedings* 148, no. 9 (2022), https://www.usni.org/magazines/proceedings/2022/september/trust-verify-building-aviator-relationships-ai; and Kyle Mizokami, "The U.S. Navy Wants Autonomous Fighter Jets, But First It Needs to Trust AI," Military Aviation, *Popular Mechanics*, April 13, 2023, https://www.popularmechanics.com/military/aviation/a43541414/the-navy-wants-ai-powered-autonomous-fighter-jets/.

47. See Heather M. Roff and David Danks, "'Trust but Verify': The Difficulty of Trusting Autonomous Weapons Systems," *Journal of Military Ethics* 17, no. 1 (2018): 2–20, https://doi.org/10.1080/15027570.2018.1481907.

48. An overview of cultural opposition within militaries to technological advancements is provided in Matt Hegarty, "Military Culture and Resistance to Technical Innovation," in *Australian Perspectives on Global Air and Space Power*, ed. Nicole Townsend et al. (Routledge, 2023).

49. On the unique world of infantry troops, consult Anthony King, *The Combat Soldier: Infantry Tactics and Cohesion in the Twentieth and Twenty-First Centuries* (Oxford University Press, 2013).

50. US Army, *Field Manual FM 3–21.8 (FM 7–8) The Infantry Rifle Platoon and Squad* (Headquarters, Department of the Army, 2007), 1–1.

51. The slogan can be traced to the following statement made by General Alfred M. Gray, the 29th Commandant of the Marine Corps: "Every Marine is, first and foremost, a rifleman. All other conditions are secondary." The quote is included as the final passage in US Marine Corps, "Annual Rifle Training Databook (NAVMC 11660 REV 02–12)," 2012, https://www.trngcmd.marines.mil/Portals/207/Docs/wtbn/ART%20NAVMC%2011660%20REV%2002-12.pdf.

52. Nick Brunetti-Lihach, "Cyber War Requires Cyber Marines," *U.S. Naval Institute Proceedings* 144, no. 11 (2018): 18–23. For more on the technical skills now needed by Marines, see Carl Forsling, "Every Marine a . . . What, Exactly?," *U.S. Naval Institute Proceedings* 151, no. 4 (2025), https://www.usni.org/magazines/proceedings/2025/april/every-marine-what-exactly.

53. Karl Flynn, "Make Every Marine a Drone Killer," *U.S. Naval Institute Proceedings* 149, no. 11 (2023), https://www.usni.org/magazines/proceedings/2023/november/make-every-marine-drone-killer; and Ken Hampshire, "Every Marine a Data Scientist?," *U.S. Naval Institute Proceedings* 145, no. 6 (2019), https://www.usni.org/magazines/proceedings/2019/june/every-marine-data-scientist.

54. For a brief essay on these challenges, see Patrick Tucker, "Pentagon Planning Huge Experiment for Its Connect-Everything Concept," *Defense One*, August 7, 2024, https://www.defenseone.com/technology/2024/08/pentagon-planning-huge-experiment-its-connect-everything-concept/398618/. A more thorough analysis of one Defense Department initiative in this manner is provided in Travis Sharp, *Ready Player None? An End-to-End Assessment of the Air Force Collaborative Combat Aircraft Program* (Center for Strategic and Budgetary Assessments, 2025), https://csbaonline.org/research/publications/ready-player-none-an-end-to-end-assessment-of-the-air-force-collaborative-combat-aircraft-program.

55. On these points, see Nina Kollars, "Genius and Mastery in Military Innovation," *Survival* 59, no. 2 (2017): 125–38, https://doi.org/10.1080/00396338.2017.1302193. For

a call to expand such efforts to middle-tier systems, see Jon Reisher, "Talent and Tech: Fielding and Wielding New Systems Requires the Right People," Modern War Institute (at West Point), December 20, 2024, https://mwi.westpoint.edu /talent-and-tech-fielding-and-wielding-new-systems-requires-the-right-people/.

56. On the cultural dynamics at play in military organizations, consult Peter R. Mansoor and Williamson Murray, eds., *The Culture of Military Organizations* (Cambridge University Press, 2019).

57. Randy George and Ryan Evans, "A Conversation with Gen. Randy George, Chief of Staff of the U.S. Army," February 5, 2024, https://warontherocks.com/2024/02/a -conversation-with-gen-randy-george-chief-of-staff-of-the-u-s-army/.

58. The broader dimensions of "transformation in contact" are outlined in James E. Rainey, "Continuous Transformation: Transformation in Contact," *Military Review: The Professional Journal of the U.S. Army*, online exclusive (August 2024): 1–7, https://www.armyupress.army.mil/journals/military-review/online-exclusive /2024-ole/Transformation-in-Contact/.

59. Rainey's quote is from Christopher Hurd, "Army Futures Command General Lays Out Continuous Transformation Plan," US Army website, April 2, 2024, https://www.army.mil/article/275040/army_futures_command_general_lays _out_continuous_transformation_plan.

60. On this point, see Vikram Mittal, "Why Military Exoskeletons Will Remain Science Fiction," Aerospace & Defense, *Forbes*, August 17, 2020, https://www.forbes .com/sites/vikrammittal/2020/08/17/military-exoskeletons-science-fiction-or -science-reality/.

61. See Samantha Murphy Kelly, "Lithium-Ion Battery Fires Are Happening More Often. Here's How to Prevent Them," CNN, March 9, 2023, https://www.cnn.com /2023/03/09/tech/lithium-ion-battery-fires/index.html.

62. On China's dominant role in the supply chains for lithium, see Philip Murray, "Power Play: Charging Up Strategic Competition over Lithium Battery Value Chains," *Journal of Advanced Military Studies* 16, no. 1 (2025): 9–25, https://doi.org /10.21140/mcuj.20251601001.

63. See Jairo I. Patino, "Better Living (and Fighting) Through Batteries," *U.S. Naval Institute Proceedings* 147, no. 2 (2021), https://www.usni.org/magazines /proceedings/2021/february/better-living-and-fighting-through-batteries.

64. Such changes could, in theory, stretch across the entire defense establishment. For an example of such a plan, see Jahara Matisek and James Micciche, "DoD 3.0: Rebooting the Pentagon for the Next War—Modern War Institute," Modern War Institute (at West Point), June 6, 2025, https://mwi.westpoint.edu/dod -3-0-rebooting-the-pentagon-for-the-next-war/. Historical examples of such undertakings are outlined in John A. Bonin and James D. Scudieri, "Change and Innovation in the Institutional Army from 1860–2020," *The US Army War College Quarterly: Parameters* 53, no. 2 (2023): 95–120, https://doi.org/10.55540/0031-1723 .3225.

65. US Army Public Affairs, "Transcript: Media Roundtable with Gen. Charles Flynn with USARPAC Update September 13, 2023," US Army website, September 18, 2023, https://www.army.mil/article/270022/transcript_media_roundtable_with_gen _charles_flynn_with_usarpac_update_september_13_2023.

66. The proposal for creation of an Army Drone Corps can be found in H.R. 8070—Servicemember Quality of Life Improvement and National Defense Authorization Act for Fiscal Year 2025, US House Subcommittee on Tactical Air and Land Forces. Accessed May 30, 2024. https://docs.house.gov/meetings/AS/AS00/20240522/117296/BILLS-118HR8070ih-TAL.pdf. For greater context, see also Joshua Suthoff, "Imagining a US Army Drone Corps," *Modern War Institute* (at West Point), December 19, 2024, https://mwi.westpoint.edu/imagining-a-us-army-drone-corps/.

67. On the debate regarding where the organizational focus should be on drones within the US Army, consult Neil Hollenbeck, "How to Transform the Army for Drone Warfare," *War Room*, January 9, 2025, https://warroom.armywarcollege.edu/articles/transform-for-drones/.

68. US Secretary of Defense, "Unleashing U.S. Military Drone Dominance," July 10, 2025, https://media.defense.gov/2025/Jul/10/2003752117/-1/-1/1/UNLEASHING-U.S.-MILITARY-DRONE-DOMINANCE.pdf. For a useful overview of the directive, see Joseph Trevithick, "Pentagon Just Made a Massive, Long Overdue Shift to Arm Its Troops with Thousands of Drones," *The War Zone (at The Drive)*, July 10, 2025, https://www.twz.com/air/pentagon-just-made-a-massive-long-overdue-shift-to-arm-its-troops-with-thousands-of-drones.

69. See Sam Skove, "Wartime Need for Drones Would Outstrip US Production. There's a Way to Fix That," *Defense One*, August 7, 2024, https://www.defenseone.com/policy/2024/08/wartime-need-drones-would-outstrip-us-production-theres-way-fix/398642/. For a sophisticated analysis tying together design, production, tactics, and procedures regarding unmanned systems, see Robert E. Price, "Defining Swarm: A Critical Step Toward Harnessing the Power of Autonomous Systems," *Military Review: The Professional Journal of the U.S. Army*, online exclusive (May 2025): 1–12, https://www.armyupress.army.mil/journals/military-review/online-exclusive/2025-ole/defining-swarm/. Weaknesses in the broader US defense industrial base are outlined in Becca Wasser and Philip Sheers, *From Production Lines to Front Lines* (Center for a New American Security, 2025), https://www.cnas.org/publications/reports/from-production-lines-to-front-lines.

70. The contents of the announcement can be found here: US Department of Defense, "Deputy Secretary of Defense Kathleen Hicks Keynote Address: 'The Urgency to Innovate,'" August 28, 2023, https://www.defense.gov/News/Speeches/Speech/Article/3507156/deputy-secretary-of-defense-kathleen-hicks-keynote-address-the-urgency-to-innov/.

71. An overview of the current state of the US acquisition system is provided in *FY2025 NDAA: Department of Defense Acquisition Policy*, no. IN12397, CRS Insight (Congressional Research Service, 2025), https://www.congress.gov/crs-product/IN12397.

72. Such an argument was powerfully made by the Ukrainian deputy minister of defense for digital development in the following op-ed: Kateryna Chernohorenko, "Digitize or Die: Ukraine's War Is a Wake-up Call for 20th Century Militaries," *Breaking Defense*, July 11, 2025, https://breakingdefense.com/2025/07/digitize-or-die-ukraines-war-is-a-wake-up-call-for-20th-century-militaries/.

73. See Jim Perkins and Mike McGinley, "Without Talent Agility, America May Lose," *War on the Rocks*, August 14, 2024, https://warontherocks.com/2024/08/without-talent-agility-america-may-lose/.

74. *Management Advisory: Army's Future Soldier Preparatory Course Places Trainees at Increased Risk of Adverse Health Effects*, Management Advisory DODIG-2025-069 (Inspector General, US Department of Defense, 2025).

75. See Steve Beynon and Konstantin Toropin, "Army and Navy Prep Courses May Be a Recruiting Crisis Silver Bullet. But Other Services Aren't Interested.," Military.com, October 31, 2024, https://www.military.com/daily-news/2024/10/30/no-thanks-air-force-and-marine-corps-say-pre-boot-camp-prep-courses-arent-them-despite-successes.html.

76. On this idea, see Charlie Phelps, "Leadership, Lethality, and (Data) Literacy: Three Keys to Prepare the Army for the Data-Driven, AI-Enabled Future of War," Modern War Institute (at West Point), May 7, 2025, https://mwi.westpoint.edu/leadership-lethality-and-data-literacy-three-keys-to-prepare-the-army-for-the-data-driven-ai-enabled-future-of-war/; and Jenna Hillhouse, "From Data to Decisions: Building a Culture of Information Literacy," *U.S. Naval Institute Proceedings* 151, no. 6 (2025), https://www.usni.org/magazines/proceedings/2025/june/data-decisions-building-culture-information-literacy.

77. Kevin Bradley, "Fighting with Robots: The Time to Prepare Is Now—Modern War Institute," Modern War Institute (at West Point), July 10, 2025, https://mwi.westpoint.edu/fighting-with-robots-the-time-to-prepare-is-now/, https://mwi.westpoint.edu/fighting-with-robots-the-time-to-prepare-is-now/.

78. On the challenges of managing these networks, see Jon R. Lindsay, *Information Technology and Military Power*, Cornell Studies in Security Affairs (Cornell University Press, 2020). Consult the following works for greater detail on the future of war: Christian Brose, *The Kill Chain: Defending America in the Future of High-Tech Warfare* (Hachette Books, 2020); John F. Antal, *Next War: Reimagining How We Fight* (Casemate Publishers, 2023); Mick Ryan, *War Transformed: The Future of Twenty-First-Century Great Power Competition and Conflict* (Naval Institute Press, 2022).

79. In August 2021, then Chairman of the Joint Chiefs of Staff General Mark Milley reflected upon the urgency of the situation. "The country that masters those technologies [at the heart of the emerging character of war], combines them with their doctrine, develops more talent . . . that country is likely to have a significant and perhaps decisive advantage at the beginning of the next war. If we fail to adapt, we fail to change. And we will be condemning a future generation." Quote from Meredith Roaten, "Milley Compares Urgency of Modernization to Interwar Period," *National Defense*, August 2, 2021, https://www.nationaldefensemagazine.org/articles/2021/8/2/milley-compares-urgency-of-modernization-to-interwar-period.

Bibliography

Ackerman, Elliot, and James Stavridis. *2034: A Novel of the Next World War*. Penguin Press, 2021.

Ackerman, Elliot, and James Stavridis. *2054: A Novel*. Penguin Press, 2024.

Act to Amend Title 10, United States Code, to Simplify Laws Relating to Members of the Army, Navy, Air Force, and Marine Corps, and for Other Purposes., Pub. L. No. Public Law 90–235, 753 (1968). https://www.congress.gov/90/statute/STATUTE-81/STATUTE-81-Pg753.pdf.

Adams, Jon Robert. *Male Armor: The Soldier-Hero in Contemporary American Culture*. University of Virginia Press, 2008.

Adamsky, Dima. *The Culture of Military Innovation: The Impact of Cultural Factors on the Revolution in Military Affairs in Russia, the US, and Israel*. Stanford University Press, 2010.

Albon, Courtney. "Space Force's Digital Push Focuses on 'Spaceverse.'" *Defense News*, October 5, 2022. https://www.defensenews.com/battlefield-tech/space/2022/10/05/space-forces-digital-push-focuses-on-spaceverse/.

Aleccia, Jonel. "Pandemic Pounds Push 10,000 U.S. Army Soldiers into Obesity." AP News, April 2, 2023. https://apnews.com/article/military-obesity-pandemic-army-covid-404bbc1a67408d390a7d462f1ecbef75.

Alexander, Nicole, and Lyla Kohistany. "Dispelling the Myth of Women in Special Operations." Center for a New American Security, March 19, 2019. https://www.cnas.org/publications/commentary/dispelling-the-myth-of-women-in-special-operations.

Alford, Chad A., and Sean Chang. "Net Benefit of Performance-Enhancing Drugs Within U.S. Army Special Forces." Master's thesis, Naval Postgraduate School, 2020.

Al-Husseini, Mahdi, Samuel J. Diehl, and Samuel L. Fricks. "Bridging Sky and Sea: Joint Strategies for Medical Evacuation in the Indo-Pacific." *The US Army War College Quarterly: Parameters* 55, no. 1 (2025): 117–32. https://doi.org/10.55540/0031-1723.3334.

Allen, Gregory C. "One Key Challenge for Diplomacy on AI: China's Military Does Not Want to Talk." Center for Strategic and International Studies, May 20, 2022. https://www.csis.org/analysis/one-key-challenge-diplomacy-ai-chinas-military-does-not-want-talk.

Allen, Robertson. *America's Digital Army: Games at Work and War*. University of Nebraska Press, 2017.

Altman, Howard. "Top Ukraine War Lessons from USAF's Commander in Europe." *The War Zone (at The Drive)*, August 8, 2023. https://www.thedrive.com/the-war-zone/top-ukraine-war-lessons-from-usafs-commander-in-europe.

Altman, Howard. "Video Games Helped Ukrainian Bradley Gunner Win Duel with Russian T-90M Tank." *The War Zone (at The Drive)*, January 20, 2024. https://www.thedrive.com/the-war-zone/video-games-helped-ukrainian-bradley-gunner-win-duel-with-russian-t-90m-tank.

Anderson, Michael G. "A People Problem: Learning from Russia's Failing Efforts to Reconstitute Its Depleted Units in Ukraine." Modern War Institute (at West Point), January 26, 2023. https://mwi.westpoint.edu/a-people-problem-learning-from-russias-failing-efforts-to-reconstitute-its-depleted-units-in-ukraine/.

Anderson, Nash, Diana Gai Robinson, Evert Verhagen, et al. "Under-Representation of Women Is Alive and Well in Sport and Exercise Medicine: What It Looks like and What We Can Do About It." *BMJ Open Sport & Exercise Medicine* 9, no. 2 (2023): e001606. https://doi.org/10.1136/bmjsem-2023-001606.

Anderson-Colon, Jessica. "Marine Corps Boot Camp during World War II: The Gateway to the Corps' Success at Iwo Jima." *Marine Corps History* 7, no. 1 (2021): 46–63. https://doi.org/10.35318/mch.2021070103.

Andreas, Peter. *Killer High: A History of War in Six Drugs*. Oxford University Press, 2020. http://ebookcentral.proquest.com/lib/utxa/detail.action?docID=6006832.

Andrews, Lena S. *Valiant Women: The Extraordinary American Servicewomen Who Helped Win World War II*. HarperCollins, 2023.

Andrews, Travis M. "The Navy's Adding a New Piece of Equipment to Nuclear Submarines: Xbox Controllers." *Washington Post*, September 25, 2017. https://www.washingtonpost.com/news/morning-mix/wp/2017/09/25/the-navys-adding-a-new-piece-of-a-equipment-to-nuclear-submarines-xbox-controllers/.

Anonymous. "Baptism by Fire." Special Report Warfare after Ukraine. *The Economist* 448, no. 9354 (2023): S12.

Antal, John F. *Next War: Reimagining How We Fight*. Casemate Publishers, 2023.

Appelboom, Thierry, Christine Rouffin, and Eric Fierens. "Sport and Medicine in Ancient Greece." *The American Journal of Sports Medicine* 16, no. 6 (1988): 594–96. https://doi.org/10.1177/036354658801600607.

Arhirova, Hanna. "Ukraine Hails Teen Drone Operator Who Spied Russian Armor." AP News, June 12, 2022. https://apnews.com/article/russia-ukraine-kyiv-politics-1115558b2d4db5a1146a2bc65ec8a275.

Army Technology. "German Military Recruitment." March 18, 2022. https://www.army-technology.com/comment/german-military-recruitment/.

Arnett, Eric H. "Welcome to Hyperwar." *Bulletin of the Atomic Scientists* 48, no. 7 (1992): 14–21. https://doi.org/10.1080/00963402.1992.11460097.

Arnold, Joseph "Jay." "Build Teams with Board Games." *U.S. Naval Institute Proceedings* 144, no. 1 (2018). https://www.usni.org/magazines/proceedings/2018/january/build-teams-board-games.

Arnold, Joseph "Jay." "Littoral Commander: Indo-Pacific." *U.S. Naval Institute Proceedings* 149, no. 7 (2023). https://www.usni.org/magazines/proceedings/2023/july/littoral-commander-indo-pacific.

Arutunyan, Anna. *Hybrid Warriors: Proxies, Freelancers and Moscow's Struggle for Ukraine.* Hurst, 2022.

Associated Press. "First Woman Completes Navy Special Warfare Training." NBC News, July 15, 2021. https://www.nbcnews.com/politics/politics-news/first-woman -completes-navy-special-warfare-training-n1274125.

Atkinson, Rick. *Crusade: The Untold Story of the Persian Gulf War.* Houghton Mifflin, 1993.

Atlantic Council. "Atlantic Council Announces New Art of Future Warfare Project." *Atlantic Council,* November 19, 2014. https://www.atlanticcouncil.org/news/press -releases/atlantic-council-announces-new-art-of-future-warfare-project/.

Atlantic Council. "A Conversation with Minister of Defense of Ukraine Oleksii Reznikov." July 19, 2022. https://www.atlanticcouncil.org/event/a-conversation-with -oleksii-reznikov/.

Aune, Clayton J. *Building the Hyper-Capable Operator: Should the Military Enhance Its Special Operations Warriors?* Research paper. Naval War College, n.d. Accessed August 23, 2023. https://apps.dtic.mil/sti/citations/AD1120842.

Australian Army. *Future Land Warfare Report 2014.* Modernisation and Strategic Planning Division—Australian Army Headquarters, 2014. https://researchcentre.army. gov.au/library/other/future-land-warfare-report-2014.

Auyeung, Tung Wai, Timothy Kwok, Jason Leung, et al. "Sleep Duration and Disturbances Were Associated with Testosterone Level, Muscle Mass, and Muscle Strength—A Cross-Sectional Study in 1274 Older Men." *Journal of the American Medical Directors Association* 16, no. 7 (2015): 630.e1–630.e6. https://doi.org/10.1016 /j.jamda.2015.04.006.

Bächle, Thomas Christian, and Jascha Bareis. "'Autonomous Weapons' as a Geopolitical Signifier in a National Power Play: Analysing AI Imaginaries in Chinese and US Military Policies." *European Journal of Futures Research* 10, no. 1 (2022). https://doi .org/10.1186/s40309-022-00202-w.

Bae, Sebastian J., ed. *Forging Wargamers: A Framework for Wargaming Education.* Marine Corps University Press, 2022.

Bae, Sebastian J. "Put Educational Wargaming in the Hands of the Warfighter." *War on the Rocks,* July 13, 2023. https://warontherocks.com/2023/07/put-educational-warga ming-in-the-hands-of-the-warfighter/.

Bae, Sebastian J., and Ian T. Brown. "Promise Unfulfilled: A Brief History of Educational Wargaming in the Marine Corps." *Journal of Advanced Military Studies* 12, no. 2 (2021): 45–80.

Baker, Deane. *Should We Ban Killer Robots?* Polity Press, 2022.

Baldor, Lolita C. "Army Sees Safety, Not 'Wokeness,' as Top Recruiting Obstacle." *Army Times,* February 12, 2023. https://www.armytimes.com/news/your-army/2023/02/12 /army-sees-safety-not-wokeness-as-top-recruiting-obstacle/.

Baldor, Lolita C. "US Army Is Slashing Thousands of Posts in Major Revamp to Prepare for Future Wars." AP News, February 27, 2024. https://apnews.com/article/army-cut s-soldiers-recruiting-shortfall-9f2f41cbe512f6330ce6008709e3435b.

Baldor, Lolita C. "U.S. Military Recruiting Rebounds After Several Tough Years, but Challenges Remain." *Los Angeles Times,* September 26, 2024. https://www.latimes.

com/world-nation/story/2024-09-26/u-s-military-recruiting-rebounds-after-sever
al-tough-years-but-challenges-remain.

Balmer, Nigel, Pascoe Pleasence, and Alan Nevill. "Evolution and Revolution: Gauging the Impact of Technological and Technical Innovation on Olympic Performance." *Journal of Sports Sciences* 30, no. 11 (2012): 1075–83. https://doi.org/10.1080/0264041 4.2011.587018.

Barbee, George A. "The Strategic Survivability Triad: The Future of Military Medicine in Support of Combat Power." *Joint Force Quarterly* 107 (October 2022): 102–15.

Barno, David, and Nora Bensahel. "Addressing the U.S. Military Recruiting Crisis." *War on the Rocks*, March 10, 2023. https://warontherocks.com/2023/03/addressing -the-u-s-military-recruiting-crisis/.

Barno, David, and Nora Bensahel. "Dumb and Dumber: The Army's New PT Test." *War on the Rocks*, October 16, 2018. https://warontherocks.com/2018/10/dumb-and -dumber-the-armys-new-pt-test/.

Barno, David, and Nora Bensahel. "Learning from Real Wars: Gaza and Ukraine." *War on the Rocks*, December 6, 2023. https://warontherocks.com/2023/12/learning-from -real-wars-gaza-and-ukraine/.

Barno, David, and Nora Bensahel. "The Other Big Lessons That the U.S. Army Should Learn from Ukraine." *War on the Rocks*, June 27, 2022. https://warontherocks.com /2022/06/the-other-big-lessons-that-the-u-s-army-should-learn-from-ukraine/.

Beals, Ginger E. "Women Marines in Counterinsurgency Operations: Lioness and Female Engagement Teams." Master's thesis, United States Marine Corps Command and Staff College, Marine Corps University, 2010. https://apps.dtic.mil/sti /pdfs/ADA604399.pdf.

Beauchamp, Zack. "Why the First Few Days of War in Ukraine Went Badly for Russia." *Vox*, February 28, 2022. https://www.vox.com/22954833/russia-ukraine -invasion-strategy-putin-kyiv.

Beehner, Lionel, and John Spencer. "To Win at War, Make America Fit Again." *The Hill*, January 12, 2017. https://thehill.com/blogs/congress-blog/healthcare/313866-to-win -at-war-make-america-fit-again/.

Belkin, Aaron. *Bring Me Men: Military Masculinity and the Benign Facade of American Empire, 1898–2001.* Oxford University Press, 2012.

Bell, Richard, Elizabeth Goldsmith, Robert Martinez, and Donghyun Lee. "Stop Talking to Yourself: Military Recruiting in the Modern Age." *Joint Force Quarterly* 115, no. 3 (2024): 53–56.

Benjamin, James. "The Hidden Battlefield: Diet and Lifestyle in Military Readiness." *U.S. Naval Institute Proceedings* 151, no. 2 (2025). https://www.usni.org/magazines /proceedings/2025/february/hidden-battlefield-diet-and-lifestyle-military-readiness.

Bennett, Brian, and Simon Shuster. "Exclusive: Ukraine's Secret Effort to Train for U.S. Jets." *Time*, August 19, 2022. https://time.com/6207115/ukraine-train -fighter-pilots-russia/.

Berger, David H. "Recruiting Requires Bold Changes." *U.S. Naval Institute Proceedings* 148, no. 11 (2022). https://www.usni.org/magazines/proceedings/2022/november /recruiting-requires-bold-changes.

Berger, David H. "Talent Management 2030." Commandant, US Marine Corps, November 2021. https://www.hqmc.marines.mil/Portals/142/Users/183/35/4535/Talent

%20Management%202030_November%202021.pdf?ver=E88HXGUdUQoiB
-edNPKOaA%3d%3d.

Berthelot, Geoffroy, Adrien Sedeaud, Adrien Marck, et al. "Has Athletic Performance Reached Its Peak?" *Sports Medicine* 45, no. 9 (2015): 1263–71. https://doi.org/10.1007/s40279-015-0347-2.

Best, Natalia. "Integrating Women into the Long Blue Line." *U.S. Naval Institute Proceedings* 148, no. 7 (2022). https://www.usni.org/magazines/proceedings/2022/july/integrating-women-long-blue-line.

Bevins, James E. "Incentivizing Innovation: Promoting Technical Competency to Win Future Wars." *Air & Space Operations Review* 1, no. 3 (2022): 22–37.

Beynon, Steve. "Army National Guard Can't Retain Enough Soldiers, Even as Active Duty Meets Goals." Military.com, June 22, 2023. https://www.military.com/daily-news/2023/06/22/army-national-guard-cant-retain-enough-soldiers-even-active-duty-meets-goals.html.

Beynon, Steve. "The Army Needs Hundreds of Officers to Leave Combat Arms." Military.com, December 2, 2024. https://www.military.com/daily-news/2024/12/02/army-needs-hundreds-of-officers-leave-combat-arms.html.

Beynon, Steve. "Debate on Army Fitness Test Takes New Turn with Chief Nominee Testimony." Military.com, July 12, 2023. https://www.military.com/daily-news/2023/07/12/armys-next-top-officer-wants-higher-gender-neutral-fitness-standards.html.

Beynon, Steve. "Recruits, Especially from the South, Are Getting Injured at Alarming Rates in Basic Training." Military.com, April 4, 2023. https://www.military.com/daily-news/2023/04/04/recruits-especially-south-are-getting-injured-alarming-rates-basic-training.html.

Beynon, Steve, and Konstantin Toropin. "Army and Navy Prep Courses May Be a Recruiting Crisis Silver Bullet. But Other Services Aren't Interested." Military.com, October 31, 2024. https://www.military.com/daily-news/2024/10/30/no-thanks-air-force-and-marine-corps-say-pre-boot-camp-prep-courses-arent-them-despite-successes.html.

Bickford, Andrew. *Chemical Heroes: Pharmacological Supersoldiers in the US Military.* Duke University Press, 2020.

Bickford, Andrew. "The 'Superman' Solution: 'Super Soldiers' and 'Superheroes' in the United States Military." *Anthropology Today* 36, no. 5 (2020): 14–17. https://doi.org/10.1111/1467-8322.12605.

Biddle, Stephen. "Back in the Trenches." *Foreign Affairs*, August 10, 2023. https://www.foreignaffairs.com/ukraine/back-trenches.

Biddle, Stephen. *Military Power: Explaining Victory and Defeat in Modern Battle.* Princeton University Press, 2004. https://doi.org/10.1515/9781400837823.

Biggs, Adam T., and Dale W. Russell. "The Challenges of Wearable Technology." *U.S. Naval Institute Proceedings* 149, no. 20 (2023). https://www.usni.org/magazines/proceedings/2023/october/challenges-wearable-technology.

Billing, Daniel C., Graham R. Fordy, Karl E. Friedl, and Henriette Hasselstrøm. "The Implications of Emerging Technology on Military Human Performance Research Priorities." *Journal of Science and Medicine in Sport* 24, no. 10 (2021): 947–53. https://doi.org/10.1016/j.jsams.2020.10.007.

Binnendijk, Anika, Timothy Marler, and Elizabeth M. Bartels. *Brain-Computer Interfaces: U.S. Military Applications and Implications, An Initial Assessment*. RR-2996-RC. RAND Corporation, 2020. https://www.rand.org/pubs/research_reports/RR2996.html.

Biondich, Amy Sue, and Jeremy David Joslin. "Coca: The History and Medical Significance of an Ancient Andean Tradition." *Emergency Medicine International* (April 2016): e4048764. https://doi.org/10.1155/2016/4048764.

Blake, Charles, Christopher W. Boyer, and David R. Hourani. "The Musculoskeletal Imperative: Enhancing Combat Capability through Effective Injury Management." *Military Review: The Professional Journal of the U.S. Army* 104, no. 5 (2024): 129–40.

Blank, Rick, and Tyler Patterson. "Relearning the Timeless Lessons of Land Operations in Asia." Modern War Institute (at West Point), February 28, 2025. https://mwi.westpoint.edu/relearning-the-timeless-lessons-of-land-operations-in-asia/.

Blanton, DeAnne, and Lauren M. Cook. *They Fought Like Demons: Women Soldiers in the American Civil War*. Louisiana State University Press, 2002.

"Володимир Зеленський on Twitter." Twitter. August 11, 2023. https://twitter.com/ZelenskyyUa/status/1689973671476023297.

Boffey, Daniel. "'Fighting Two Enemies': Ukraine's Female Soldiers Decry Harassment." *The Guardian*, August 4, 2023. https://www.theguardian.com/world/2023/aug/04/fighting-two-enemies-ukraine-female-soldiers-decry-harassment.

Bondar, Kateryna. *Ukraine's Future Vision and Current Capabilities for Waging AI-Enabled Autonomous Warfare*. Center for Strategic and International Studies (CSIS), 2025. https://www.csis.org/analysis/ukraines-future-vision-and-current-capabilities-waging-ai-enabled-autonomous-warfare.

Bondar, Kateryna. *Understanding the Military AI Ecosystem of Ukraine*. Center for Strategic and International Studies (CSIS), 2024. https://www.csis.org/analysis/understanding-military-ai-ecosystem-ukraine.

Bonin, John A., and James D. Scudieri. "Change and Innovation in the Institutional Army from 1860–2020." *The US Army War College Quarterly: Parameters* 53, no. 2 (2023): 95–120. https://doi.org/10.55540/0031-1723.3225.

Bonné, Jon. "'Go Pills': A War on Drugs?" NBC News, January 13, 2003. https://www.nbcnews.com/id/wbna3071789.

Boot, Max. "The New American Way of War." *Foreign Affairs* 82, no. 4 (2003): 41–58. https://doi.org/10.2307/20033648.

Borin, Elliot. "The U.S. Military Needs Its Speed." *Wired*, February 10, 2003. https://www.wired.com/2003/02/the-u-s-military-needs-its-speed/.

Bornstein, Daniel B., George L. Grieve, Morgan N. Clennin, et al. "Which US States Pose the Greatest Threats to Military Readiness and Public Health? Public Health Policy Implications for a Cross-Sectional Investigation of Cardiorespiratory Fitness, Body Mass Index, and Injuries Among US Army Recruits." *Journal of Public Health Management and Practice* 25, no. 1 (2019): 36–44. https://doi.org/10.1097/PHH.0000000000000778.

Bourque, Stephen A. "'An Incredible Degree of Rugged and Realistic Training': The 4th Infantry Division's Preparation for D-Day." *Military Review: The Professional Journal of the U.S. Army* 104, no. 3 (2024): 20–32.

Bowers, Matthew T., and Thomas M. Hunt. "The President's Council on Physical Fitness and the Systematisation of Children's Play in America." *International Journal of the History of Sport* 28, no. 11 (2011): 1496–511. https://doi.org/10.1080/09523367.2011.586789.

Bowsher, Herbert. "Air Denial Lessons from Ukraine." *U.S. Naval Institute Proceedings* 149, no. 9 (2023). https://www.usni.org/magazines/proceedings/2023/september/air-denial-lessons-ukraine.

Boydston, Keith R. "Fort Benning Graduates First Women Armor Officers." Army.mil, December 2, 2016. https://www.army.mil/article/179006/fort_benning_graduates_first_women_armor_officers.

Boyle, Annette. "DHA Adds New Weapons in the Military's War on Obesity in Servicemembers." US Medicine, June 18, 2018. https://www.usmedicine.com/clinical-topics/obesity/dha-adds-new-weapons-in-the-militarys-war-on-obesity-in-servicemembers/.

Boyle, Michael J. *The Drone Age: How Drone Technology Will Change War and Peace.* Oxford University Press, 2020.

Bradley, Kevin. "Fighting with Robots: The Time to Prepare Is Now—Modern War Institute." Modern War Institute (at West Point), July 10, 2025. https://mwi.westpoint.edu/fighting-with-robots-the-time-to-prepare-is-now/, https://mwi.westpoint.edu/fighting-with-robots-the-time-to-prepare-is-now/.

Bradley, Omar Nelson, and Clay Blair. *A General's Life: An Autobiography.* Simon and Schuster, 1983.

Brands, Hal, ed. *War in Ukraine: Conflict, Strategy, and the Return of a Fractured World.* Johns Hopkins University Press, 2024.

Breaking Defense. "Better User Interface and Experience Create Faster Decision-Making in Multi-Domain Operations." October 28, 2024. http://breakingdefense.com/2024/10/better-user-interface-and-experience-create-faster-decision-making-in-multi-domain-operations/.

Brendel, Hadiyah. "MIRROR Project Creates Body-Worn Sensor Technology Predictive of Musculoskeletal Injury." Defense Visual Information Distribution Service (DVIDS), February 23, 2023. https://www.dvidshub.net/news/439242/mirror-project-creates-body-worn-sensor-technology-predictive-musculoskeletal-injury.

Britton, Rick. "How Gettysburg Inspired Modern War Gaming." HistoryNet, October 20, 2022. https://www.historynet.com/how-gettysburg-inspired-modern-war-gaming/.

Britzky, Haley. "The Army Is Developing an Alternate Combat Fitness Test for Soldiers with Permanent Injuries." *Task & Purpose*, May 21, 2019. https://taskandpurpose.com/news/army-combat-fitness-test-alternative/.

Britzky, Haley. "The Army's New Pre-Basic-Training Boot Camp Might Actually Build Better Soldiers." *Task & Purpose*, October 19, 2022. https://taskandpurpose.com/news/army-future-soldier-prep-course-recruitment/.

Britzky, Haley. "The Military Says It's Finally Designing Body Armor for Women for the Umpteenth Time This Decade." *Task & Purpose*, October 16, 2020. https://taskandpurpose.com/news/women-combat-gear-problem/.

Broderick, Ryan. *The Most Online War of All Time Until the Next One.* February 24, 2022. https://www.thecontentmines.com/p/the-most-online-war-of-all-time-until-a13.

Bronk, Justin, and Jack Watling. *Mass Precision Strike: Designing UAV Complexes for Land Forces.* Occasional paper. Royal United Services Institute for Defence and Security Studies, 2024.

Brooks, Drew. "The Army's Next Crisis: Americans Aren't Fit Enough to Fight." *Task & Purpose,* February 27, 2018. https://taskandpurpose.com/news/army-physical-fitness-crisis/.

Brose, Christian. *The Kill Chain: Defending America in the Future of High-Tech Warfare.* Hachette Books, 2020.

Brown, Melissa T. *Enlisting Masculinity: The Construction of Gender in US Military Recruiting Advertising During the All-Volunteer Force.* Oxford Studies in Gender and International Relations. Oxford University Press, 2012.

Brown, Steve. "Ukraine Trained AI for Its 'Spiders Web' Airfield Drone Attacks at Aviation Museum." *Kyiv Post,* June 2, 2025. https://www.kyivpost.com/post/53784.

Brunetti-Lihach, Nick. "Cyber War Requires Cyber Marines." *U.S. Naval Institute Proceedings* 144, no. 11 (2018): 18–23.

Bruno, Mikala K., Sean M. Boulanger, Christian V. Olivarez, and Jason E. Hartman. "Taking Holistic Health and Fitness to the Fight: A Case Example of One of the First H2F Teams to Deploy on Rotation to the Republic of Korea." *Military Review: The Professional Journal of the U.S. Army* Online exclusive (April 2025): 1–11. https://www.armyupress.army.mil/Journals/Military-Review/Online-Exclusive/2025-OLE/Holistic-Health-and-Fitness/.

Bryant, Susan F., and Andrew Harrison. *Finding Ender: Exploring the Intersections of Creativity, Innovation, and Talent Management in the U.S. Armed Forces.* No. 31. Strategic Perspectives, edited by Thomas F. Lynch III. Center for Strategic Research, Institute for National Strategic Studies, National Defense University, 2019.

Burgess, Scott, Robert Walton, and P. Michael Politano. "Characteristics of Helicopter Accidents Involving Male and Female Pilots." *International Journal of Aviation, Aeronautics, and Aerospace* 5, no. 2 (2018). https://doi.org/10.15394/ijaaa.2018.1216.

Burke, Ashley. "Canadian Military Reports Sagging Recruitment as NATO Ramps Up Deployment in Eastern Europe." CBC News, March 23, 2022. https://www.cbc.ca/news/politics/canadian-armed-forces-staff-shortfall-1.6395131.

Burke, Crispin. "The Pentagon Should Adjust Standards for Cyber Soldiers—as It Has Always Done." *War on the Rocks,* January 24, 2018. https://warontherocks.com/2018/01/pentagon-adjust-standards-cyber-soldiers-always-done/.

Buzzard, Curtis A., Thomas M. Feltey, John M. Nimmons, Austin T. Schwartz, and Robert S. Cameron. "The Tank Is Dead . . . Long Live the Tank: The Persistent Value of Armored Combined Arms Teams in the 21st Century." *Military Review: The Professional Journal of the U.S. Army* 103, no. 6 (2023): 22–30.

Caddell, Joseph. "Rewind and Reconnoiter: Joseph Caddell on the Evolving Historiography of National Security Events." *War on the Rocks,* November 16, 2023. https://warontherocks.com/2023/11/rewind-and-reconnoiter-joseph-caddell-on-the-evolving-historiography-of-national-security-events/.

Caffrey, Matthew B., Jr. *On Wargaming: How Wargames Have Shaped History and How They May Shape the Future.* Naval War College Newport Papers 43. US Naval War College Press, 2019.

Caldwell, John, Lynn Caldwell, Jennifer Smith, et al. *The Efficacy of Modafinil for Sustaining Alertness and Simulator Flight Performance in F-117 Pilots During 37 Hours of Continuous Wakefulness.* United States Air Force Research Laboratory, 2004.

"Call to Action: Ender's Game." *Marine Corps Association* (blog), June 10, 2019. https:// mca-marines.org/blog/2019/06/10/call-to-action-enders-game/.

Campbell, James D. *"The Army Isn't All Work": Physical Culture in the Evolution of the British Army, 1860–1920.* Ashgate Publishing Company, 2012.

Cancian, Mark. "Blue-Haired Soldiers? Just Say No." *War on the Rocks*, January 18, 2018. https://warontherocks.com/2018/01/blue-haired-soldiers-just-say-no/.

Cancian, Mark F., Matthew Cancian, and Eric Heginbotham. *The First Battle of the Next War: Wargaming a Chinese Invasion of Taiwan.* A Report of the CSIS International Security Program. Center for Strategic and International Studies (CSIS), 2023. https:// www.csis.org/analysis/first-battle-next-war-wargaming-chinese-invasion-taiwan.

Carberry, Sean. "Air Force Wants to Sift Through Simulation Data to Boost Training." *National Defense* 107, no. 835 (2023): 13.

Carberry, Sean. "Siloed Simulators: Improved Training at NATO a Bureaucracy, Not Technology Problem." *National Defense* 107, no. 835 (2023): 24–25.

Card, Orson Scott. *Ender's Game.* Century Publishing, 1985.

Caron, Jean-François. *A Theory of the Super Soldier: The Morality of Capacity-Increasing Technologies in the Military.* Manchester University Press, 2018.

Carter, Ash, US Secretary of Defense. "Implementation Guidance for the Full Integration of Women in the Armed Forces." Memorandum for Secretaries of the Military Departments to Acting Under Secretary of Defense for Personnel and Readiness, Chiefs of the Military Services, and Commander, US Special Operations Command. December 3, 2015. https://dod.defense.gov/Portals/1/Documents/pubs/OSD 014303-15.pdf.

Carter, Ashton B., Deputy Secretary of Defense. "U.S. Department of Defense Directive 3000.09: Autonomy in Weapon Systems." US Department of Defense, November 21, 2012. https://www.hsdl.org/?view&did=726163.

Ceder, Riley. "18th Airborne Orders Soldiers on Staff Duty to Get 4 Hours of Sleep." *Army Times*, May 21, 2024. https://www.armytimes.com/news/your-army/2024/05/21/18th -airborne-orders-soldiers-on-staff-duty-to-get-4-hours-of-sleep/.

Ceder, Riley. "MARSOC Is Fusing Traditionally Rugged Marines with Tech-Curious Ones." *Military Times*, April 30, 2025. https://www.militarytimes.com/land/2025 /04/30/marsoc-is-fusing-traditionally-rugged-marines-with-tech-curious-ones/.

Center for a New American Security. "Virtual Fireside Chat: Recruitment, Retention, and Quality of Life in the Force with the Hon. Christine Wormuth, Secretary of the Army." With Katherine Kuzminski and Christine Wormuth. November 18, 2022. https://www.cnas.org/events/virtual-fireside-chat-recruitment-retention-and-quali ty-of-life-in-the-force-honorable-christine-wormuth-secretary-of-the-army.

Centers for Disease Control and Prevention. "Childhood Obesity Facts." April 2, 2024. https://www.cdc.gov/obesity/childhood-obesity-facts/childhood-obesity-facts .html.

Centers for Disease Control and Prevention. "New CDC Data Show Adult Obesity Prevalence Remains High." CDC Newsroom, September 12, 2024. https://www.cdc .gov/media/releases/2024/p0912-adult-obesity.html.

Centers for Disease Control and Prevention. "Unfit to Serve: Obesity and Physical Inactivity Are Impacting National Security." July 2022. https://www.cdc.gov/physi calactivity/downloads/unfit-to-serve-062322-508.pdf.

Chadwick, Andrew Lewis. *Part-Time Soldiers: Reserve Readiness Challenges in Modern Military History*. Studies in Civil-Military Relations. University Press of Kansas, 2023.

Chalk, Andy. "America's Army Is Finally Closing for Good." *PC Gamer*, February 8, 2022. https://www.pcgamer.com/americas-army-is-finally-closing-for-good/.

Chapa, Joseph O. *Is Remote Warfare Moral?: Weighing Issues of Life and Death from 7,000 Miles*. PublicAffairs, 2022.

Chávez, Kerry. "Learning on the Fly: Drones in the Russian-Ukrainian War." *Arms Control Today* (Washington, United States) 53, no. 1 (2023): 6–11.

Chávez, Kerry, and Ori Swed. "Emulating Underdogs: Tactical Drones in the Russia-Ukraine War." *Contemporary Security Policy* 44, no. 4 (2023): 592–605. https://doi.or g/10.1080/13523260.2023.2257964.

Chayka, Kyle. "Ukraine Becomes the World's 'First TikTok War.'" *The New Yorker*, March 3, 2022. https://www.newyorker.com/culture/infinite-scroll/watching-the -worlds-first-tiktok-war.

Cheney, Stephen A., and Stephen N. Xenakis. *Perspective—Obesity's Increasing Threat to Military Readiness: The Challenge to U.S. National Security*. American Security Project, 2022. https://www.jstor.org/stable/resrep46869.

Chernohorenko, Kateryna. "Digitize or Die: Ukraine's War Is a Wake-up Call for 20th Century Militaries." *Breaking Defense*, July 11, 2025. https://breakingdefense. com/2025/07/digitize-or-die-ukraines-war-is-a-wake-up-call-for-20th-century -militaries/.

Chief of Naval Operations. "NAVADMIN 205/10—ESTABLISHMENT OF THE CYBER WARFARE ENGINEER DESIGNATOR." Navy Specific Administrative Message (NAVADMIN). June 10, 2010. https://navadmin-viewer.github.io/?type =NAVADMIN&year=2010&number=252.

Christesen, Paul. "Sport and Society in Sparta." In *A Companion to Sport and Spectacle in Greek and Roman Antiquity*, edited by Paul Christesen and Donald G. Kyle. John Wiley & Sons, Ltd., 2013. https://doi.org/10.1002/9781118609965.ch17.

Claburn, Thomas. "US Mulls Drafting Gray-Haired Hackers During Times of Crisis." *The Register*, March 21, 2018. https://www.theregister.com/2018/03/21/uncle _sam_mulls_drafting_grayhaired_hackers_during_times_of_crisis/.

Clancy, Tom. *Red Storm Rising*. G.P. Putnam Sons, 1986.

Clark, Ian, and Kyle Atkinson. "Gen-Z Will Fight: But First, They Need to Know Why." *U.S. Naval Institute* 149, no. 1 (2023). https://www.usni.org/magazines/pro ceedings/2023/january/gen-z-will-fight-first-they-need-know-why.

Clark, Matt. "The Army Has a Physical Fitness Problem, Part 1: Eight Myths That Weaken Combat Readiness." Modern War Institute (at West Point), January 25, 2020. https://mwi.usma.edu/army-physical-fitness-problem-part-1-eight-myths-we aken-combat-readiness/.

Clark, Matt. "The Army Has a Physical Fitness Problem, Part 2: Toward a More Combat-Ready Force." Modern War Institute (at West Point), January 31, 2020. https://mwi .usma.edu/army-physical-fitness-problem-part-2-toward-combat-ready-force/.

Clemis, Martin. "The Enduring Lessons of Vietnam: Implications for US Strategy and Policy." *The US Army War College Quarterly: Parameters* 55, no. 2 (2025): 117–33. https://doi.org/10.55540/0031-1723.3347.

CNN. "Chinese Online Game Lets Players Fight Japan over Disputed Islands." August 5, 2013. https://www.cnn.com/2013/08/05/world/asia/china-online-game-fight-japan-diaoyu-senkaku.

CNN. "Hezbollah Video Game: War with Israel." August 16, 2007. http://edition.cnn.com/2007/WORLD/meast/08/16/hezbollah.game.reut/.

CNN International. "Web Video Game Aim: 'Kill' Bush Characters." September 18, 2006. http://edition.cnn.com/2006/WORLD/meast/09/18/bush.game/.

Cobbs, Elizabeth. *The Hello Girls: America's First Women Soldiers*. Harvard University Press, 2017.

Cohen, Eliot A., and Wesley Clark. "Transcript: Reflections on the Ukraine War." Transcript. Center for Strategic and International Studies, February 20, 2024. https://www.csis.org/analysis/reflections-ukraine-war.

Cohen, Rachel. "Can the Air Force Train New Pilots Without Planes?" *Air Force Times*, November 16, 2022. https://www.airforcetimes.com/news/your-air-force/2022/11/16/can-the-air-force-train-new-pilots-without-planes/.

Coker, Christopher. *Future War*. John Wiley & Sons, 2015.

Cole, August, ed. *The Atlantic Council Art of Future Warfare Project: War Stories from the Future*. Ebook. Atlantic Council, 2015. https://www.atlanticcouncil.org/in-depth-research-reports/books/war-stories-from-the-future/.

Collins, Liam, Michael Kofman, and John Spencer. "The Battle of Hostomel Airport: A Key Moment in Russia's Defeat in Kyiv." *War on the Rocks*, August 10, 2023. https://warontherocks.com/2023/08/the-battle-of-hostomel-airport-a-key-moment-in-russias-defeat-in-kyiv/.

Colomb, P. H. "Broadside Fire, and a Naval War Game." *Royal United Services Institution. Journal* 23, no. 101 (1879): 507–35. https://doi.org/10.1080/03071847909417112.

Comprehensive Review of Recent Surface Force Incidents. US Fleet Forces Command, United States Navy, 2017. https://s3.documentcloud.org/documents/4172114/Comprehensive-Review-of-Recent-Surface-Force.pdf.

Connable, Ben. *Ground Combat: Puncturing the Myths of Modern War*. Georgetown University Press, 2025.

Connell, Michael, Brooke Lennox, and Paul Schwartz. *Training in the Russian Armed Forces*. Center for Naval Analyses, 2023. https://www.cna.org/reports/2023/09/training-in-the-russian-armed-forces.

Consensus Conference Panel. "Recommended Amount of Sleep for a Healthy Adult: A Joint Consensus Statement of the American Academy of Sleep Medicine and Sleep Research Society." *Sleep* 38, no. 6 (2015): 843–44. https://doi.org/10.5665/sleep.4716.

Consumer Technology Association. "CTA Research: Exploring Gen Z Views and Preferences in Technology." February 20, 2024. https://www.cta.tech/Resources/Newsroom/Media-Releases/2024/February/CTA-Research-Exploring-Gen-Z-Views-and-Preferences.

Cook, Ellie. "How Russia and Ukraine's Losses Compare." *Newsweek*, February 23, 2024. https://www.newsweek.com/russia-ukraine-losses-casualties-tanks-death-toll-anniversary-1864726.

Cook, Shawn M., Andrew Hall, and Todd Spanton. "Developing Alternative Manning Strategies to Maintain the Combat Effectiveness of the Joint Force." *Joint Force Quarterly* 117, no. 2 (2025): 58–63.

Copp, Tara. "Goodbye, Tape Test? Sweeping DoD Fitness Review Underway." *Military Times*, February 7, 2018. https://www.militarytimes.com/news/your-military/2018/02/05/goodbye-tape-test-sweeping-dod-fitness-review-underway/.

Cordle, John. "Physical Fitness Standards Should Be Tied to the Mission." *U.S. Naval Institute Proceedings* 151, no. 7 (2025). https://www.usni.org/magazines/proceedings/2025/july/physical-fitness-standards-should-be-tied-mission.

Cordle, John P. "Fatigue Is the Navy's Black Lung Disease." *U.S. Naval Institute Proceedings* 146, no. 1 (2020). https://www.usni.org/magazines/proceedings/2020/january/fatigue-navys-black-lung-disease.

Cordle, John, and Robert Sweetman. "Human Factors Meets New Technology in 2025." Center for International Maritime Security, August 30, 2021. https://cimsec.org/human-factors-meets-new-technology-in-2025/.

Cornum, Rhonda, John Caldwell, and Kory Cornum. "Stimulant Use in Extended Flight Operations." *Airpower Journal* (Maxwell AFB, United States) 11, no. 1 (1997): 53–58.

Correll, Diana. "Navy Prep Course Aims to Launch Hundreds of Recruits into the Fleet." *Navy Times*, July 3, 2023. https://www.navytimes.com/news/your-navy/2023/07/03/navy-prep-course-aims-to-launch-hundreds-of-recruits-into-the-fleet/.

Correll, Diana. "Navy's New 'Project Avenger' Flight Training Program Aims to Produce Stronger Aviators." *Navy Times*, May 26, 2021. https://www.navytimes.com/news/your-navy/2021/05/25/navys-new-project-avenger-flight-training-program-aims-to-produce-stronger-aviators/.

Courtwright, David T. *Forces of Habit: Drugs and the Making of the Modern World.* Harvard University Press, 2009.

Cox, Matthew. "Men and Women Seeing Different Failure Rates on Army's Gender-Neutral Fitness Test." Military.com, October 15, 2019. https://www.military.com/daily-news/2019/10/15/men-and-women-seeing-different-failure-rates-armys-gender-neutral-fitness-test.html.

Creveld, Martin Van. *Technology and War: From 2000 B.C. to the Present.* Free Press, 1989.

Creveld, Martin Van. *Wargames: From Gladiators to Gigabytes.* Cambridge University Press, 2013.

Cruickshank, Iain. "An AI-Ready Military Workforce." *Joint Force Quarterly* 110 (Quarter 2023): 46–53.

Cummings, Bobby. "Video Games Can Enhance Warrior Cognitive Performance." US Navy website, February 17, 2022. https://www.navy.mil/Press-Office/News-Stories/Article/2938975/video-games-can-enhance-warrior-cognitive-performance/.

Curley, Gregg. "The Provision of Cyber Manpower: Creating a Virtual Reserve." *Marine Corps University Journal* 9, no. 1 (2018): 191–217.

Curthoys, Kathleen. "Army Offers Direct Commissions to Boost Cyber Force." *Army Times*, December 6, 2017. https://www.armytimes.com/news/your-army/2017/12/06/army-offers-direct-commissions-to-boost-cyber-force/.

Daddis, Gregory A. "Understanding Fear's Effect on Unit Effectiveness." *Military Review: The Professional Journal of the U.S. Army* 84, no. 5 (2004): 22–27.

Danby, Nick. "Carrier Strike Groups Should Be Ready to Go Dark in Conflict." *War on the Rocks*, August 29, 2023. https://warontherocks.com/2023/08/carrier-strike-groups-should-be-ready-to-go-dark-in-conflict/.

Darbasie, Bailee A. "New Direction for Female-Specific Flight Equipment." Air Force website, April 1, 2019. https://www.af.mil/News/Article-Display/Article/1801680/new-direction-for-female-specific-flight-equipment/.

Dayan, Leore, and Noah Smith. "A New Israeli Tank Features Xbox Controllers, AI Honed by 'StarCraft II' and 'Doom.'" *Washington Post*, July 28, 2020. https://www.washingtonpost.com/video-games/2020/07/28/new-israeli-tank-features-xbox-controllers-ai-honed-by-starcraft-ii-doom/.

De Backer, Jan. "The Physical Fitness of Soldiers: A Critical Factor in Modern Warfare." *Surfers and Chess Players: Everything About Healthcare and Technology* (blog), December 13, 2023. https://www.linkedin.com/pulse/physical-fitness-soldiers-critical-factor-modern-qs9nf/.

De Pauw, Linda Grant. "Women in Combat: The Revolutionary War Experience." *Armed Forces and Society* 7, no. 2 (1981): 209–26.

Defense Advanced Research Projects Agency (DARPA). "DARPA Exploring Ways to Assess Ethics for Autonomous Weapons." December 19, 2024. https://www.darpa.mil/news/2024/asimov-approaches.

Defense Advisory Committee on Women in the Services. *Executive Summary.* Defense Advisory Committee on Women in the Services, 2018. https://dacowits.defense.gov/Portals/48/Documents/Reports/2018/Annual%20Report/DACOWITS%20ES%202018.pdf.

Deibert, Ronald J. "The Autocrat in Your iPhone: How Mercenary Spyware Threatens Democracy." *Foreign Affairs* 102, no. 1 (2023): 72–88.

Delpero, Walter T., Hugh O'Neill, Evanne Casson, and Jeff Hovis. "Aviation-Relevant Epidemiology of Color Vision Deficiency." *Aviation, Space, and Environmental Medicine* 76, no. 2 (2005): 127–33.

Deuster, Patricia A., Francis G. O'Connor, Kurt A. Henry, et al. "Human Performance Optimization: An Evolving Charge to the Department of Defense." *Military Medicine* 172, no. 11 (2007): 1133–37. https://doi.org/10.7205/milmed.172.11.1133.

Devine, Kieran. "Ukraine War: Mobile Networks Being Weaponised to Target Troops on Both Sides of Conflict." Sky News, January 4, 2023. https://news.sky.com/story/ukraine-war-mobile-networks-being-weaponised-to-target-troops-on-both-sides-of-conflict-12577595.

Dickerson, Kelly M. "Protecting Posterity: Women and the Military Selective Service Act." *Military Review: The Professional Journal of the U.S. Army* 102, no. 5 (2022): 52–60.

Dimeo, Paul. *A History of Drug Use in Sport: 1876–1976: Beyond Good and Evil.* Routledge, 2008.

Dobkin, Adin. "The Video Game That Could Shape the Future of War." *The Atlantic*, October 26, 2017. https://www.theatlantic.com/technology/archive/2017/10/operation-overmatch/544062/.

Donham, Benjamin P. "It's Not Just About the Algorithm: Development of a Joint Medical Artificial Intelligence." *Joint Force Quarterly* 111 (2023): 50–57.

Donnelly, John M. "F-35 Ejection Seats Could Endanger Many Pilots." *Roll Call*, October 16, 2015. https://www.rollcall.com/2015/10/16/exclusive-f-35-ejection-seats -could-endanger-many-pilots/.

Dougherty, Chris. *More Than Half the Battle*. Center for a New American Security, 2021. https://www.cnas.org/publications/reports/more-than-half-the-battle.

Dowd, Alan W. "Drone Wars: Risks and Warnings." *The US Army War College Quarterly: Parameters* 42/43, no. 4/1 (2012): 7–16.

Dowd, Alan W. "Ukraine Reminds Us Necessity Is Still the Mother of Invention." RealClearDefense, October 7, 2023. https://www.realcleardefense.com/articles /2023/10/07/ukraine_reminds_us_necessity_is_still_the_mother_of_invention _984556.html.

Dragomir, Cristina-Ioana. *Making the Immigrant Soldier: How Race, Ethnicity, Class, and Gender Intersect in the US Military*. University of Illinois Press, 2023.

Duffie, Warren, Jr. "Special Delivery: New Autonomous Flight Technology Means Rapid Resupply for Marines." Office of Naval Research, December 13, 2017. https:// www.nre.navy.mil/media-center/news-releases/special-delivery-new-autonomous -flight-technology-means-rapid-resupply.

Dumenil, Lynn. *The Second Line of Defense: American Women and World War I*. University of North Carolina Press, 2017.

Dyreson, Mark S. "Optimizing Human Performance—A Brief History of Macro and Micro Perspectives." *Kinesiology Review* 9, no. 1 (2020): 4–12.

Dyson, Tom, and Yuriy Pashchuk. "Organisational Learning During the Donbas War: The Development of Ukrainian Armed Forces Lessons-Learned Processes." *Defence Studies* 22, no. 2 (2022): 141–67. https://doi.org/10.1080/14702436.2022.2037427.

East, Whitfield B. *Fit to Serve: A History of US Army Physical Readiness*. Army University Press, US Army Combined Arms Center, 2024.

East, Whitfield B. *A Historical Review and Analysis of Army Physical Readiness Training and Assessment*. Combat Studies Institute, US Army Command and General Staff College, 2013. https://apps.dtic.mil/sti/citations/ADA622014.

Eastwood, Clint, dir. *Firefox*. Produced by Clint Eastwood. Warner Bros., 1982. https:// www.imdb.com/title/tt0083943/.

Eckstein, Megan. "Fleet Finding New Sleep-Sensitive Watch Schedules Boosts Crew Performance, Efficiency." *U.S. Naval Institute News*, July 15, 2019. https://news.usni .org/2019/07/15/fleet-finding-new-sleep-sensitive-watch-schedules-boosts-crew -performance-efficiency.

Eckstein, Megan. "US Navy, Marines Look for Training Systems with Accurate Adversaries, Ability to Track Individual Performance." *Defense News*, November 30, 2021. https://www.defensenews.com/naval/2021/11/30/navy-marines-looking-for-training -systems-with-accurate-adversaries-ability-to-track-individuals-performance/.

Edmonds, Jeffrey, and Samuel Bendett. *Russia's Use of Uncrewed Systems in Ukraine*. Center for Naval Analyses, 2023. https://www.cna.org/reports/2023/05/russias -use-of-drones-in-ukraine.

Egnell, Robert, and Mayesha Alam. *Women and Gender Perspectives in the Military: An International Comparison*. Georgetown University Press, 2019.

Egozi, Arie. "IAI Wins Israeli Carmel Land Vehicle Competition; Company Wants American Partner." *Breaking Defense*, October 11, 2021. https://breakingdefense. com/2021/10/iai-wins-israeli-carmel-land-vehicle-competition-company-wants -american-partner/.

Eisenhower, Dwight D. "Executive Order 10673—Fitness of American Youth." *Federal Register* 21, no. 138 (1956): 5341.

El Khoury, Joseph. "The Use of Stimulants in the Ranks of Islamic State: Myth or Reality of the Syrian Conflict." *Studies in Conflict & Terrorism* 43, no. 8 (2020): 679–87. https://doi.org/10.1080/1057610X.2018.1495291.

El Sirgany, Sarah, Ben Wedeman, and Kostyantin Gak. "Captured Russian Soldiers Tell of Low Morale, Disarray and Horrors of Trench Warfare." CNN, July 6, 2023. https://www.cnn.com/2023/07/06/europe/captured-russian-soldiers-ukraine-intl-cmd/index.html.

Elkus, Adam. "War Is a Video Game, and We're Losing." *War on the Rocks*, March 20, 2014. https://warontherocks.com/2014/03/war-is-a-video-game-and-were-losing/.

Emery, John R. "Moral Choices Without Moral Language: 1950s Political-Military Wargaming at the RAND Corporation." *Texas National Security Review* 4, no. 4 (2021): 12–31. https://doi.org/10.26153/tsw/17528.

Emery, John R., and Hadley Biggs. "Human, All Too Human: Drones, Ethics, and the Psychology of Military Technologies." *Political Psychology* 43, no. 3 (2022): 605–13. https://doi.org/10.1111/pops.12809.

Emmerich, Samuel D., Cheryl D. Fryar, Bryan Stierman, and Cynthia L. Ogden. "Obesity and Severe Obesity Prevalence in Adults: United States, August 2021–August 2023." NCHS Data Briefs No. 508. National Center for Health Statistics, 2024. https://doi .org/10.15620/cdc/159281.

Emonson, David L., and Rodger D. Vanderbeek. "The Use of Amphetamines in U.S. Air Force Tactical Operations During Desert Shield and Storm." *Aviation, Space, and Environmental Medicine* (US) 66, no. 3 (1995): 260–63.

Estes, Adam Clark. "Syrian Rebels Now Have a Tank Powered by a Playstation Controller." *The Atlantic*, December 10, 2012. https://www.theatlantic.com/international/archive /2012/12/syrian-rebels-now-have-tank-powered-playstation-controller/320687/.

Etkind, Alexander. *Russia Against Modernity*. Polity Press, 2023.

Etter, Gregg W., Sr., and David H. McElreath. "Why the Ukrainians Fight: The Holodomor (1932–33)." *Journal of Advanced Military Studies* 16, no. 1 (2025): 96–108. https:// doi.org/10.21140/mcuj.20251601005.

Etzioni, Amitia, and Oren Etzioni. "Pros and Cons of Autonomous Weapons Systems." *Military Review: The Professional Journal of the U.S. Army* 97, no. 3 (2017): 72–81.

Fabian, Sandor. "The Illusion of Conventional War: Europe Is Learning the Wrong Lessons from the Conflict in Ukraine." Modern War Institute (at West Point), April 23, 2024. https://mwi.westpoint.edu/the-illusion-of-conventional-war-europe -is-learning-the-wrong-lessons-from-the-conflict-in-ukraine/.

Fair, Jeffrey J. "America's Cyber Auxiliary: Building Capacity and Future Operators." *The Cyber Defense Review* 7, no. 2 (2022): 57–66.

Fandre, Matthew. "Medical Changes Needed for Large-Scale Combat Operations: Observations from Mission Command Training Program Warfighter Exercises." *Military Review: The Professional Journal of the U.S. Army* 100, no. 3 (2020): 36–45.

Fazal, Tanisha M. "Ukraine's Military Medicine Is a Critical Advantage." *Foreign Policy*, October 31, 2022. https://foreignpolicy.com/2022/10/31/ukraine-military -medicine-russia-war/.

Feldstein, Steven. "Disentangling the Digital Battlefield: How the Internet Has Changed War." *War on the Rocks*, December 7, 2022. https://warontherocks.com/2022/12 /disentangling-the-digital-battlefield-how-the-internet-has-changed-war/.

Ferguson, Michael P. "Ghost in the Machine: Coming to Terms with the Human Core of Unmanned War." *Texas National Security Review* 8, no. 2 (2025): 27–46.

Final Report: National Security Commission on Artificial Intelligence. National Security Commission on Artificial Intelligence's (NSCAI), 2021. https://www.nscai.gov/wp -content/uploads/2021/03/Full-Report-Digital-1.pdf.

Finkenstadt, Daniel J., and Erik Helzer. "Gamified Learning Can Be Effective." *U.S. Naval Institute Proceedings* 149, no. 3 (2023). https://www.usni.org/magazines/proceedings /2023/march/gamified-learning-can-be-effective.

Fish, Lauren, and Paul Scharre. *Soldier Protection Today*. Super Soldiers. Center for a New American Security, 2018. https://www.cnas.org/publications/reports/super -soldiers-1.

Fish, Lauren, and Paul Scharre. *The Soldier's Heavy Load*. Super Soldiers. Center for a New American Security, 2018. https://www.cnas.org/publications/reports/the -soldiers-heavy-load-1.

Fitzgerald, David. *Uncertain Warriors: The United States Army between the Cold War and the War on Terror*. Cambridge University Press, 2023.

Fitzsimonds, James R., and Jan Van Tol. "Revolutions in Military Affairs." *Joint Force Quarterly*, Spring 1994, 24–31.

Flam, Faye. "Obese Kids: How Did We Get Here?" *Washington Post*, September 27, 2013. https://www.washingtonpost.com/postlive/how-childhood-obesity-became- a-crisis/2013/09/26/b2f87652-1708-11e3-804b-d3a1a3a18f2c_story.html.

Flynn, Karl. "Make Every Marine a Drone Killer." *U.S. Naval Institute Proceedings* 149, no. 11 (2023). https://www.usni.org/magazines/proceedings/2023/november/make -every-marine-drone-killer.

Foley, Jordan. "The Supreme Court Should End Male-Only Selective Service Registration." *U.S. Naval Institute Proceedings* 147, no. 5 (2021). https://www.usni.org /magazines/proceedings/2021/may/supreme-court-should-end-male-only-selective -service-registration.

Folse, Mark R. "'The Cleanest and Strongest of Our Young Manhood': Marines, Belleau Wood, and the Test of American Manliness." *Marine Corps History* 4, no. 1 (2018): 6–22.

Folse, Mark Ryland. *The Globe and Anchor Men: U.S. Marines and American Manhood in the Great War Era*. Modern War Studies. University Press of Kansas, 2024.

Ford, Matthew. *War in the Smartphone Age: Conflict, Connectivity and the Crises at Our Fingertips*. Oxford University Press, 2025.

Forney, Andrew J. "Davy Crockett and the Boy Scouts: The Korean War and Mismanaging Protracted Conflict." *Texas National Security Review* 8, no. 1 (2024): 31–48.

Forsling, Carl. "Every Marine a . . . What, Exactly?" *U.S. Naval Institute Proceedings* 151, no. 4 (2025). https://www.usni.org/magazines/proceedings/2025/april/every-marine -what-exactly.

Fox, Collin. "Restore MAVNI for Legal Aliens to Enter the Military." *U.S. Naval Institute Proceedings* 147, no. 8 (2021). https://www.usni.org/magazines/proceedings/2021/august/restore-mavni-legal-aliens-enter-military.

Frankenberg, Sydney, and Hallie Lucas. "Draft Us Too, America." *U.S. Naval Institute Proceedings* 147, no. 6 (2021). https://www.usni.org/magazines/proceedings/2021/june/draft-us-too-america.

Freedberg, Sydney J., Jr. "Mattis Upguns Infantry: Task Force to Invest Over $1B." *Breaking Defense*, February 21, 2018. https://breakingdefense.com/2018/02/mattis-upguns-infantry-close-combat-lethality-task-force/.

Freedberg, Sydney J., Jr. "SOCOM Tests Sarcos Exoskeleton (No, It Isn't 'Iron Man')." *Breaking Defense*, March 18, 2019. https://breakingdefense.com/2019/03/socom-tests-sarcos-exoskeleton-no-it-isnt-iron-man/.

Friedl, Karl E. "Biomedical Research on Health and Performance of Military Women: Accomplishments of the Defense Women's Health Research Program (DWHRP)." *Journal of Women's Health* 14, no. 9 (2005): 764–802. https://doi.org/10.1089/jwh.2005.14.764.

Fry, Brian J. "Taking the Brakes Off Uniformed Scientists and Engineers." *Air & Space Operations Review* 1, no. 1 (2022): 18–33.

Fuentes, Gidget. "Latest Surface Navy Sleep Policy Aims for Better-Rested, More Alert, Healthier Crews." *U.S. Naval Institute News*, January 28, 2021. https://news.usni.org/2021/01/28/latest-surface-navy-sleep-policy-aims-for-better-rested-more-alert-healthier-crews.

Fuentes, Gidget. "Pilot Course Aims to Build Marines' Skills as Communicators for the Future Fight." *US Naval Institute (USNI) News*, February 28, 2023. https://news.usni.org/2023/02/28/pilot-course-aims-to-build-marines-skills-as-communicators-for-the-future-fight.

FY2025 NDAA: Department of Defense Acquisition Policy. No. IN12397. CRS Insight. Congressional Research Service, 2025. https://www.congress.gov/crs-product/IN12397.

Gady, Franz Stefan. "How an Army of Drones Changed the Battlefield in Ukraine." *Foreign Policy*, December 6, 2023. https://foreignpolicy.com/2023/12/06/ukraine-russia-war-drones-stalemate-frontline-counteroffensive strategy/.

Galliott, Jai, and Mianna Lotz. *Super Soldiers: The Ethical, Legal and Social Implications.* Routledge, 2016.

Garamone, Jim. "Training Key to Ukrainian Advantages in Defending Nation." US Department of Defense, September 6, 2022. https://www.defense.gov/News/News-Stories/Article/article/3149975/training-key-to-ukrainian-advantages-in-defending-nation/.

Garcia, Juan. "Bring the Navy's Physical Fitness Assessment into the 21st Century." *U.S. Naval Institute Proceedings* 145, no. 7 (2019). https://www.usni.org/magazines/proceedings/2019/july/bring-navys-physical-fitness-assessment-21st-century.

Gasca, Frank, Ryan Voneida, and Ken Goedecke. "Unique Capabilities of Women in Special Operations Forces." *Special Operations Journal* 1, no. 2 (2015): 105–11. https://doi.org/10.1080/23296151.2015.1070613.

Gatzemeyer, Garrett. *Bodies for Battle: US Army Physical Culture and Systematic Training, 1885–1957.* University Press of Kansas, 2021.

GBD 2021 US Obesity Forecasting Collaborators. "National-Level and State-Level Prevalence of Overweight and Obesity Among Children, Adolescents, and Adults in the USA, 1990–2021, and Forecasts up to 2050." *The Lancet* 404, no. 10469 (2024): 2278–98. https://doi.org/10.1016/S0140-6736(24)01548-4.

Gearhart, Grant. "The Knight and War: Alternative Displays of Masculinity in El Passo Honroso de Suero de Quiñones, El Victorial, and the Historia de los Hechos Del Marqués de Cádiz." PhD diss., University of North Carolina at Chapel Hill, 2015.

George, Randy, and Ryan Evans. "A Conversation with Gen. Randy George, Chief of Staff of the U.S. Army." February 5, 2024. https://warontherocks.com/2024/02/a-conversation-with-gen-randy-george-chief-of-staff-of-the-u-s-army/.

Germano, Kate, and Kelly Kennedy. *Fight Like a Girl: The Truth Behind How Female Marines Are Trained.* Prometheus Books, 2018.

Gerras, Stephen, and Andrew A. Hill. "Lucy in the Chocolate Factory: On the Inevitability of Killer Machines." *War Room*, September 12, 2024. https://warroom.army warcollege.edu/articles/the-chocolate-factory/.

Gertler, Jeremiah. *U.S. Unmanned Aerial Systems.* CRS Report for Congress. Congressional Research Service, 2012. https://sgp.fas.org/crs/natsec/R42136.pdf.

Gilliland, C. Herbert. "Dan Lenson Comes Home: An Appreciation of David Poyer's The Academy." *U.S. Naval Institute Proceedings* 150, no. 3 (2024). https://www.usni.org/magazines/proceedings/2024/march/dan-lenson-comes-home-appreciation-david-poyers-academy.

Gilman, Luke S. "Endurance and Executive Function: Implications for Military Education from a Study of Marine Officer Fitness and Cognition." *International Perspectives on Military Education* 2 (2025): 127–46.

Glick, Stephen P., and L. Ian Charters. "War, Games, and Military History." *Journal of Contemporary History* 18, no. 4 (1983): 567–82.

GlobalData Thematic Intelligence. "The Metaverse Is Preparing Ukrainian Fighter Pilots for Combat." *Airforce Technology* (blog), May 30, 2023. https://www.airforce-technology.com/comment/metaverse-ukrainian-fighter-pilots/.

Goldfarb, Avi, and Jon R. Lindsay. "Prediction and Judgment: Why Artificial Intelligence Increases the Importance of Humans in War." *International Security* 46, no. 3 (2021): 7–50. https://doi.org/10.1162/isec_a_00425.

Goldman, Patrick A. "In the Digital Age, Make Ships Go Dark." *U.S. Naval Institute Proceedings* 148, no. 8 (2022). https://www.usni.org/magazines/proceedings/2022/august/digital-age-make-ships-go-dark.

Gonzalez, Kris. "Soldiers Bridge Cultural Gap: Female Interaction Critical to Deploying Units' Success." US Army website, March 11, 2010. https://www.army.mil/article/35644/soldiers_bridge_cultural_gap_female_interaction_critical_to_deploying_units_success.

Good, Cameron H., Allison J. Brager, Vincent F. Capaldi, and Vincent Mysliwiec. "Sleep in the United States Military." *Neuropsychopharmacology* 45, no. 1 (2020): 1. https://doi.org/10.1038/s41386-019-0431-7.

Goodley, Héloïse. *Pharmacological Performance Enhancement and the Military: Exploring an Ethical and Legal Framework for "Supersoldiers."* Research paper. Chatham House, The Royal Institute of International Affairs, 2020.

https://policycommons.net/artifacts/1423207/pharmacological-performance-enhancement-and-the-military/2037472/.

Gozzi, Laura. "How Ukraine Carried Out Daring 'Spider Web' Attack on Russian Bombers." BBC News, June 2, 2025. https://www.bbc.com/news/articles/cq69qnvj6nlo.

Grady, John. "Marines Considering Autonomous Systems for Almost Everything, General Says." *US Naval Institute (USNI) News*, September 6, 2023. https://news.usni.org/?p=105549.

Grady, John. "Navy Sees 'Difficult Times' with Recruiting Goals for Nuclear, Cyber Sailors." *US Naval Institute (USNI) News*, February 15, 2018. https://news.usni.org/2018/02/15/navy-wants-congressional-permission-bring-new-cyber-sailors-higher-ranks.

Greenburg, Jennifer. *At War with Women: Military Humanitarianism and Imperial Feminism in an Era of Permanent War.* Cornell University Press, 2023.

Grier, Tyson, Timothy Benedict, Olivia Mahlmann, Latoya Goncalves, and Bruce H. Jones. "Physical and Behavioral Characteristics of Soldiers Acquiring Recommended Amounts of Sleep per Night." *Sleep Health* 9, no. 5 (2023): 626–33. https://doi.org/10.1016/j.sleh.2023.03.003.

Grinspan, Jon. "How Coffee Fueled the Civil War." Opinionator. *New York Times*, July 9, 2014. https://archive.nytimes.com/opinionator.blogs.nytimes.com/2014/07/09/how-coffee-fueled-the-civil-war/.

Grisdale, David J. "Unfit to Fight: Medical Readiness Lessons Learned from the Draft from 1940 To1947, and Why the United States Should Classify Registrants of the Selective Service." Master's thesis, US Army Command and General Staff College, 2019. https://apps.dtic.mil/sti/citations/AD1106237.

Gross, Mary E., Stacie E. Taylor, and Kathleen M Robinette. *Flight Suit Sizes for Women.* Final Report AFRL-HE-WP-TR-2000-0072. United States Air Force Research Laboratory, 2020.

Guthart, Andrew. "Twitter Trackers Jeopardize Military Aircraft." *U.S. Naval Institute Proceedings* 148, no. 9 (2022). https://www.usni.org/magazines/proceedings/2022/september/twitter-trackers-jeopardize-military-aircraft-o.

Haberman, Ryan, and Michael Pollard. "The Army Should Be Looking for a Few Older Soldiers." *Defense One*, April 7, 2023. https://www.defenseone.com/ideas/2023/04/army-should-be-looking-few-older-soldiers/384875/.

Hacker, Tyler. "How the US Army Can Close Its Dangerous—and Growing—Small Drone Gap." Modern War Institute (at West Point), March 6, 2024. https://mwi.westpoint.edu/how-the-us-army-can-close-its-dangerous-and-growing-small-drone-gap/.

Hackett, General Sir John. *The Third World War: A Future History.* Sidgwick and Jackson, Limited, 1978.

Hackett, Michael T., and John A. Nagl. "A Long, Hard Year: Russia-Ukraine War Lessons Learned 2023." *The US Army War College Quarterly: Parameters* 54, no. 3 (2024): 41–52. https://doi.org/10.55540/0031-1723.3302.

Hadley, Greg. "Everything You Need to Know About the Space Force's Fitness Tracker PT Study." *Air & Space Forces Magazine*, May 25, 2023. https://www.airandspaceforces.com/space-forces-fitness-tracker-pt-study/.

Haldeman, Joe. *The Forever War.* St. Martin's Press, 1974.

Halem, Harry. "Ukraine's Lessons for Future Combat: Unmanned Aerial Systems and Deep Strike." *The US Army War College Quarterly: Parameters* 53, no. 4 (2023): 19–32. https://doi.org/10.55540/0031-1723.3252.

Hall, Matthew Thayer. "Drones Can Speed Medical Care, Search and Rescue." *U.S. Naval Institute Proceedings* 147, no. 2 (2021). https://www.usni.org/magazines/proceedings/2021/february/drones-can-speed-medical-care-search-and-rescue.

Halter, Ed. *From Sun Tzu to XBox: War and Video Games.* PublicAffairs, 2006.

Hambling, David. "Are Drugs Making Russian Soldiers Act Like Zombies?" *Forbes*, February 2, 2023. https://www.forbes.com/sites/davidhambling/2023/02/02/are-drugs-making-russian-soldiers-act-like-zombies/.

Hambling, David. "Game Controllers Driving Drones, Nukes." *Wired*, July 19, 2008. https://www.wired.com/2008/07/wargames/.

Hampshire, Ken. "Every Marine a Data Scientist?" *U.S. Naval Institute Proceedings* 145, no. 6 (2019). https://www.usni.org/magazines/proceedings/2019/june/every-marine-data-scientist.

Hardison, Chaitra M., Eyal Aharoni, Christopher Larson, Steven Trochlil, and Alexander C. Hou. *Stress and Dissatisfaction in the Air Force's Remotely Piloted Aircraft Community: Focus Group Findings.* Research Reports. RAND Corporation, 2017. https://www.rand.org/pubs/research_reports/RR1756.html.

Hardison, Chaitra M., Paul W. Mayberry, Heather Krull, et al. *Independent Review of the Army Combat Fitness Test: Summary of Key Findings and Recommendations.* RAND Corporation, 2022. https://www.rand.org/pubs/research_reports/RRA1825-1.html.

Hardy, Brad. "Citizen Candidates: Cold War Naturalization, Military Service, and the Lodge Act of 1950." *Journal of Military History* 87, no. 1 (2023): 169–88.

Haring, Ellen. "Marines' Requirements for Infantry Officers Are Unrealistic, Army Colonel Says." *Marine Corps Times*, October 15, 2016. https://www.marinecorpstimes.com/opinion/2016/10/15/marines-requirements-for-infantry-officers-are-unrealistic-army-colonel-says/.

Haring, Ellen L. "What Women Bring to the Fight." *The US Army War College Quarterly: Parameters* 43, no. 2 (2013): 27–32.

Harkins, Gina. "The Military Is Overhauling Troops' Chow as Obesity Rates Soar." Military.com, August 19, 2018. https://www.military.com/daily-news/2018/08/19/military-overhauling-troops-chow-obesity-rates-soar.html.

Harper, Jon. "Army Creates Video Game to Prototype New Weapons." *National Defense* 102, no. 769 (2017): 35–37.

Harper, Jon. "Army Taps General Dynamics, American Rheinmetall for Next Phases of Optionally Manned Fighting Vehicle Program." *DefenseScoop*, June 26, 2023. https://defensescoop.com/2023/06/26/army-taps-general-dynamics-american-rheinmetall-for-next-phases-of-optionally-manned-fighting-vehicle-program/.

Harper, Jon. "Gen. Milley Anticipates New 'Joint Futures' Organization Will Come to Fruition." *DefenseScoop*, June 30, 2023. https://defensescoop.com/2023/06/30/gen-milley-anticipates-new-joint-futures-organization-will-come-to-fruition/.

Harrison, Y., and J. A. Horne. "One Night of Sleep Loss Impairs Innovative Thinking and Flexible Decision Making." *Organizational Behavior and Human Decision Processes* 78, no. 2 (1999): 128–45. https://doi.org/10.1006/obhd.1999.2827.

Hass, Alex. "Vitamin D Supplementation Program." *Marine Corps Gazette* 108, no. 8 (2024): 59–61.

Hatfield, Bradley. "2019 Annual Meeting to Focus on Optimization of Human Performance." *National Academy of Kinesiology Newsletter* 40, no. 2 (2019): 4–5.

Haydock, Thomas, and Jack Meeker. "Lessons in Reconstitution from the Russia-Ukraine War: Gaining Asymmetric Advantage Through Transformative Reconstitution." *Military Review: The Professional Journal of the U.S. Army* 105, no. 1 (2025): 26–41.

Headquarters Marine Corps. "1956 US Marine Corps PFT Order." Headquarters, United States Marine Corps, August 9, 1956. http://archive.org/details/1956UsMarine CorpsPftOrder.

Heckmann, Laura. "Upgraded Kevlar to Lighten Loads for Troops." *National Defense* 107, no. 835 (2023): 10.

Heffernan, Teresa. "Autonomous Weapons in Fiction and the Fiction of Autonomous Weapons." In *The Realities of Autonomous Weapons*, edited by Thomas Christian Bächle and Jascha Bareis. Bristol University Press, 2025.

Hefling, Kimberly. "Army Tests New Body Armor for Women." NBC News, April 22, 2011. https://www.nbcnews.com/id/wbna42721218.

Hefti, Michael L. *Training the Professional Soldier: Bridging Inexperience and Sophisticated Warfighting Technologies*. School of Advanced Military Studies Monograph. School of Advanced Military Studies, US Army Command and General Staff College, 2021.

Hegarty, Matt. "Military Culture and Resistance to Technical Innovation." In *Australian Perspectives on Global Air and Space Power*, edited by Nicole Townsend, Kus Pandey, and Jarrod Pendlebury. Routledge, 2023.

Heinlein, Robert Anson. *Starship Troopers*. Putnam, 1959.

Henkin, Yagil. "The 'Big Three' Revisited: Initial Lessons from 200 Days of War in Ukraine." *Expeditions with Marine Corps University Press* (November 2022). https:// doi.org/10.36304/ExpwMCUP.2022.13.

Herlihy, Daniel. "Cognitive Performance Enhancement for Multi-Domain Operations." *The US Army War College Quarterly: Parameters* 52, no. 4 (2022): 75–95. https://doi.org/10.55540/0031-1723.3188.

Herr, Andrew. "Will Humans Matter in the Wars of 2030?" *Joint Force Quarterly* 77 (2nd Quarter 2015): 76–83.

Herr, Christopher, and Dennis M Allen. *Video Games as a Training Tool to Prepare the Next Generation of Cyber Warriors*. Software Engineering Institute, Carnegie Mellon University, 2015.

Hessert, M. Josephine. "Needed: Body Armor Built for Women." *U.S. Naval Institute Proceedings* 149, no. 6 (2023). https://www.usni.org/magazines/proceedings/2023 /june/needed-body-armor-built-women.

Hester, Brian A., Dennis Doyle, and Ronan A. Sefton. "Techcraft on Display in Ukraine." *War on the Rocks*, May 16, 2024. https://warontherocks.com/2024/05/tech craft-on-display-in-ukraine/.

Higgins, A. J. "PL04 from Ancient Greece to Modern Athens: 3000 Years of Doping in Competition Horses." *Journal of Veterinary Pharmacology and Therapeutics* 29, no. s1 (2006): 4–8. https://doi.org/10.1111/j.1365-2885.2006.00770_4.x.

Hillhouse, Jenna. "From Data to Decisions: Building a Culture of Information Literacy." *U.S. Naval Institute Proceedings* 151, no. 6 (2025). https://www.usni.org/magazines /proceedings/2025/june/data-decisions-building-culture-information-literacy.

Hlad, Jennifer. "In the Pacific, Army Leaders Expect Today's Fiction to Be Near-Term Reality." *Defense One*, July 25, 2025. https://www.defenseone.com/technology/2025/07 /pacific-army-leaders-expect-todays-fiction-be-near-term-reality/407001/.

Hoberman, John. *Dopers in Uniform: The Hidden World of Police on Steroids.* University of Texas Press, 2017.

Hodgdon, James A. *A History of the U.S. Navy Physical Readiness Program From 1976 to 1999:* Technical Document No. 99–6F. Naval Health Research Center, 1999. https:// doi.org/10.21236/ADA375322.

Hodge, Nathan. "Future Warbot Powered by Xbox Controller." *Wired*, June 12, 2009. https://www.wired.com/2009/06/future-warbot-powered-by-xbox-controller/.

Hoffman, Frank. "American Defense Priorities After Ukraine." *War on the Rocks*, January 2, 2023. https://warontherocks.com/2023/01/american-defense-priorities-after -ukraine/.

Holdridge, Michael. "Mission Command Means Emissions Control." *U.S. Naval Institute Proceedings* 147, no. 5 (2021). https://www.usni.org/magazines/proceedings /2021/may/mission-command-means-emissions-control.

Holland, T. J. "Decoding Lethality: Measuring What Matters." *Military Review: The Professional Journal of the U.S. Army.* Online Exclusive (October 2024): 1–8.

Holland, T. J. "Technology at the Point of Contact: Shaping the Future of Warfighting." *NCO Journal.* Muddy Boots Online Collection (November 2024): 1–4.

Hollenbeck, Neil. "How to Transform the Army for Drone Warfare." *War Room*, January 9, 2025. https://warroom.armywarcollege.edu/articles/transform-for-drones/.

Holley, Peter. "The Tiny Pill Fueling Syria's War and Turning Fighters into Superhuman Soldiers." *Washington Post*, November 19, 2015.

Hope, Gregory C. *"Army Training, Sir": The Impact of the World War I Experience on the Evolution of Training Doctrine in the US Army.* Art of War Papers. Army University Press, 2021.

Horbyk, Roman. "'The War Phone': Mobile Communication on the Frontline in Eastern Ukraine." *Digital War* 3, no. 1 (2022): 9–24. https://doi.org/10.1057/s42984 -022-00049-2.

House, Jonathan M. *A Military History of the New World Disorder, 1989–2022.* University of Oklahoma Press, 2025.

House of Commons Committee of Public Accounts. *Skill Shortages in the Armed Forces.* Fifty-Ninth Report of Session 2017–19. House of Commons of the United Kingdom, 2018.

Howell, Sam. "The United States' Quantum Talent Shortage Is a National Security Vulnerability." *Foreign Policy*, July 31, 2023. https://foreignpolicy.com/2023/07/31/us -quantum-technology-china-competition-security/.

H.R. 2670: National Defense Authorization Act for Fiscal Year 2024, § 577 (2023). https:// www.govinfo.gov/content/pkg/BILLS-118hr2670enr/uslm/BILLS-118hr2670enr.xml.

H.R. 8070—Servicemember Quality of Life Improvement and National Defense Authorization Act for Fiscal Year 2025, US House Subcommittee on Tactical Air and Land Forces. Accessed May 30, 2024. https://docs.house.gov/meetings/AS/AS00/20240522/117296/BILLS-118HR8070ih-TAL.pdf.

Hudson, John, and Kostiantyn Khudov. "The War in Ukraine Is Spurring a Revolution in Drone Warfare Using AI." *Washington Post*, July 26, 2023. https://www.washingtonpost.com/world/2023/07/26/drones-ai-ukraine-war-innovation/.

Hukovskyy, Oleh, James C. West, Joshua C. Morganstein, et al. "The Combat Path: Sustaining Mental Readiness in Ukrainian Soldiers." *The US Army War College Quarterly: Parameters* 54, no. 2 (2024): 21–36. https://doi.org/10.55540/0031-1723.3285.

Human Performance. Technical Report JSR-07-625. Jason Program Office, The MITRE Corporation, 2008. https://apps.dtic.mil/sti/citations/ADA480029.

Humr, Scott. "Bridging the Marine Corps's Digital Chasm." Modern War Institute (at West Point), November 20, 2024. https://mwi.westpoint.edu/bridging-the-marine-corpss-digital-chasm/.

Humr, Scott A. "Take the Conn! Steering a Course for Technical Talent in Modern Naval Warfare." Center for International Maritime Security, September 3, 2024. https://cimsec.org/take-the-conn-steering-a-course-for-technical-talent-in-modern-naval-warfare/.

Hunt, Thomas M. "The Digital Games That Militaries Play." Stratfor.com, January 8, 2018. https://worldview.stratfor.com/article/digital-games-militaries-play.

Hunt, Thomas M. *Drug Games: The International Olympic Committee and the Politics of Doping, 1960–2008*. University of Texas Press, 2011.

Hunter, Jamie. "Fighter Pilots Warn of Newly Trained Pilots' Lack of Actual Flying Experience." *The War Zone (at The Drive)*, July 22, 2020. https://www.thedrive.com/the-war-zone/35033/fighter-pilots-warn-of-lack-of-new-aviators-flying-experience-as-force-remodels-training.

Hurd, Christopher. "Army Futures Command General Lays Out Continuous Transformation Plan." US Army website, April 2, 2024. https://www.army.mil/article/275040/army_futures_command_general_lays_out_continuous_transformation_plan.

Ignatius, David. "How the Algorithm Tipped the Balance in Ukraine." Opinion. *Washington Post*, December 19, 2022. https://www.washingtonpost.com/opinions/2022/12/19/palantir-algorithm-data-ukraine-war/.

Ingram, Jonas H. "Physical Fitness: A Necessary Requisite for War-Time Leadership and Longer Life." *U.S. Naval Institute Proceedings* 68, no. 1 (1942): 9–17.

Insinna, Valerie. "To Get More Female Pilots, the Air Force Is Changing the Way It Designs Weapons." *Air Force Times*, August 19, 2020. https://www.airforcetimes.com/news/your-air-force/2020/08/19/to-get-more-female-pilots-the-air-force-is-changing-the-way-it-designs-weapons/.

Intelligence Section, the General Staff of the American Expeditionary Forces. *Candid Comment on the American Soldier of 1917–1918 and Kindred Topics by the Germans: Soldiers, Priests, Women, Village Notables, Politicians, and Statesmen*. Headquarters, General Staff of the American Expeditionary Forces, 1919. https://irp.fas.org/agency/army/wwi-soldiers.pdf.

ITV News, dir. *Robots Join Ukraine's Soldiers in the Battle to Rescue the Dead and Shield the Living*. 2023. 4:02. https://www.youtube.com/watch?v=ye_OlAatH8I.

Jacobsen, Mark. "Ukraine's Drone Strikes Are a Window into the Future of Warfare." Atlantic Council, September 14, 2023. https://www.atlanticcouncil.org/blogs/new-atlanticist/ukraines-drone-strikes-are-a-window-into-the-future-of-warfare/.

Jahner, Kyle. "Start-Up May Revolutionize Fit of Body Armor, Other Gear." *Army Times*, December 10, 2014. https://www.armytimes.com/news/your-army/2014/12/10/start-up-may-revolutionize-fit-of-body-armor-other-gear/.

Janco, Andrew P. "Training in the Amusements of Mars: Peter the Great, War Games and the Science of War, 1673–1699." *Russian History* 30, no. 1/2 (2003): 35–112.

Jarvis, Christina S. *The Male Body at War: American Masculinity During World War II.* Northern Illinois University Press, 2004.

Jbeily, Mark. "Trust, but Verify: Building Aviator Relationships with AI." *U.S. Naval Institute Proceedings* 148, no. 9 (2022). https://www.usni.org/magazines/proceedings/2022/september/trust-verify-building-aviator-relationships-ai.

Jecker, Nancy S., and Andrew Ko. "Brain-Computer Interfaces Could Allow Soldiers to Control Weapons with Their Thoughts and Turn Off Their Fear—but the Ethics of Neurotechnology Lags Behind the Science." *The Conversation*, December 2, 2022. http://theconversation.com/brain-computer-interfaces-could-allow-soldiers-to-control-weapons-with-their-thoughts-and-turn-off-their-fear-but-the-ethics-of-neurotechnology-lags-behind-the-science-194017.

Jeffers, Jasper. "AN41." Modern War Institute (at West Point), May 6, 2019. https://mwi.westpoint.edu/an41-2/.

Jensen, Benjamin M., Christopher Whyte, and Scott Cuomo. *Information in War: Military Innovation, Battle Networks, and the Future of Artificial Intelligence.* Georgetown University Press, 2022.

Jensen, Rebecca. "Opening Marine Infantry to Women: A Civil-Military Crisis?" *Marine Corps University Journal.* Special Issue: Gender Integration and the Military (2018): 132–55.

Jernigan, Michael J. "Marine Doom." *Marine Corps Gazette* (Quantico, United States) 81, no. 8 (1997): 19–20.

Johnson, David. "The Tank Is Dead: Long Live the Javelin, the Switchblade, the . . . ?" *War on the Rocks*, April 18, 2022. https://warontherocks.com/2022/04/the-tank-is-dead-long-live-the-javelin-the-switchblade-the/.

Johnson, Irik, Alan Kolackovsky, and Ken Kryszyn. "How Well Is an Officer Handling the Pressure of Battle? Wearables May Be Able to Tell." US Naval Institute, accessed October 10, 2023. https://www.usni.org/magazines/proceedings/sponsored/how-well-officer-handling-pressure-battle-wearables-may-be-able.

Jones, Seth G., Riley McCabe, and Alexander Palmer. *Ukrainian Innovation in a War of Attrition.* CSIS Briefs. Center for Strategic and International Studies (CSIS), 2023. https://www.csis.org/analysis/ukrainian-innovation-war-attrition.

Joshi, Shashank. "Ypres with AI." *The Economist*, July 8, 2023 (Special Report: Warfare after Ukraine), S3–S5.

Journal of Advanced Military Studies 12, no. 2 (2021). https://www.usmcu.edu/Outreach/Marine-Corps-University-Press/MCU-Journal/JAMS-Vol-12-No-2/.

Journeyman Pictures, dir. *The Drug Fueling Conflict in Syria.* BBC Arabic, 2015. 28:56. https://www.youtube.com/watch?v=7ke13JNlpBQ.

Jumper, John P. "Global Strike Task Force: A Transforming Concept, Forged by Experience." *Aerospace Power Journal* 15, no. 1 (2001): 24–33.

Jung, Hyokwon. "A Glimpse into the Future Battlefield with AI-Embedded Wargames." *U.S. Naval Institute Proceedings* 150, no. 6 (2024). https://www.usni.org/magazines/proceedings/2024/june/glimpse-future-battlefield-ai-embedded-wargames.

Kaempf, Artur, and Sebastian Gruszczak, eds. *Routledge Handbook of the Future of Warfare*. Routledge, 2023. https://doi.org/10.4324/9781003299011.

Kamienski, Lukasz. "The Drugs That Built a Super Soldier." *The Atlantic*, April 8, 2016. https://www.theatlantic.com/health/archive/2016/04/the-drugs-that-built-a-super-soldier/477183/.

Kamienski, Lukasz. *Shooting Up: A Short History of Drugs and War*. Oxford University Press, 2016.

Kania, Elsa B. "Minds at War: China's Pursuit of Military Advantage Through Cognitive Science and Biotechnology." *PRISM* 8, no. 3 (2019): 82–101.

Kania, Elsa, and Ian Burns McCaslin. *Learning Warfare from the Laboratory: China's Progression in Wargaming and Opposing Force Training*. Military Learning and The Future of War Project. Institute for the Study of War, 2021. https://www.understandingwar.org/report/learning-warfare-laboratory-china%E2%80%99s-progression-wargaming-and-opposing-force-training.

Kanz, Fabian. "Dying in the Arena: The Osseous Evidence from Ephesian Gladiators." *Roman Amphitheatres and Spectacula, a 21st-Century Perspective: Papers from an International Conference Held at Chester, 16th–18th February*, 2009, 19. https://dialnet.unirioja.es/servlet/articulo?codigo=4164418.

Katzmarzyk, Peter T., Kara D. Denstel, Kim Beals, et al. "Results from the United States 2018 Report Card on Physical Activity for Children and Youth." *Journal of Physical Activity and Health* 15, no. s2 (2018): S422–24. https://doi.org/10.1123/jpah.2018-0476.

Keeley, Hunter. "Envisioning a Hellscape: Ukrainian Lessons for a Taiwan Drone Strategy." *U.S. Naval Institute Proceedings* 151, no. 4 (2025): 54–59.

Keenan, Kevin G., Jonathon W. Senefeld, and Sandra K. Hunter. "Girls in the Boat: Sex Differences in Rowing Performance and Participation." *PLOS ONE* 13, no. 1 (2018): e0191504. https://doi.org/10.1371/journal.pone.0191504.

Kelland, Kate. "Ancient Dopers Got Their Kicks from Raw Testicles." *Reuters*, August 1, 2012. https://www.reuters.com/article/us-oly-doping-history-day5-idUSBRE8700YC20120801.

Keller, Jared. "Soldiers Are Finally Getting a 'Robotic Mule' to Haul All Their Heavy Sh*t for Them." *Task & Purpose*, November 6, 2022. https://taskandpurpose.com/tech-tactics/army-s-met-robotic-mule-fielding/.

Keller, Jared. "This Video of a Drone Airdropping an Armed Robot Dog Is the Stuff of Dystopian Nightmares." *Task & Purpose*, October 27, 2022. https://taskandpurpose.com/tech-tactics/robot-dog-machine-gun-drone-airdrop-kestrel-defense-video/.

Keller, Jared. "The US Military Will Fight the Next Big War with Xbox-Style Video Game Controllers." *Task & Purpose*, March 22, 2023. https://taskandpurpose.com/tech-tactics/us-military-video-game-controllers-war/.

Keller, John. "Boeing Starts Producing MQ-25 Stingray Unmanned Tanker Aircraft to Support Navy Aircraft Carrier Operations." Military Aerospace,

November 8, 2022. https://www.militaryaerospace.com/sensors/article/14285399/unmanned-tanker-aircraft-carrier.

Keller-Lynn, Carrie. "Israel Is Putting More Women on the Front Line to Help Fix Its Manpower Problem." *Wall Street Journal*, June 8, 2025. https://www.wsj.com/world/middle-east/israel-is-putting-more-women-on-the-front-line-to-help-fix-its-manpower-problem-a962fe79.

Kelly, Samantha Murphy. "Lithium-Ion Battery Fires Are Happening More Often. Here's How to Prevent Them." CNN, March 9, 2023. https://www.cnn.com/2023/03/09/tech/lithium-ion-battery-fires/index.html.

Kenagy, David N., Christopher T. Bird, Christopher M. Webber, and Joseph R. Fischer. "Dextroamphetamine Use During B-2 Combat Missions." *Aviation, Space, and Environmental Medicine* 75, no. 5 (2004): 381–86.

Kennedy, John F. "The Soft American." *Sports Illustrated* 13, no. 26 (1960): 15–17.

Kennedy, Merrit. "First Female Marine Completes Grueling Infantry Officer Course." *NPR*, September 25, 2017. https://www.npr.org/sections/thetwo-way/2017/09/25/553494258/first-female-marine-completes-grueling-infantry-officer-course.

Kerg, Brian. "Maneuver Chess: Integrating Think-and-Fight Drills into Military Training." *U.S. Naval Institute Proceedings* 150, no. 11 (2024). https://www.usni.org/magazines/proceedings/2024/november/maneuver-chess-integrating-think-and-fight-drills-military.

Kernstine, Kelsey. "US Army Recruiting: Could Female Soldiers Close the Gap?" *NewsNation*, April 17, 2023. https://www.newsnationnow.com/us-news/military/us-army-recruiting-could-female-soldiers-close-the-gap/.

Kheel, Rebecca. "Navy Follows Army in Offering Prep Courses to Recruits Who Don't Meet Fitness, Academic Standards." Military.com, March 22, 2023. https://www.military.com/daily-news/2023/03/22/navy-follows-army-offering-prep-courses-recruits-who-dont-meet-fitness-academic-standards.html.

Khona, Jatin G., Michael A. Insler, and Roger D. Little. "Retention of Women in the Naval Officer Corps." *U.S. Naval Institute* 148, no. 3 (2022). https://www.usni.org/magazines/proceedings/2022/march/retention-women-naval-officer-corps.

Kim, Jeong Soo. "Beyond Force Design 2030: Preparing for Fourth-Generation Combat." *U.S. Naval Institute Proceedings* 148, no. 11 (2022). https://www.usni.org/magazines/proceedings/2022/november/beyond-force-design-2030-preparing-fourth-generation-combat.

Kim, Lucian. "How U.S. Military Aid Has Helped Ukraine Since 2014." *NPR*, December 18, 2019. https://www.npr.org/2019/12/18/788874844/how-u-s-military-aid-has-helped-ukraine-since-2014.

Kindervater, Katharine Hall. "The Emergence of Lethal Surveillance: Watching and Killing in the History of Drone Technology." *Security Dialogue* 47, no. 3 (2016): 223–38.

King, Anthony. "On Combat Effectiveness in the Infantry Platoon: Beyond the Primary Group Thesis." *Security Studies* 25, no. 4 (2016): 699–728. https://doi.org/10.1080/09636412.2016.1220205.

King, Anthony. *The Combat Soldier: Infantry Tactics and Cohesion in the Twentieth and Twenty-First Centuries*. Oxford University Press, 2013.

King, James. "The Overweight Infantryman." Modern War Institute (at West Point), January 10, 2017. https://mwi.usma.edu/the-overweight-infantryman/.

Kleisner, Theodore W. "Tactical TikTok for Great Power Competition: Applying the Lessons of Ukraine's IO Campaign to Future Large-Scale Conventional Operations." *Military Review: The Professional Journal of the U.S. Army* 102, no. 4 (2022): 23–43.

Knapik, Joseph J., and Whitfield B. East. "History of United States Army Physical Fitness and Physical Readiness Testing." *U.S. Army Medical Department Journal*, US Army Medical Department Center & School, April 1, 2014, 5–20.

Knight, Mariya, Olga Voitovych, Andrew Carey, Tim Lister, and Josh Pennington. "Russian Commander Killed While Jogging May Have Been Tracked on Strava App." CNN, July 11, 2023. https://www.cnn.com/2023/07/11/europe/russian-submarine -commander-killed-krasnador-intl/index.html.

Koehler, Jeff. "In WWI Trenches, Instant Coffee Gave Troops a Much-Needed Boost." Food History & Culture. *NPR*, April 6, 2017. https://www.npr.org/sections/the salt/2017/04/06/522071853/in-wwi-trenches-instant-coffee-gave-troops-a-much-nee ded-boost.

Koistinen, Paul A. C. *Arsenal of World War II: The Political Economy of American Warfare, 1940–1945.* University Press of Kansas, 2004.

Kollars, Nina. "Genius and Mastery in Military Innovation." *Survival* 59, no. 2 (2017): 125–38. https://doi.org/10.1080/00396338.2017.1302193.

Kraus, Hans, and Ruth P. Hirschland. "Minimum Muscular Fitness Tests in School Children." *Research Quarterly. American Association for Health, Physical Education and Recreation* 25, no. 2 (1954): 178–88. https://doi.org/10.1080/10671188.1954.106249 57.

Krepinevich, Andrew F. *The Military-Technical Revolution: A Preliminary Assessment.* Office of Net Assessment, 1992. https://csbaonline.org/research/publications/the -military-technical-revolution-a-preliminary-assessment/publication/1.

Krepinevich, Andrew F. *The Origins of Victory: How Disruptive Military Innovation Determines the Fates of Great Powers.* Yale University Press, 2023.

Kreps, Sarah E. *Drones: What Everyone Needs to Know.* Oxford University Press, 2016.

Kreuzer, Michael P. *Drones and the Future of Air Warfare: The Evolution of Remotely Piloted Aircraft.* Routledge, 2016.

Krulak, Charles. "Marine Corps Order 1500.55." April 12, 1997. https://www.marines. mil/Portals/1/Publications/MCO%201500.55.pdf.

Kube, Courtney, and Molly Boigon. "Every Branch of the Military Is Struggling to Make Its 2022 Recruiting Goals, Officials Say." NBC News, June 27, 2022. https:// www.nbcnews.com/news/military/every-branch-us-military-struggling-meet-202 2-recruiting-goals-officia-rcna35078.

Kuhn, Scott. "Soldiers Maintain Readiness Playing Video Games." Defense Visual Information Distribution Service (DVIDS), April 29, 2020. https://www.dvidshub. net/news/368773/soldiers-maintain-readiness-playing-video-games.

Kuntz, Carol. *Genomes: The Era of Purposeful Manipulation Begins.* A Report of the CSIS Strategic Technologies Program. Center for Strategic and International Studies (CSIS), 2022. https://www.csis.org/analysis/genomes-era-purposeful-manipulation-begins.

Kuntzman, Allison. "Mad Scientist FY17: A Retrospective." Mad Scientist Laboratory, January 8, 2018. https://madsciblog.tradoc.army.mil/18-mad-scientist-fy17-a-retrospective/.

Kuzminski, Katherine L. "The Military Needs to Make Human-Performance Optimization Part of Daily Ops." Center for a New American Security, August 1, 2024. https://www.cnas.org/publications/commentary/the-military-needs-to-make-human-performance-optimization-part-of-daily-ops.

Kuzminski, Katherine. "The Pentagon's New Opportunity to Boost Readiness Among Female Troops." *Military Times*, November 22, 2023. https://www.militarytimes.com/opinion/2023/11/22/the-pentagons-new-opportunity-to-boost-readiness-among-female-troops/.

Kuzminski, Katherine L., and Taren Dillon Sylvester. *Back to the Drafting Board: U.S. Draft Mobilization Capability for Modern Operational Requirements*. Center for a New American Security, 2024. https://www.cnas.org/publications/reports/back-to-the-drafting-board.

Labrenz, Robert. "Drill Emission Control as a Main Battery." *U.S. Naval Institute Proceedings* 149, no. 6 (2023). https://www.usni.org/magazines/proceedings/2023/june/drill-emission-control-main-battery.

Laco, Kelly. "Republicans Move to Strike Dem Proposal to Include Women in the Military Draft." Fox News, October 12, 2022. https://www.foxnews.com/politics/republicans-move-kill-dem-proposal-include-women-military-draft.

Laird, Robbin. *The Coming of Maritime Autonomous Systems: Empowering and Enhancing the Kill Web Force*. Independently published, 2024.

Lake, Daniel R. *The Pursuit of Technological Superiority and the Shrinking American Military*. Springer, 2019.

Lamothe, Dan. "How Big Is Opposition to Women in Combat Units Among Marines? This Report Explains." *Washington Post*, March 10, 2016. https://www.washingtonpost.com/news/checkpoint/wp/2016/03/10/survey-details-depth-of-opposition-in-marine-corps-to-allowing-women-in-combat-jobs/.

Lamothe, Dan. "Marine Experiment Finds Women Get Injured More Frequently, Shoot Less Accurately Than Men." *Washington Post*, September 10, 2015. https://www.washingtonpost.com/news/checkpoint/wp/2015/09/10/marine-experiment-finds-women-get-injured-more-frequently-shoot-less-accurately-than-men/.

Landen, Xander. "Putin Mobilizing Medically, Mentally 'Unfit' Troops to Ukraine: General." *Newsweek*, September 22, 2022. https://www.newsweek.com/vladimir-putin-mobilizing-medically-mentally-unfit-troops-ukraine-general-jack-keane-1745492.

Larose, Stephen W. "A View from the Trenches on the Debate Wracking the Marine Corps." *War on the Rocks*, May 6, 2022. https://warontherocks.com/2022/05/a-view-from-the-trenches-on-the-debate-wracking-the-marine-corps/.

Larter, David. "'Performance Enhancing Drugs' Considered for Special Operations Soldiers." *Defense News*, May 16, 2017. https://www.defensenews.com/digital-show-dailies/sofic/2017/05/16/performance-enhancing-drugs-considered-for-special-operations-soldiers/.

Latham, Andrew. "The Canadian Armed Forces Recruitment Crisis." RealClearDefense, December 4, 2024. https://www.realcleardefense.com/articles/2024/12/04/the_canadian_armed_forces_recruitment_crisis_1076308.html.

Latiff, Robert H. *Future Peace: Technology, Aggression, and the Rush to War.* University of Notre Dame Press, 2022.

Latiff, Robert H. *Future War: Preparing for the New Global Battlefield.* Alfred A. Knopf, 2017.

Lawrence, Quil. "Ukrainian Soldiers Benefit from U.S. Prosthetics Expertise but Their War Is Different." *NPR*, February 7, 2023. https://www.npr.org/2023/02/07/1153472827/ukrainian-soldiers-benefit-from-u-s-prosthetics-expertise-but-their-war-is-diffe.

Leese, Bryan. "The Cold War Computer Arms Race." *Journal of Advanced Military Studies* 14, no. 2 (2023): 102–20. https://doi.org/10.21140/mcuj.20231402006.

Legros, Joe. "Sky Soldier Makes History as First Active-Duty Female Army Sniper." US Army website, December 5, 2023. https://www.army.mil/article/272115/sky_soldier_makes_history_as_first_active_duty_female_army_sniper.

Lemmon, Gayle Tzemach. *Ashley's War: The Untold Story of a Team of Women Soldiers on the Special Ops Battlefield.* HarperCollins, 2015.

Lemmon, Gayle Tzemach. "No Turning Back?: First Woman Makes Army's Elite 75th Ranger Regiment, Big Step for Women in Combat." *Defense One*, January 19, 2017. https://www.defenseone.com/ideas/2017/01/no-turning-back-first-woman-makes-armys-elite-75th-ranger-regiment-big-step-women-combat/134691/.

Lemmon, Gayle Tzemach. "The Women of the Army Rangers' Cultural Support Teams." World. *At War, Notes from the Front Lines—New York Times* (blog), September 14, 2015. https://archive.nytimes.com/atwar.blogs.nytimes.com/2015/09/14/army-rangers-cultural-support-teams/.

Lemoine, C. W., dir. *How Do Fighter Pilots Prepare for Long Sorties/Ocean Crossing?* 2023. 9:23. https://www.youtube.com/watch?v=342Qie3FPjo.

Leshchinskiy, Brandon, and Andrew Bowne. "Digital Transformation Is a Cultural Problem, Not a Technological One." *War on the Rocks*, May 17, 2022. https://warontherocks.com/2022/05/digital-transformation-is-a-cultural-problem-not-a-technological-one/.

Letendre, Linell. "Women Warriors: Why the Robotics Revolution Changes the Combat Equation." *PRISM: The Journal of Complex Operations* 6, no. 1 (2016): 91–103.

Lewis, David H. "Build This, Not That." *U.S. Naval Institute Proceedings* 149, no. 10 (2023). https://www.usni.org/magazines/proceedings/2023/october/build-not.

Lin, Patrick, Maxwell Mehlman, and Keith Abney. *Enhanced Warfighters: Risk, Ethics, and Policy.* Greenwell Foundation, 2013. https://ethics.calpoly.edu/Greenwall_report.pdf.

Lindsay, Jon R. *Information Technology and Military Power.* Cornell Studies in Security Affairs. Cornell University Press, 2020.

Lin-Greenberg, Erik. "Wargame of Drones: Remotely Piloted Aircraft and Crisis Escalation." *Journal of Conflict Resolution* 66, no. 10 (2022): 1737–65. https://dspace.mit.edu/handle/1721.1/148674.

Linn, Brian McAllister. *The Echo of Battle: The Army's Way of War.* Harvard University Press, 2007.

Linn, Brian McAllister. "A Historical Perspective on Today's Recruiting Crisis." *The US Army War College Quarterly: Parameters* 53, no. 3 (2023): 7–20. https://doi.org/10.55540/0031-1723.3239.

Linn, Brian McAllister. *Real Soldiering: The US Army in the Aftermath of War, 1815–1980*. Modern War Studies. University Press of Kansas, 2023.

Lockheed Martin. "Gaming Technology: The Future of Training." November 28, 2022. https://www.lockheedmartin.com/en-us/news/features/2022/gaming-technology -the-future-of-training.html.

Lofgren, Stephen J. "Unready for War: The Army and World War I." *Army History*, no. 22 (1992): 11–19.

Londoño, Ernesto. "A Decade into War, Body Armor Gets Curves." *Washington Post*, September 20, 2012. https://www.washingtonpost.com/world/national-security/a -decade-into-war-body-armor-gets-curves/2012/09/20/ffaa06c4-0353-11e2-91e7 -2962c74e7738_story.html.

Lopez, German. "Captagon, ISIS's Favorite Amphetamine, Explained." *Vox*, November 20, 2015. https://www.vox.com/world/2015/11/20/9769264/captagon-isis-drug.

Lord, Alfred F., and Thomas R. Kunish. "Lessons from the War at Sea." In *A Call to Action: Lessons from Ukraine for the Future Force*, by John A. Nagl and Katie Crombie. US Army War College Press, 2024. https://press.armywarcollege.edu/monographs/968.

LoRusso, Nick. "Maintain the Standards: Why Physical Fitness Will Always Matter in War." Modern War Institute (at West Point), February 14, 2017. https://mwi.usma .edu/maintain-standards-physical-fitness-will-always-matter-war/.

Losey, Stephen. "Air Force Meets Recruitment Goals, Eyes 20% Increase in 2025." *Defense News*, September 17, 2024. https://www.defensenews.com/news/your-air -force/2024/09/17/air-force-meets-recruitment-goals-eyes-20-increase-in-2025/.

Lowe, Miranda Summers. "'Swat the Kaiser' and Stork Stands: The History of Army Physical Fitness." *New York Times Magazine*, March 28, 2019. https://www.nytimes. com/2019/03/28/magazine/army-physical-fitness-test.html.

Lowther, Adam, and Mahbube K. Siddiki. "Combat Drones in Ukraine." *Air & Space Operations Review* 1, no. 4 (2022): 3–13.

Lynch, Justin. "A Great Power War Could Require More Troops, and More Quickly, Than the United States Can Generate—Even with a Draft." Modern War Institute (at West Point), October 1, 2020. https://mwi.westpoint.edu/a-great-power-war-could -require-more-troops-and-more-quickly-than-the-united-states-can-generate-eve n-with-a-draft/.

MacCalman, Alex, Jeff Grubb, Joe Register, and Mike McGuire. "The Hyper-Enabled Operator." *Small Wars Journal*, June 6, 2019. https://smallwarsjournal.com/jrnl/art /hyper-enabled-operator.

MacDonald, Heather. "Women Don't Belong in Combat Units." Opinion. *Wall Street Journal*, January 13, 2019. https://www.wsj.com/articles/women-dont-belong-in -combat-units-11547411638.

Magnuson, Stew. "Why the Coast Guard Lags When It Comes to Unmanned Systems." *National Defense*, July 7, 2023. https://www.nationaldefensemagazine.org/articles /2023/7/7/why-the-coast-guard-lags-when-it-comes-to-unmanned-systems.

Mahnken, Thomas G. *Technology and the American Way of War*. Columbia University Press, 2008.

Malsin, Jared. "Ukraine's Sea Drones Alter Balance of Power in Black Sea." *Wall Street Journal*, August 11, 2023. https://www.wsj.com/articles/ukraines-sea-drones-alter -balance-of-power-in-black-sea-391cebee.

Malyasov, Dylan. "Russia Follows Ukraine in Creating Drone Forces." Defence Blog, December 16, 2024. https://defence-blog.com/russia-follows-ukraine-in-creating-drone-forces/.

Mamiche, Mona. "Adapting Military Equipment to Account for Gender Differences." NATO Association of Canada, October 10, 2019. https://natoassociation.ca/adapting-military-equipment-to-account-for-gender-differences/.

Management Advisory: Army's Future Soldier Preparatory Course Places Trainees at Increased Risk of Adverse Health Effects. Management Advisory DODIG-2025-069. Inspector General, US Department of Defense, 2025.

Manning, Courtney. "Combating Military Obesity: Stigma's Persistent Impact on Operational Readiness." White Paper. American Security Project, 2023. https://www.americansecurityproject.org/wp-content/uploads/2023/10/Ref-0286-Combating-Military-Obesity.pdf.

Mansoor, Peter R., and Williamson Murray, eds. *The Culture of Military Organizations.* Cambridge University Press, 2019.

"MARADMIN 601/12: Announcement of High Intensity Tactical Training Program." United States Marine Corps, October 17, 2012. https://www.marines.mil/News/Messages/Messages-Display/Article/895087/announcement-of-high-intensity-tactical-training-program/.

Marine Corps Headquarters. "Expeditionary Advanced Base Operations (EABO)." US Marine Corps website, August 2, 2021. https://www.marines.mil/News/News-Display/Article/2708120/expeditionary-advanced-base-operations-eabo/.

Marion, Forrest L. *Standing Up Space Force: The Road to the Nation's Sixth Armed Service.* US Naval Institute Press, 2023. https://www.usni.org/press/books/standing-space-force.

Marshall, Andrew W. "Some Thoughts on Military Revolutions—Second Version." ONA Memorandum for the Record, August 23, 1993. https://stacks.stanford.edu/file/druid:yx275qm3713/yx275qm3713.pdf.

Marson, James. "The Nerdy Gamers Who Became Ukraine's Deadliest Drone Pilots." *Wall Street Journal,* November 5, 2024.

Martin, Richard. "It's Wake-Up Time." *Wired,* November 1, 2003. https://www.wired.com/2003/11/sleep/.

Martinez, Christine. "Women, Peace, and Security: An Underutilized Tool in Countering the People's Republic of C." *Journal of Indo-Pacific Affairs* 6, no. 7 (2023): 110–16.

Martinson, Ryan D. "The PLA Navy's Blue Team Center Games for War." *U.S. Naval Institute* 150, no. 5 (2024). https://www.usni.org/magazines/proceedings/2024/may/pla-navys-blue-team-center-games-war.

Marx, Patricia. "Is the Army's New Tactical Bra Ready for Deployment?" *The New Yorker,* June 19, 2023. https://www.newyorker.com/magazine/2023/06/26/is-the-armys-new-tactical-bra-ready-for-deployment.

Masih, Niha. "Biden Nominates Adm. Lisa Franchetti as First Woman to Lead U.S. Navy." *Washington Post,* July 22, 2023. https://www.washingtonpost.com/national-security/2023/07/22/lisa-franchetti-navy-biden-nomination/.

Massicot, Dara. "The Russian Military's Looming Personnel Crises of Retention and Veteran Mental Health." *The RAND Blog, The RAND Corporation,* June 1, 2023.

https://www.rand.org/blog/2023/06/the-russian-militarys-looming-personnel-crises-of-retention.html.

Matisek, Jahara. "Change Physical Standards to Recognize Most Jobs Don't Require 'Combat Fitness.'" *Task & Purpose*, November 17, 2016. https://taskandpurpose.com/news/change-physical-standards-recognize-jobs-dont-require-combat-fitness/.

Matisek, Jahara. "Physical Fitness Is Not the Key to Winning America's Future Wars." Modern War Institute (at West Point), February 7, 2017. https://mwi.usma.edu/physical-fitness-not-key-winning-americas-future-wars/.

Matisek, Jahara, and James Micciche. "DoD 3.0: Rebooting the Pentagon for the Next War—Modern War Institute." Modern War Institute (at West Point), June 6, 2025. https://mwi.westpoint.edu/dod-3-0-rebooting-the-pentagon-for-the-next-war/.

Matthews, Luke J., Mary Lee, Brandon De Bruhl, Daniel Elinoff, and Christopher A. Eusebi. *Plagues, Cyborgs, and Supersoldiers: The Human Domain of War*. Research Report RR-A2520-1. RAND Corporation, 2024. https://www.rand.org/pubs/research_reports/RRA2520-1.html.

Matthews, Owen. *Overreach: The Inside Story of Putin's War Against Ukraine*. Mudlark, 2022.

Mattis, Jim. *Summary of the 2018 National Defense Strategy*. US Department of Defense, 2018.

Mayer, Holly A. "Bearing Arms, Bearing Burdens: Women Warriors, Camp Followers and Home-Front Heroines of the American Revolution." In *Gender, War and Politics: Transatlantic Perspectives, 1775–1830*, edited by Karen Hagemann, Gisela Mettele, and Jane Rendall. War, Culture and Society, 1750–1850. Palgrave Macmillan UK, 2010. https://doi.org/10.1057/9780230283046_9.

Mazanov, Jason. "Anti-Doping in Sport and Human Enhancing Technologies in Army." *Land Power Forum, Australian Army Research Centre* (blog), October 13, 2017. https://researchcentre.army.gov.au/library/land-power-forum/anti-doping-sport-and-human-enhancing-technologies-army.

Mazzara, Jennifer L. *Shared Experience: Organizational Culture and Ethos at the U.S. Marine Corps Basic School, 1924–1941*. Marine Corps University Press, 2023. https://doi.org/10.56686/9798986259406.

McArdle, Jennifer, and Caitlin Dohrman. "The Full Potential of a Military Metaverse." *War on the Rocks*, February 18, 2022. https://warontherocks.com/2022/02/the-full-potential-of-a-military-metaverse/.

McClure, Jim. "Flight of the Phoenix." *U.S. Naval Institute Proceedings* 148, no. 9 (2022). https://www.usni.org/magazines/proceedings/2022/september/flight-phoenix.

McConnell, Richard A., and Mark T. Gerges. "Seeing the Elephant: Improving Leader Visualization Skills Through Simple War Games." *Military Review: The Professional Journal of the U.S. Army*. Online exclusive (October 2018): 1–9. https://www.armyupress.army.mil/Journals/Military-Review/Online-Exclusive/2018-OLE/Oct/Seeing-the-Elephant/.

McElroy, Damien. "Mumbai Attacks: Terrorists Took Cocaine to Stay Awake During Assault." *The Telegraph*, December 2, 2008. https://www.telegraph.co.uk/news/worldnews/asia/india/3540964/Mumbai-attacks-Terrorists-took-cocaine-to-stay-awake-during-assault.html.

McGee, Will. "Social Media Puts EABO at Risk." *U.S. Naval Institute Proceedings* 148, no. 4 (2022). https://www.usni.org/magazines/proceedings/2022/april/social-media-puts-eabo-risk.

McGeehan, Timothy. "Web Storm Rising." *U.S. Naval Institute Proceedings* 148, no. 3 (2022). https://www.usni.org/magazines/proceedings/2022/march/web-storm-rising.

McHugh, Francis J. "Eighty Years of War Gaming." *Naval War College Review* 21, no. 7 (1969): 88–90.

McKeown, Robert. "Assessing Military Capability: More Than Just Counting Guns." *U.S. Naval Institute Proceedings* 148, no. 12 (2022). https://www.usni.org/magazines/proceedings/2022/december/assessing-military-capability-more-just-counting-guns.

McMahon, Christopher, and Colin Bernard. "Storm Clouds on the Horizon—Challenges and Recommendations for Military Recruiting and Retention." *Naval War College Review* 72, no. 3 (2019): 84–100.

"MDM 22—Tomahawk Robotics RAID Plate." *Soldier Systems: An Industry Daily.* May 11, 2022. https://soldiersystems.net/2022/05/11/mdm-22-tomahawk-robotics-raid-plate/.

Mead, Corey. *War Play: Video Games and the Future of Armed Conflict.* Houghton Mifflin Harcourt, 2013.

Mehlhorn, Tom, and Allen Garner. "Electric Plasmas Can Extend the Golden Hour." *U.S. Naval Institute Proceedings* 149, no. 4 (2023). https://www.usni.org/magazines/proceedings/2023/april/electric-plasmas-can-extend-golden-hour.

Meyer, Josh, and Kim Hjelmgaard. "Were the Hamas Attacks on Israel So Brutal Because the Killers Were High on the Drug Captagon?" *USA Today*, November 2, 2023. https://www.usatoday.com/story/news/world/2023/11/02/hamas-captagon-drug-use-idf-claim/71288873007/.

Meyer, Leisa D. *Creating GI Jane: Sexuality and Power in the Women's Army Corps During World War II.* Columbia University Press, 1996.

Meyers, C. D. "GI, Robot: The Ethics of Using Robots in Combat." *Public Affairs Quarterly* 25, no. 1 (2011): 21–36.

Michaels, Daniel. "A Million People Play This Video Wargame. So Does the Pentagon." *Wall Street Journal*, October 27, 2024. https://www.wsj.com/politics/national-security/a-million-people-play-this-video-wargame-so-does-the-pentagon-e6388f50?st=ZkysKC.

Michaels, Daniel. "The Secret of Ukraine's Military Success: Years of NATO Training." *Wall Street Journal*, April 13, 2022. https://www.wsj.com/articles/ukraine-military-success-years-of-nato-training-11649861339.

Milburn, Andrew. "Time Is Not on Kyiv's Side: Training, Weapons, and Attrition in Ukraine." Modern War Institute (at West Point), June 27, 2022. https://mwi.westpoint.edu/time-is-not-on-kyivs-side-training-weapons-and-attrition-in-ukraine/.

Milley, Mark A. "Strategic Inflection Point: The Most Historically Significant and Fundamental Change in the Character of War Is Happening Now—While the Future Is Clouded in Mist and Uncertainty." *Joint Force Quarterly* 110 (Quarter 2023): 6–15.

Mills, Walker, and Timothy Heck. "War Books—Preparing for Great-Power Competition: A Fiction Reading List." Modern War Institute (at West Point), May 13, 2020. https://mwi.westpoint.edu/war-books-preparing-great-power-competition-fiction-reading-list/.

Mills, Walker J., and Christopher Booth. "Marines Need a Few Good Mules." *U.S. Naval Institute Proceedings* 148, no. 4 (2022). https://www.usni.org/magazines/proceedings/2022/april/marines-need-few-good-mules.

Milner, Allison, Anna J. Scovelle, Belinda Hewitt, Humaira Maheen, Leah Ruppanner, and Tania L. King. "Shifts in Gender Equality and Suicide: A Panel Study of Changes over Time in 87 Countries." *Journal of Affective Disorders* 276 (2020): 495–500. https://doi.org/10.1016/j.jad.2020.07.105.

Minzenberg, Michael J., and Cameron S. Carter. "Modafinil: A Review of Neurochemical Actions and Effects on Cognition." *Neuropsychopharmacology* 33, no. 7 (2008): 7. https://doi.org/10.1038/sj.npp.1301534.

Mittal, Vikram. "Why Military Exoskeletons Will Remain Science Fiction." *Forbes*, August 17, 2020. https://www.forbes.com/sites/vikrammittal/2020/08/17/military-exoskeletons-science-fiction-or-science-reality/.

Mittal, Vikram, and Andrew Davidson, "Combining Wargaming with Modeling and Simulation to Project Future Military Technology Requirements," *IEEE Transactions on Engineering Management* 68, no. 4 (2021): 1195–1207, https://doi.org/10.1109/TEM.2020.3017459.

Mizokami, Kyle. "The U.S. Navy Wants Autonomous Fighter Jets, but First It Needs to Trust AI." *Popular Mechanics*, April 13, 2023. https://www.popularmechanics.com/military/aviation/a43541414/the-navy-wants-ai-powered-autonomous-fighter-jets/.

Modern War Institute staff. "War Books: Science Fiction and Modern War." Modern War Institute (at West Point), May 12, 2023. https://mwi.westpoint.edu/war-books-science-fiction-and-modern-war/.

Molenda, Patrick. "Winning the Next War—Virtually." *U.S. Naval Institute Proceedings* 149, no. 7 (2023). https://www.usni.org/magazines/proceedings/2023/july/winning-next-war-virtually.

Monahan, Evelyn, and Rosemary Neidel-Greenlee. *A Few Good Women: America's Military Women from World War I to the Wars in Iraq and Afghanistan.* Knopf Doubleday Publishing Group, 2010.

Mongilio, Heather. "Marine Corps Personnel Change Was Key to New Force Design, Says CMC Berger." *US Naval Institute (USNI) News*, July 6, 2023. https://news.usni.org/2023/07/06/marine-corps-personnel-change-was-key-to-new-force-design-says-cmc-berger.

Mongilio, Heather. "Military Services Competing Over the Same Recruiting Pool of Less Than 500,000." News & Analysis. *US Naval Institute (USNI) News*, February 14, 2023. https://news.usni.org/2023/02/14/military-services-competing-over-the-same-recruiting-pool-of-less-than-500000.

Mongilio, Heather. "Navy Medicine Is Preparing for the Future of Expeditionary Combat." *US Naval Institute (USNI) News*, February 22, 2023. https://news.usni.org/2023/02/22/navy-medicine-is-preparing-for-the-future-of-expeditionary-combat.

Mongilio, Heather. "Navy Needs to Fill About 9,000 At-Sea Billets in More Than a Dozen Ratings, Says Personnel Command." *US Naval Institute (USNI) News*, October 25, 2022. https://news.usni.org/2022/10/25/navy-needs-to-fill-about-9000-at-sea-billets-in-more-than-a-dozen-ratings-says-personnel-command.

Mongilio, Heather. "USNI News Video: Naval Aviation Breaks Out the Tape Measure to Size Up a New Generation of Pilots." *US Naval Institute (USNI) News*, May 17,

2023. https://news.usni.org/2023/05/17/usni-news-video-naval-aviation-breaks-out-the-tape-measure-to-size-up-a-new-generation-of-pilots.

Mon-López, Daniel, Carlos M. Tejero-González, and Santiago Calero. "Recent Changes in Women's Olympic Shooting and Effects in Performance." *PLOS ONE* 14, no. 5 (2019): e0216390. https://doi.org/10.1371/journal.pone.0216390.

Mon-López, Daniel, Carlos M. Tejero-González, Alfonso de la Rubia Riaza, and Jorge Lorenzo Calvo. "Pistol and Rifle Performance: Gender and Relative Age Effect Analysis." *International Journal of Environmental Research and Public Health* 17, no. 4 (2020): 1365. https://doi.org/10.3390/ijerph17041365.

Moore, Emma. "The ACFT and the Problems with the Military's Cult of Physical Fitness." Military.com, December 16, 2019. https://www.military.com/daily-news/2019/12/16/acft-and-problems-militarys-cult-physical-fitness.html.

Moore-Carrillo, Jaime. "Air Force, Space Force Raise Max Enlistment Age to 42." *Air Force Times*, October 30, 2023. https://www.airforcetimes.com/news/your-air-force/2023/10/30/air-force-space-force-raise-max-enlistment-age-to-42/.

Moran, Nick, and Arnel P. David. "Why Gamers Will Win the Next War." Modern War Institute (at West Point), June 30, 2022. https://mwi.usma.edu/why-gamers-will-win-the-next-war/.

Moran, Rachel Louise. *Governing Bodies: American Politics and the Shaping of the Modern Physique.* University of Pennsylvania Press, 2018.

Motyl, Alexander. "Ukraine's Military Advantage? How Quick It Treats Its Wounded." *EUobserver* (Brussels), April 20, 2023. https://euobserver.com/opinion/156933.

MSNBC, dir. *Eric Schmidt: The War in Ukraine Is the First Broadband War.* MSNBC, 2023. 2:51. https://www.youtube.com/watch?v=G3ksdKxDsdk.

Mulchandani, Nand, and Lieutenant General (Ret.) John N. T. "Jack" Shanahan. *Software-Defined Warfare: Architecting the DOD's Transition to the Digital Age.* A Report of the CSIS Strategic Technologies Program. Center for Strategic and International Studies, 2022. https://www.csis.org/analysis/software-defined-warfare-architecting-dods-transition-digital-age.

Mulrine, Anna. "How Can Army Keep Soldiers Fighting Fit After Afghanistan? Avatars." *Christian Science Monitor*, May 3, 2012. https://www.csmonitor.com/USA/Military/2012/0503/How-can-Army-keep-soldiers-fighting-fit-after-Afghanistan-Avatars.

Murphy, M. P., and G. A. Gates. "Hearing Loss: Does Gender Play a Role?" *Medscape Women's Health* 2, no. 10 (1997): 2.

Murphy, Neil. "We Need Soldiers, Not Robots: British Army's Plea for New Recruits." *The National*, July 1, 2022. https://www.thenationalnews.com/world/uk-news/2022/07/01/we-need-soldiers-not-robots-british-armys-plea-for-new-recruits/.

Murray, Philip. "Power Play: Charging Up Strategic Competition over Lithium Battery Value Chains." *Journal of Advanced Military Studies* 16, no. 1 (2025): 9–25. https://doi.org/10.21140/mcuj.20251601001.

Myers, Meghann. "America's Obesity Is Threatening National Security, According to This Study." *Army Times*, October 10, 2018. https://www.armytimes.com/news/your-army/2018/10/10/americas-obesity-is-threatening-national-security-according-to-this-study/.

Myers, Meghann. "Army Scrambles to Improve Drone Training." *Defense One*, May 30, 2025. https://www.defenseone.com/technology/2025/05/army-scrambles-improve-drone-training/405706/.

Myers, Meghann. "The Army's New Combat Readiness Test Is Part of a Holistic Push for Soldier Health and Nutrition." *Defense News*, October 8, 2017. https://www.defensenews.com/news/your-army/2017/10/08/the-armys-new-combat-readiness-test-is-part-of-a-holistic-push-for-soldier-health-and-nutrition/.

Myers, Meghann. "Experts, Data Point to Women as Best Military Recruiting Pool." *Military Times*, January 26, 2023. https://www.militarytimes.com/news/your-military/2023/01/26/experts-data-point-to-women-as-best-military-recruiting-pool/.

Myers, Meghann. "Get Ready for Dietitians, Physical Therapists and More in Every Army Battalion." *Defense News*, October 9, 2018. https://www.defensenews.com/news/your-army/2018/10/09/get-ready-for-dietitians-physical-therapists-and-more-in-every-army-battalion/.

Myers, Meghann, and Charlsy Panzino. "The Army Is Testing a New Combat Fitness Test." *Army Times*, August 15, 2017. https://www.armytimes.com/news/your-army/2017/08/15/the-army-is-testing-a-new-combat-fitness-test/.

Myers, Steven Lee. "The Old Army, It Turns Out, Was the Fitter One." *New York Times*, June 25, 2000. https://www.nytimes.com/2000/06/25/health/the-old-army-it-turns-out-was-the-fitter-one.html.

Myre, Greg. "How Ukraine Created an 'Army of Drones' to Take on Russia." Ukraine Invasion—Explained. *NPR*, June 20, 2023. https://www.npr.org/2023/06/20/1183050117/how-ukraine-created-an-army-of-drones-to-take-on-russia.

Mysliwiec, Vincent, Jessica Gill, Hyunhwa Lee, et al. "Sleep Disorders in US Military Personnel: A High Rate of Comorbid Insomnia and Obstructive Sleep Apnea." *Chest* 144, no. 2 (2013): 549–57. https://doi.org/10.1378/chest.13-0088.

Nadibaidze, Anna. "Great Power Identity in Russia's Position on Autonomous Weapons Systems." *Contemporary Security Policy* 43, no. 3 (2022): 407–35. https://doi.org/10.1080/13523260.2022.2075665.

Nagl, John A., and Katie Crombe. *A Call to Action: Lessons from Ukraine for the Future Force*. US Army War College Press, 2024. https://press.armywarcollege.edu/monographs/968.

Nanos, Janelle. "Armor All: New Body Armor Issued for Women in the Military." *Boston Magazine*, September 26, 2013. https://www.bostonmagazine.com/news/2013/09/26/new-body-armor-women-military/.

National Defense Authorization Act for Fiscal Year 2006, Pub. L. No. Public Law 109–163, 3136 (2006). https://www.congress.gov/109/plaws/publ163/PLAW-109publ163.pdf.

National Institute of Diabetes and Digestive and Kidney Diseases. "Overweight & Obesity Statistics." Accessed March 16, 2024. https://www.niddk.nih.gov/health-information/health-statistics/overweight-obesity.

Naval Air Systems Command. "NAWCTSD's Mobile NEST Expands Navy Training Capabilities." April 18, 2025. https://www.navair.navy.mil/news/NAWCTSDs-Mobile-NEST-Expands-Navy-Training-Capabilities/Fri-04182025-0959.

Nevill, A. M., G. P. Whyte, R. L. Holder, and M. Peyrebrune. "Are There Limits to Swimming World Records?" *International Journal of Sports Medicine*, May 29, 2007, 1012–17. https://doi.org/10.1055/s-2007-965088.

Nevill, Alan M., and Gregory Whyte. "Are There Limits to Running World Records?" *Medicine & Science in Sports & Exercise* 37, no. 10 (2005): 1785–88. https://doi.org/10 .1249/01.mss.0000181676.62054.79.

Newdick, Thomas. "Navy's Aviation Boss Lays Out Big Vision for Drone-Packed Carriers of the Future." *The War Zone (at The Drive)*, April 1, 2021. https://www.twz. com/40007/navys-aviation-boss-lays-out-big-vision-for-drone-packed-carrier-air -wings-of-the-future.

Newdick, Thomas. "Ukrainian MiG-29 Pilot's Front-Line Account of the Air War Against Russia." *The War Zone (at The Drive)*, April 1, 2022. https://www.thedrive. com/the-war-zone/45019/fighting-russia-in-the-sky-mig-29-pilots-in-depth- account-of-the-air-war-over-ukraine.

"News Notes: Future Soldier-Human Tank." *Armor* 70, no. 3 (1961): 54–55.

Nieberg, Patty. "Air Force Pilots Get a New Way to Pee at 30,000 Feet." *Task & Purpose*, April 17, 2025. https://taskandpurpose.com/news/air-force-pilots-pee-while-flying/.

Nieberg, Patty. "The Army Is Planning for a New Robotics Technician MOS." *Task & Purpose*, October 31, 2024. https://taskandpurpose.com/news/robot-tech-warrant-officer/.

Nieberg, Patty. "Army Ups Recruiting Goal to 61,000 Soldiers in 2025, an 11% Jump." *Task & Purpose*, October 14, 2024. https://taskandpurpose.com/news/army-new -recruiting-goal-61000/.

Nieberg, Patty. "Blood Brothers: In the Pacific, US Medics May Soon Use Blood from Allies." *Task & Purpose*, November 6, 2023. https://taskandpurpose.com/history /blood-agreements-china-allies/.

Nikolova, Vania. "American Runners Have Never Been Slower (Mega Study)." RunRepeat, September 21, 2021. https://runrepeat.com/american-runners-have-never-been -slower-mega-study.

Nindl, Bradley C. *Strategies for Enhancing Military Physical Readiness in the 21st Century*. United States Army War College, 2012. https://apps.dtic.mil/sti/citations/ADA 561612.

Nindl, Bradley C., Bruce H. Jones, Stephanie J. Van Arsdale, Karen Kelly, and William J. Kraemer. "Operational Physical Performance and Fitness in Military Women: Physiological, Musculoskeletal Injury, and Optimized Physical Training Considerations for Successfully Integrating Women Into Combat-Centric Military Occupations." *Military Medicine* 181, no. suppl_1 (2016): 50–62. https://doi.org /10.7205/MILMED-D-15-00382.

N-iX. "Education That Makes Developers in Ukraine the Best in the World." September 9, 2019. https://www.n-ix.com/tech-education-developers-in-ukraine/.

Nixon, Dave. "A Lack of Sleep Is Breaking the US Military." *Military Times*, December 12, 2023. https://www.militarytimes.com/opinion/2023/12/12/a-lack-of-sleep-is -breaking-the-us-military/.

Nixon, David, and Porter Riley. "Sleep and Performance: Why the Army Must Change Its Sleepless Culture." *Military Review: The Professional Journal of the U.S. Army* 103, no. 6 (2023): 147–57.

Norris, Michele. "Roles for Women in U.S. Army Expand." *NPR*, October 1, 2007. https://www.npr.org/2007/10/01/14869648/roles-for-women-in-u-s-army-expand.

Novelly, Thomas. "Even More Young Americans Are Unfit to Serve, a New Study Finds. Here's Why." Military.com, September 28, 2022. https://www.military.com /daily-news/2022/09/28/new-pentagon-study-shows-77-of-young-americans-are -ineligible-military-service.html.

Novelly, Thomas. "Female Air Force Pilots Would Be Able to Safely Pee In-Flight During Long Missions with New Tech Being Tested." Military.com, March 20, 2023. https://www.military.com/daily-news/2023/03/20/female-air-force-pilots-would-be -able-safely-pee-flight-during-long-missions-new-tech-being-tested.html.

Nutton, Vivian. *Ancient Medicine*. Series of Antiquity. Routledge, 2004.

Nye, Logan. "5 Video Games Worth Using as 'Hip-Pocket Training.'" Military.com, July 14, 2025. https://www.military.com/off-duty/games/5-video-games-worth -using-hip-pocket-training.html.

de Oca, Jeffrey Montez. "The 'Muscle Gap': Physical Education and US Fears of a Depleted Masculinity, 1954–1963." In *East Plays West: Sport and the Cold War*, edited by Stephen Wagg and David Andrews. Routledge, 2012. http://www.taylorfrancis. com/books/e/9781134241682/chapters/10.4324%2F9780203007112-13.

O'Connell, Aaron B. *Underdogs: The Making of the Modern Marine Corps*. Harvard University Press, 2012.

O'Connor, Ema, and Vera Bergengruen. "Military Doctors Told Them It Was Just 'Female Problems' Weeks Later, They Were in the Hospital." BuzzFeed News, March 8, 2019. https://www.buzzfeednews.com/article/emaoconnor/woman-military -doctors-female-problems-health-care.

Odom, William E. *America's Military Revolution: Strategy and Structure After the Cold War*. American University Press, 1993.

O'Hanlon, Michael E. *Technological Change and the Future of Warfare*. Brookings Institution Press, 2000.

Ohler, Norman. *Blitzed: Drugs in Nazi Germany*. Translated by Shaun Whiteside. Houghton Mifflin, 2017.

Ohnesorge, Hendrik W. "Profile in Vigor: John F. Kennedy and the Quest for Athletic Excellence." In *Sports and the American Presidency: From Theodore Roosevelt to Donald Trump*, edited by Adam Burns and Rivers Gambrell. Edinburgh University Press, 2022. https://www.cambridge.org/core/books/sports-and-the-american -presidency/profile-in-vigor-john-f-kennedy-and-the-quest-for-athletic-excellence /2177B6913CE98991AB3048EBB57566B6.

O'Keefe, Megan. "Grow, Borrow, Recruit, and Reorganize: How the Military Can Get the Personnel It Needs for Digital War." Modern War Institute (at West Point), June 9, 2022. https://mwi.westpoint.edu/grow-borrow-recruit-and-reorganize-how -the-military-can-get-the-personnel-it-needs-for-digital-war/.

Okonofua, Benjamin A., Nicole Laster-Loucks, and Andrew Johnson. "'Will to Fight': Twenty-First-Century Insights from the Russo-Ukrainian War." *Military Review: The Professional Journal of the U.S. Army* 104, no. 3 (2024): 34–39.

"Oleksii Reznikov on Twitter." Twitter. August 5, 2023. https://twitter.com/oleksi ireznikov/status/1687859613289349126.

Oppermann, Brenda. "Women and Gender in the U.S. Military: A Slow Process of Integration." In *Women and Gender Perspectives in the Military: An International Comparison*, edited by Robert Egnell and Mayesha Alam. Georgetown University Press, 2019.

O'Rourke, Brian. "Wargaming at MCU: A Small Step for Marines, a Giant Leap for the Marine Corps." *U.S. Naval Institute Proceedings* 148, no. 11 (2022). https://www.usni.org/magazines/proceedings/2022/november/wargaming-mcu.

Owen, Peter F. *To the Limit of Endurance: A Battalion of Marines in the Great War.* Texas A&M University Press, 2007.

Pallas, Ryan. "How to Boil a Frog: The Dangers of Downsizing in the U.S. Military." *War on the Rocks*, August 26, 2024. https://warontherocks.com/2024/08/how-to-boil-a-frog-the-dangers-of-downsizing-in-the-u-s-military/.

Parra-Orlandoni, M. Alejandra. "Women in the Ranks, but Not in the Clear." *War on the Rocks*, May 29, 2025. https://warontherocks.com/2025/05/women-in-the-ranks-but-not-in-the-clear/.

Parsons, Dan. "Ukraine Situation Report: Defense Chief Wants Advanced Weapons Testing Against Russians." *The War Zone (at The Drive)*, July 19, 2022. https://www.thedrive.com/the-war-zone/ukraine-situation-report-defense-chief-wants-advanced-weapons-testing-against-russians.

Pashakhanlou, Arash Heydarian. "AI, Autonomy, and Airpower: The End of Pilots?" *Defence Studies* 19, no. 4 (2019): 337–52. https://doi.org/10.1080/14702436.2019.1676156.

Patino, Jairo I. "Better Living (and Fighting) Through Batteries." *U.S. Naval Institute Proceedings* 147, no. 2 (2021). https://www.usni.org/magazines/proceedings/2021/february/better-living-and-fighting-through-batteries.

Patterson, Erin. "Ship Collisions: Address the Underlying Causes, Including Culture." *U.S. Naval Institute Proceedings* 143, no. 8 (2017). https://www.usni.org/magazines/proceedings/2017/august/ship-collisions-address-underlying-causes-including-culture.

Patterson, Sarah E. "'Beauty Isn't Prerequisite for Girl Marines': Images of Female Marines During World War II." *Marine Corps History* 8, no. 1 (2022): 5–20. https://doi.org/10.35318/mch.2022080101.

Patton, Erik M., Christopher L. Chapman, and Gabrielle E. W. Giersch. "Preparing for Hot Conflicts: Army Training and Operations in a Warming World." *Military Review: The Professional Journal of the U.S. Army* 105, no. 1 (2025): 132–44.

Patton, Erik, Antonio Salinas, and Dagomar Degroot. "Fremen Stillsuits and Heated Bubble Gyms: Preparing the Army to Fight and Win on a Rapidly Warming Battlefield." Modern War Institute (at West Point), June 19, 2024. https://mwi.westpoint.edu/fremen-stillsuits-and-heated-bubble-gyms-preparing-the-army-to-fight-and-win-on-a-rapidly-warming-battlefield/.

Patton, George S. *War as I Knew It.* Houghton Mifflin Harcourt, 1947.

Pawlyk, Oriana. "Here's How the Air Force Plans to Recruit Teenage Gamers." Military.com, May 25, 2018. https://www.military.com/defensetech/2018/05/25/heres-how-air-force-plans-recruit-teenage-gamers.html.

Pearcy, Aimee. "Meet Gen Alpha, the 'Mini-Millennials' Who Are Poised to Take Over the Internet." *Business Insider*, November 7, 2023. https://www.businessinsider.com/gen-alpha-explained-technology-views-mental-health-2023-10.

Peduzzi, Olivia. "The Promising Virtual Mental Health Pilot." *U.S. Naval Institute Proceedings* 147, no. 2 (2021). https://www.usni.org/magazines/proceedings/2021/february/promising-virtual-mental-health-pilot.

People's Project.Com—Ukraine's Military and Civil Crowdfunding. "Sabre Remote Weapon Station." Accessed May 8, 2023. https://www.peoplesproject.com/en/sabre-remote-weapon-station/.

Pérez-Peña, Richard, and Matthew Rosenberg, "Strava Fitness App Can Reveal Military Sites, Analysts Say," *New York Times*, January 29, 2018, https://www.nytimes.com/2018/01/29/world/middleeast/strava-heat-map.html.

Perkins, Jim, and Mike McGinley. "Without Talent Agility, America May Lose." *War on the Rocks*, August 14, 2024. https://warontherocks.com/2024/08/without-talent-agility-america-may-lose/.

Pete Hegseth—Secretary of Defense Nominee | SRS #143. Shawn Ryan Show. 2024. YouTube video, 0:15. https://www.youtube.com/watch?v=DoN5ovwB8s4.

Peters, Andrea M., Michael A. Washington, Lolita Burrell, and James Ness. "Rethinking Female Urinary Devices for the US Army." *The US Army War College Quarterly: Parameters* 52, no. 1 (2022): 41–56. https://doi.org/10.55540/0031-1723.3128.

Pettyjohn, Stacie L. "America's Eroding Airpower: Washington Must Upgrade Its Fleet of Planes, Drones, and Missiles." *Foreign Affairs*, March 10, 2025. https://www.foreignaffairs.com/united-states/americas-eroding-airpower.

Pettyjohn, Stacie, Becca Wasser, and Chris Dougherty. *Dangerous Straits: Wargaming a Future Conflict over Taiwan*. Center for a New American Security, 2022. https://www.cnas.org/publications/reports/dangerous-straits-wargaming-a-future-conflict-over-taiwans.

Pew Research Center. *From Businesses and Banks to Colleges and Churches: Americans' Views of U.S. Institutions*. Pew Research Center, 2024. https://www.pewresearch.org/politics/2024/02/01/the-u-s-military/.

Pfaff, C. Anthony. "Soldier Enhancement Ethics and the Lessons of World War I." In *A Persistent Fire: The Strategic Ethical Impact of World War I on the Global Profession of Arms*, edited by Timothy S. Mallard and Nathan H. White. National Defense University Press, 2019.

Pfaltzgraf, Daniel, and Gary S. Insch. "Digitally Native, yet Technologically Illiterate: Methods to Prepare Business Students to Create Versus Consume." *Journal of Applied Business and Economics* 23, no. 2 (2021): 25–34. https://articlearchives.co/index.php/JABE/article/view/2181.

Phelps, Charlie. "Leadership, Lethality, and (Data) Literacy: Three Keys to Prepare the Army for the Data-Driven, AI-Enabled Future of War." Modern War Institute (at West Point), May 7, 2025. https://mwi.westpoint.edu/leadership-lethality-and-data-literacy-three-keys-to-prepare-the-army-for-the-data-driven-ai-enabled-future-of-war/, https://mwi.westpoint.edu/leadership-lethality-and-data-literacy-three-keys-to-prepare-the-army-for-the-data-driven-ai-enabled-future-of-war/.

Phelps, Wayne. *On Killing Remotely: The Psychology of Killing with Drones*. Little, Brown and Co., 2021.

Philipps, Dave. "Death in Navy SEAL Training Exposes a Culture of Brutality, Cheating and Drugs." *New York Times*, August 30, 2022. https://www.nytimes.com/2022/08/30/us/navy-seal-training-death.html.

Philipps, Dave. "Navy Will Start Testing SEALs for Illicit Drug Use." *New York Times*, September 29, 2023. https://www.nytimes.com/2023/09/29/us/navy-seal-drug-testing .html.

Philipps, Dave. "With Few Able and Fewer Willing, U.S. Military Can't Find Recruits." *New York Times*, July 14, 2022. https://www.nytimes.com/2022/07/14/us/us-military -recruiting-enlistment.html.

Philipps, Dave, and Tim Arango. "Who Signs Up to Fight? Makeup of U.S. Recruits Shows Glaring Disparity." *New York Times*, January 10, 2020. https://www.nytimes .com/2020/01/10/us/military-enlistment.html.

Phillips-Levine, Trevor, and Walker D. Mills. "Outgunned in the Drone Fight: The U.S. Military Is Failing to Adopt the Next Machine Gun." *War on the Rocks*, March 6, 2024. https://warontherocks.com/2024/03/outgunned-in-the-drone-fight-the-u-s -military-is-failing-to-adopt-the-next-machine-gun/.

Pickrell, Ryan. "US Army Tankers Are Playing Video Games Online to Train for Tank Warfare During the Coronavirus Pandemic." *Business Insider*, May 1, 2020. https:// www.businessinsider.com/coronavirus-us-army-tank-crews-play-video-games-to -train-2020-4.

Pike, Travis. "Energy Drinks—The Unsung Hero of the Global War on Terror." *Sandboxx News*, June 7, 2023. https://www.sandboxx.us/blog/energy-drinks-the -unsung-hero-of-the-global-war-on-terror/.

Plichta, Marcel. "You Think It's a Game? What Video Games Can—and Can't—Teach About Strategy and History." Modern War Institute (at West Point), October 8, 2021. https://mwi.westpoint.edu/you-think-its-a-game-what-video-games-can-and-cant -teach-about-strategy-and-history/.

Pomerleau, Mark. "Army Trying to Expose Entire Force to Electromagnetic War- fare During Training." *DefenseScoop*, August 17, 2023. https://defensescoop.com /2023/08/17/army-trying-to-expose-entire-force-to-electromagnetic-warfare -during-training/.

Pons, Chip. "Flying Training Reimagined as First PTN Class Graduates." Air Force website, August 6, 2018. https://www.af.mil/News/Article-Display/Article/1594573 /flying-training-reimagined-as-first-ptn-class-graduates/.

Posey, Misty. "Physical Fitness Fuels Cognitive Power." *U.S. Naval Institute Proceedings* 146, no. 11 (2020). https://www.usni.org/magazines/proceedings/2020/november /physical-fitness-fuels-cognitive-power.

Poyer, David. "The War with China, as I Saw It." *Air Force Times*, November 27, 2020. https://www.airforcetimes.com/opinion/commentary/2020/11/27/the-war-with -china-as-i-saw-it/.

Poznyakov, Andrey, and Frances Lopez. "Ukraine War: How Have Western Weapons Per- formed in Combat?" *Euronews*, July 13, 2023. https://www.euronews.com/2023/07/13 /war-in-ukraine-how-have-western-weapons-performed-in-combat.

Prados, John. "Waging War with Cardboard Navies." *Naval History* 20, no. 4 (2006). https://www.usni.org/magazines/naval-history-magazine/2006/august/waging -war-cardboard-navies.

Pranshu, Verma. "Inside the Pentagon's Long Debate: Do Gamers Make Good Troops?" *Washington Post*, June 10, 2022. https://www.washingtonpost.com/video-games /2022/06/10/military-gaming/.

Price, Jay. "To Keep Up with Modern Combat, Marines Add Drone Operators to Infantry Units." WUNC 91.5, North Carolina Public Radio, June 4, 2018. https://www.wunc.org/military/2018-06-04/to-keep-up-with-modern-combat-marines-add-drone-operators-to-infantry-units.

Price, Robert E. "Defining Swarm: A Critical Step Toward Harnessing the Power of Autonomous Systems." *Military Review: The Professional Journal of the U.S. Army.* Online exclusive (May 2025): 1–12. https://www.armyupress.army.mil/journals/military-review/online-exclusive/2025-ole/defining-swarm/.

Pugh, James. "'Not . . . Like a Rum-Ration': Amphetamine Sulphate, the Royal Navy, and the Evolution of Policy and Medical Research During the Second World War." *War in History* 24, no. 4 (2017): 498–519. https://doi.org/10.1177/0968344516643348.

Pugh, James. "The Royal Air Force, Bomber Command and the Use of Benzedrine Sulphate: An Examination of Policy and Practice During the Second World War." *Journal of Contemporary History* 53, no. 4 (2018): 740–61. https://doi.org/10.1177/0022009416652717.

Rainey, James. "The Questionable Training of the AEF in World War I." *The US Army War College Quarterly: Parameters* 22, no. 1 (1992): 89–103. https://doi.org/10.55540/0031-1723.1638.

Rainey, James E. "Continuous Transformation: Transformation in Contact." *Military Review: The Professional Journal of the U.S. Army.* Online exclusive (August 2024): 1–7. https://www.armyupress.army.mil/journals/military-review/online-exclusive/2024-ole/Transformation-in-Contact/.

Ramani, Samuel. *Putin's War on Ukraine: Russia's Campaign for Global Counter-Revolution.* Oxford University Press, 2023.

Ramsey, Benjamin, Ann Bednash, and John Folks. "Retaining Female Leaders: A Key Readiness Issue." *Joint Force Quarterly* 104 (Quarter 2022): 81–88.

Raska, Michael. "The Sixth RMA Wave: Disruption in Military Affairs?" *Journal of Strategic Studies* 44, no. 4 (2021): 456–79. https://doi.org/10.1080/01402390.2020.1848818.

Rasmussen, Nicolas. "Medical Science and the Military: The Allies' Use of Amphetamine During World War II." *Journal of Interdisciplinary History* 42, no. 2 (2011): 205–33. https://doi.org/10.1162/JINH_a_00212.

Raum, Mary. "Warrior Women: 3,000 Years in the Fight." *Joint Force Quarterly* 93 (Quarter 2019): 47–54.

Rehman, Iskander. *Planning for Protraction: A Historically Informed Approach to Great-Power War and Sino-US Competition.* Routledge, 2023.

Reisher, Jon. "Talent and Tech: Fielding and Wielding New Systems Requires the Right People." Modern War Institute (at West Point), December 20, 2024. https://mwi.westpoint.edu/talent-and-tech-fielding-and-wielding-new-systems-requires-the-right-people/.

Rempfer, Kyle. "Citing 13 Brigades-Worth of Non-Deployable Troops, Army Looks to Holistic Health Solutions." *Defense News,* September 30, 2020. https://www.defensenews.com/news/your-army/2020/09/30/citing-13-brigades-worth-of-non-deployable-troops-army-looks-to-holistic-health-solutions/.

Rennolds, Nathan. "Ukraine Unveils Military Uniforms for Women After Complaints That Ill-Fitting Men's Clothes Hindered Them on the Frontline." *Business Insider,*

August 6, 2023. https://www.businessinsider.com/ukraine-russia-army-approves-female-military-uniforms-first-time-2023-8.

"Research Guides: CMC Professional Reading Program 2020: Home." Accessed September 2, 2022. https://grc-usmcu.libguides.com/usmc-reading-list-2020/home.

Reuther, Zachary. "The Coast Guard Must Adopt the Physical Fitness Test." *U.S. Naval Institute Proceedings* 149, no. 1 (2023). https://www.usni.org/magazines/proceedings/2023/january/coast-guard-must-adopt-physical-fitness-test.

Rhoades, Stephen L. "Are We Too Fat to Fight? Changing Chow Hall Offerings Can Help." *Stars and Stripes*, November 21, 2023. https://www.stripes.com/opinion/2023-11-21/military-chow-hall-healthy-troops-12126933.html.

Richards, Evelyn. "Lowdown on High Tech Weapons in Gulf War." *Washington Post*, January 27, 1991. https://www.washingtonpost.com/archive/politics/1991/01/27/lowdown-on-high-tech-weapons-in-gulf-war/54219640-02e7-4270-b328-bfa05bf90974/.

Ricks, Thomas E. "Military Physical Training: It's a Problem Bigger than Obesity, with No Easy Solutions." *Foreign Policy*, August 10, 2015. https://foreignpolicy.com/2015/08/10/military-physical-training-its-a-problem-bigger-than-obesity-with-no-easy-solutions-2/.

Riddell, Rob. "Doom Goes to War." *Wired*, April 1, 1997. https://www.wired.com/1997/04/ff-doom/.

Rielage, Dale. "How We Lost the Great Pacific War." *U.S. Naval Institute Proceedings* 144, no. 5 (2018). https://www.usni.org/magazines/proceedings/2018/may/how-we-lost-great-pacific-war.

Rifkin, Jesse. "Body Armor for Females Modernization Act Would Help Military Women's Uniforms and Protective Equipment Better Fit Them." GovTrack Insider, January 13, 2020. https://govtrackinsider.com/body-armor-for-females-modernization-act-would-help-military-womens-uniforms-and-protective-23d66ea3853f.

Rinella, Michael A. *Pharmakon: Plato, Drug Culture, and Identity in Ancient Athens.* Lexington Books, 2010.

Roaten, Meredith. "Milley Compares Urgency of Modernization to Interwar Period." *National Defense*, August 2, 2021. https://www.nationaldefensemagazine.org/articles/2021/8/2/milley-compares-urgency-of-modernization-to-interwar-period.

Roaten, Meredith. "Telling Tales: Startup Helps Officers Explain Future Warfare Through Storytelling." *National Defense* 107, no. 829 (2022): 36–39.

Robb, James. "Ukraine Serves as Lesson in the Value of Training." *National Defense* CVII, no. 829 (2022): 3.

Robert, Leon L., Jr., and Carl J. Wojtaszek. "Closing the Gap: Officer Advanced Education STEM+M (Management)." *The US Army War College Quarterly: Parameters* 54, no. 2 (2024): 111–27. https://doi.org/10.55540/0031-1723.3290.

Robins, Lee N. *The Vietnam Drug User Returns.* Special Action Monograph Series A, Number 2. Special Action Office for Drug Abuse Prevention, 1973. https://prhome.defense.gov/Portals/52/Documents/RFM/Readiness/DDRP/docs/35%20Final%20Report.%20The%20Vietnam%20drug%20user%20returns.pdf.

Robinson, Amy. "Army Seeks Gamers' Input to Help Shape Future Force." US Department of Defense, August 23, 2017. https://www.defense.gov/News/News-Stories/Article/Article/1286948/army-seeks-gamers-input-to-help-shape-future-force/.

Robinson, Tim, and Stephen Bridgewater. "Highlights from the RAeS Future Combat Air & Space Capabilities Summit." Royal Aeronautical Society, May 26, 2023. https://www.aerosociety.com/news/highlights-from-the-raes-future-combat-air-space-capabilities-summit/.

Roco, Mihail C., and William Sims Bainbridge. *Converging Technologies for Improving Human Performance: Nanotechnology, Biotechnology, Information Technology and Cognitive Science.* National Science Foundation, 2002. https://obamawhitehouse.archives.gov/sites/default/files/microsites/ostp/bioecon-%28%23%20023SUPP%29%20NSF-NBIC.pdf.

Roff, Heather M., and David Danks. "'Trust but Verify': The Difficulty of Trusting Autonomous Weapons Systems." *Journal of Military Ethics* 17, no. 1 (2018): 2–20. https://doi.org/10.1080/15027570.2018.1481907.

Rogoway, Tyler. "Marines Winding Down Weapon Systems Officer Position, F/A-18Ds to Fly with Pilot Only." *The War Zone (at The Drive)*, July 19, 2019. https://www.thedrive.com/the-war-zone/29078/marines-winding-down-weapon-system-officers-position-f-a-18ds-to-fly-with-pilot-only.

Rominger, Donald W. "From Playing Field to Battleground: The United States Navy V-5 Preflight Program in World War II." *Journal of Sport History* 12, no. 3 (1985): 252–64.

Rominger, Donald William. "The Impact of the United States Government's Sports and Physical Training Policy on Organized Athletics During World War II." PhD diss., Oklahoma State University, 1976. https://www.proquest.com/docview/302843272/citation/D5F8FC91514C4E38PQ/1.

Roosevelt, Theodore. *Theodore Roosevelt: An Autobiography.* Charles Scribner's Sons, 1913.

Roque, Ashley. "Army Eyeing First Human Machine Integrated Formations in 2027, Common Controller for Robotics." *Breaking Defense*, April 14, 2025. https://breakingdefense.com/2025/04/army-eyeing-first-human-machine-integrated-formations-in-2027-common-controller-for-robotics/.

Roque, Ashley. "More Soldiers, Control Vehicles May Be Required to Maneuver Army's Early Robotic Combat Vehicles." *Breaking Defense*, October 31, 2024. https://breakingdefense.com/2024/10/more-soldiers-control-vehicles-may-be-required-to-maneuver-armys-early-remote-combat-vehicles/.

Rosa-Hernandez, Gabriela Iveliz, and Patrick Enoch. *We Need a Medic!: The Russian Military Medicine Experience in Ukraine.* Center for Naval Analyses, 2024. https://www.cna.org/reports/2024/11/we-need-a-medic.

Rosenberg, Matthew, and John Markoff. "The Pentagon's 'Terminator Conundrum': Robots That Could Kill on Their Own." *New York Times*, October 25, 2016. https://www.nytimes.com/2016/10/26/us/pentagon-artificial-intelligence-terminator.html.

Roshchina, Olena. "Russians Destroy 60–70% of Uncrewed Surface Vessels, but 30% Remains Dangerous for Russian Navy." *Ukrainska Pravda*, August 24, 2023. https://www.pravda.com.ua/eng/news/2023/08/24/7416940/.

Roth, Tanya L. *Her Cold War: Women in the U.S. Military, 1945–1980.* University of North Carolina Press, 2021.

Roza, David. "The Space Force Wants to Try Ditching Physical Fitness Tests for a Year." *Task & Purpose*, March 18, 2022. https://taskandpurpose.com/news/space-force -fitness-test/.

Rubel, Robert C. "Acey Deucey Can Teach Military Strategy." *U.S. Naval Institute Proceedings* 149, no. 3 (2023). https://www.usni.org/magazines/proceedings/2023 /march/acey-deucey-can-teach-military-strategy.

Rubel, Robert C. "Research & Debate—The Medium Is the Message: Weaving Wargaming More Tightly into the Fabric of the Navy." *Naval War College Review* 72, no. 4 (2019): 150–59.

Rushing, Bonnie L., and Kyleanne Hunter. "The Human Weapon System in Gray Zone Competition." *Journal of Advanced Military Studies* 14, no. 1 (2023): 255–71. https:// doi.org/10.21140/mcuj.20231401011.

Ryan, Mick. *The War for Ukraine: Strategy and Adaptation Under Fire*. Naval Institute Press, 2024.

Ryan, Mick. *War Transformed: The Future of Twenty-First-Century Great Power Competition and Conflict*. Naval Institute Press, 2022.

Ryan, Mick. *White Sun War: The Campaign for Taiwan*. Casemate Publishers, 2023.

Salinas, Antonio, Mark Askew, and Jason P. Levay. "The Meaning of Drone-Enabled Infantry Striking Beyond Line of Sight." *War on the Rocks*, June 23, 2025. https:// warontherocks.com/2025/06/the-meaning-of-drone-enabled-infantry-striking-bey ond-line-of-sight/.

Sanchez-Bustamante, Claudia. "Why Today's 'Gen Z' Is at Risk for Boot Camp Injuries." Defense Visual Information Distribution Service (DVIDS), February 22, 2022. https://web.archive.org/web/20220222221823/https://www.dvidshub.net/news /414340/why-todays-gen-z-risk-boot-camp-injuries.

Sandbakk, Øyvind, Guro Strøm Solli, and Hans-Christer Holmberg. "Sex Differences in World-Record Performance: The Influence of Sport Discipline and Competition Duration." *International Journal of Sports Physiology and Performance* 13, no. 1 (2018): 2–8. https://doi.org/10.1123/ijspp.2017-0196.

Sankar, Shyam. "Ukraine's Software Warrior Brigade." *Wall Street Journal*, March 9, 2023.

Sarantakes, Nicholas. "The Future-War Literature of the Reagan Era—Winning World War III in Fiction." *Naval War College Review* 76, no. 3 (2023): 1–20.

Sattler, Brian. "Recruit Medical Standards Are Out of Touch." *U.S. Naval Institute Proceedings* 150, no. 3 (2024). https://www.usni.org/magazines/proceedings/2024 /march/recruit-medical-standards-are-out-touch.

Scales, Robert H. *Scales on War: The Future of America's Military at Risk*. Naval Institute Press, 2016.

Scalzi, John. *Old Man's War*. Tor Publishing, 2005.

Scarbro, Graham. "Naval Flight Officers' Unmanned Future." *U.S. Naval Institute Proceedings* 147, no. 9 (2021). https://www.usni.org/magazines/proceedings/2021/september /naval-flight-officers-unmanned-future.

Scharre, Paul. *Army of None: Autonomous Weapons and the Future of War*. W.W. Norton, 2018.

Scharre, Paul. *Four Battlegrounds: Power in the Age of Artificial Intelligence*. W.W. Norton, 2023.

Scharre, Paul, and Lauren Fish. *Human Performance Enhancement*. Super Soldiers. Center for a New American Security, 2018.

Scharre, Paul, Lauren Fish, Katherine Kidder, and Amy Schafer. *Super Soldiers: Summary of Findings and Recommendations*. Super Soldiers. Center for a New American Security, 2018.

Schei, Tonje Hessen, dir. *Drone*. With Brandon Bryant, Zubair Rehman, and John Bellinger III. Flimmer Film, Radiator Film, Volt Film, 2014. 1h18m.

Schillinger, Nicolas. "Playing Soldiers: The War Game in Late Qing and Republican China." *Journal of Chinese Military History* 9, no. 1 (2020): 38–64. https://doi.org/10.1163/22127453-BJA10003.

Schlegel, Petr, and Adam Křehký. "Performance Sex Differences in CrossFit®." *Sports* 10, no. 11 (2022): 11. https://doi.org/10.3390/sports10110165.

Schmidt, Eric. "Innovation Power: Why Technology Will Define the Future of Geopolitics." *Foreign Affairs* 102, no. 2 (2023): 38–52.

Schmitt, Eric, Julian E. Barnes, Helene Cooper, and Thomas Gibbons-Neff. "Ukraine's Forces and Firepower Are Misallocated, U.S. Officials Say." *New York Times*, August 22, 2023. https://www.nytimes.com/2023/08/22/us/politics/ukraine-counteroffensive-russia-war.html.

Schneider, Jacquelyn. "Blue Hair in the Gray Zone." *War on the Rocks*, January 10, 2018. https://warontherocks.com/2018/01/blue-hair-gray-zone/.

Schneider, Jacquelyn. "Does Technology Win Wars?" *Foreign Affairs*, March 3, 2023. https://www.foreignaffairs.com/ukraine/does-technology-win-wars.

Schogol, Jeff. "Army Revamps Basic Training to Simulate Battlefield Stalked by Drone Swarms." *Task & Purpose*, October 17, 2024. https://taskandpurpose.com/news/army-training-drones-conceal/.

Schogol, Jeff. "The Army's Yearslong Fight Over Its Controversial New Fitness Test Isn't Over Yet." *Task & Purpose*, June 28, 2023. https://taskandpurpose.com/news/army-combat-fitness-test-problems-congress/.

Schogol, Jeff. "Every Marine a Rifleman No More?" Your Marine Corps. *Marine Corps Times*, May 7, 2017. https://www.marinecorpstimes.com/news/your-marine-corps/2017/05/07/every-marine-a-rifleman-no-more/.

Schogol, Jeff. "Here's What The Army's Proposed Gender-Neutral Combat Test Really Looks Like." *Task & Purpose*, February 22, 2018. https://taskandpurpose.com/news/army-gender-neutral-combat-test/.

Schogol, Jeff. "How Coffee Helped the Union Army Win the Civil War." *Task & Purpose*, September 30, 2020. https://taskandpurpose.com/history/civil-war-union-army-coffee/.

Schogol, Jeff. "Marine Corps Plan Calls for Some Future Marines to Skip Boot Camp." *Task & Purpose*, November 23, 2021. https://taskandpurpose.com/news/marine-corps-plan-cyber-skip-boot-camp/.

Schogol, Jeff. "Russia Jams US GPS-Guided Weapons Given to Ukraine, Leaked Info Shows." *Task & Purpose*, April 13, 2023. https://taskandpurpose.com/news/russia-ukraine-gps-weapons-jam/.

Schogol, Jeff. "US Vet on the Front Line Says Ukrainians Need Better Infantry Gear to Beat the Russians." *Task & Purpose*, June 27, 2022. https://taskandpurpose.com/news/ukraine-american-military-veteran-infantry-gear/.

Schogol, Jeff, and Patty Nieberg. "War with China 'Would Result in Large-Scale Casualties,' Army General Says." *Task & Purpose*, July 24, 2025. https://taskandpurpose.com/news/us-china-war-casualties/.

Schultz, Colin. "A Military Contractor Just Went Ahead and Used an Xbox Controller for Their New Giant Laser Cannon." Smart News, Smart News Science, Articles. *Smithsonian Magazine*, September 9, 2014. https://www.smithsonianmag.com/smart-news/military-contractor-just-went-ahead-and-used-xbox-controller-their-new-giant-laser-cannon-180952647/.

Schultz, Jaime. *Qualifying Times: Points of Change in U.S. Women's Sport*. University of Illinois Press, 2014.

Schulzke, Marcus. "Rethinking Military Gaming: America's Army and Its Critics." *Games and Culture* 8, no. 2 (2013): 59–76. https://doi.org/10.1177/1555412013478686.

Schuurman, Paul. "A Game of Contexts: Prussian-German Professional Wargames and the Leadership Concept of Mission Tactics 1870–1880." *War in History* 28, no. 3 (2021): 504–24. https://doi.org/10.1177/0968344519855104.

Schwab, Klaus. "The Fourth Industrial Revolution." *Foreign Affairs*, December 12, 2015. https://www.foreignaffairs.com/world/fourth-industrial-revolution.

Schwartz, Zachary. "Infantry Battalions as Sensor Webs for the Fleet." *U.S. Naval Institute Proceedings* 150, no. 9 (2024). https://www.usni.org/magazines/proceedings/2024/september/infantry-battalions-sensor-webs-fleet.

Scott-Skillern, Carmelia. "Army Logistics in the Pacific." New America, September 13, 2023. http://newamerica.org/future-security/reports/army-logistics-in-the-pacific/.

Seck, Hope. "MARSOC Raiders to Deploy in Smaller, Tech-Loaded Teams as Conflict Gets More Complex." *Sandboxx News*, July 17, 2023. https://www.sandboxx.us/blog/marsoc-raiders-to-deploy-in-smaller-tech-loaded-teams-as-conflict-gets-more-complex/.

Seck, Hope. "Researchers Want to Prevent Injuries in Soldiers Before They Happen." SANDBOXX News, April 5, 2023. https://www.sandboxx.us/blog/researchers-want-to-prevent-injuries-in-soldiers-before-they-happen/.

Seck, Hope Hodge. "Marine Corps Shelves Futuristic Robo-Mule Due to Noise Concerns." Military.com, December 22, 2015. https://www.military.com/daily-news/2015/12/22/marine-corps-shelves-futuristic-robo-mule-due-to-noise-concerns.html.

Seck, Hope Hodge. "Navy Developing Smart Bra, Undershirt for Fighter Pilots." *Navy Times*, December 27, 2022. https://www.navytimes.com/news/your-navy/2022/12/27/navy-developing-smart-bra-undershirt-for-fighter-pilots/.

Seck, Hope Hodge. "'Robot Marines' in Every Formation: Corps' Robotics Chief Casts Vision." *The War Zone (at The Drive)*, May 7, 2024. https://www.twz.com/sea/robot-marines-in-every-formation-corps-robotics-chief-casts-vision.

See How Gamers Are Outwitting and Helping to Kill Russian Soldiers. With Erin Burnett. CNN, 2022. 5:31. https://www.youtube.com/watch?v=b166ecyNBCw.

Sefton, JoEllen M., K. R. Lohse, and J. S. McAdam. "Prediction of Injuries and Injury Types in Army Basic Training, Infantry, Armor, and Cavalry Trainees Using a Common Fitness Screen." *Journal of Athletic Training* 51, no. 11 (2016): 849–57. https://doi.org/10.4085/1062-6050-51.9.09.

Shanahan, Patrick M. "Use of Geolocation-Capable Devices, Applications, and Services." Memorandum. August 3, 2018. https://media.defense.gov/2018/Aug/06/2001951064/-1/-1/1/GEOLOCATION-DEVICES-APPLICATIONS-SERVICES.PDF.

Sharp, Travis. *Ready Player None? An End-to-End Assessment of the Air Force Collaborative Combat Aircraft Program*. Center for Strategic and Budgetary Assessments, 2025. https://csbaonline.org/research/publications/ready-player-none-an-end-to-end-assessment-of-the-air-force-collaborative-combat-aircraft-program.

Sharp, Travis, and Tyler Hacker. *Evaluate Like We Operate: Why DOD Should Evaluate Weapons Systems as Networked Force Packages, Not Individual Platforms*. Center for Strategic and Budgetary Assessments, 2023. https://csbaonline.org/research/publications/evaluate-like-we-operate-why-dod-should-evaluate-weapons-systems-as-networked-force-packages-not-individual-platforms.

Shaw, Ian G. R. *Predator Empire: Drone Warfare and Full Spectrum Dominance*. University of Minnesota Press, 2016.

Sherlock, Ruth. "Hezbollah Designed a Video Game to Appeal to the U.S." *NPR*, July 16, 2018. https://www.npr.org/2018/07/16/629588453/hezbollah-designed-a-video-game-to-appeal-to-the-u-s.

Shimko, Keith L. *The Iraq Wars and America's Military Revolution*. Cambridge University Press, 2010.

Shunk, Dave. "Ethics and the Enhanced Soldier of the Near Future." *Military Review: The Professional Journal of the U.S. Army* 95, no. 1 (2015): 91–98.

Sicard, Sarah. "Coffee Was So Important to the Army That GI Joe Gave It His Name." *Task & Purpose*, April 7, 2017. https://taskandpurpose.com/news/coffee-important-army-gi-joe-gave-name/.

Sicard, Sarah. "The Army Is Working on a Tactical Bra." *Military Times*, August 3, 2022. https://www.militarytimes.com/off-duty/military-culture/2022/08/03/the-army-is-working-on-a-tactical-bra/.

Simpson, Leanne K. "Eight Myths About Women on the Military Frontline—and Why We Shouldn't Believe Them." *The Conversation*, April 1, 2016. http://theconversation.com/eight-myths-about-women-on-the-military-frontline-and-why-we-shouldnt-believe-them-55594.

Singer, P. W. *Wired for War: The Robotics Revolution and Conflict in the 21st Century*. Penguin, 2009.

Singer, P. W., and August Cole. *Burn-In: A Novel of the Real Robotic Revolution*. Houghton Mifflin Harcourt, 2020.

Singer, P. W., and August Cole. *Ghost Fleet: A Novel of the Next World War*. Houghton Mifflin Harcourt, 2015.

Singer, Peter Warren, and Emerson T. Brooking. *Likewar: The Weaponization of Social Media*. Houghton Mifflin Harcourt, 2018.

Sirimarco-Lang, Rebecca. "SJAFB Tests New In-Flight Bladder Relief System." Air Combat Command, March 20, 2023. https://www.acc.af.mil/News/Article-Display/Article/3334916/sjafb-tests-new-in-flight-bladder-relief-system/.

Sisk, Richard. "Dunford Pushes Services to Move Faster on Body Armor Fitted for Women." Military.com, May 14, 2018. https://www.military.com/kitup/daily-news/2018/05/14/dunford-pushes-services-move-faster-body-armor-fitted-women.html.

Sisk, Richard. "Robert Bales Among 8 Former Troops, Contractors Petitioning Trump for Pardons." Military.com, December 4, 2020. https://www.military.com/daily-news/2020/12/04/robert-bales-among-8-former-troops-contractors-petitioning-trump-pardons.html.

Skove, Sam. "Wartime Need for Drones Would Outstrip US Production. There's a Way to Fix That." *Defense One*, August 7, 2024. https://www.defenseone.com/policy/2024/08/wartime-need-drones-would-outstrip-us-production-theres-way-fix/398642/.

Slayton, Nicholas. "The Air Force Is Letting Troops Play This Wargame on Its Secure Networks." *Task & Purpose*, October 27, 2024. https://taskandpurpose.com/culture/air-force-command-video-game/.

Slayton, Nicholas. "The Marine Corps' Fitness Test from the 1950s Wasn't That Hard." *Task & Purpose*, October 30, 2022. https://taskandpurpose.com/military-life/marine-corps-1956-physical-fitness-test/.

Slayton, Rebecca. "What Is a Cyber Warrior? The Emergence of U.S. Military Cyber Expertise, 1967–2018." *Texas National Security Review* 4, no. 1 (2020): 61–96.

Slisco, Aila. "Russia's Military Has a Major Desertion Problem." *Newsweek*, December 5, 2023. https://www.newsweek.com/russias-military-has-major-desertion-problem-1849729.

Smith, Bryan. "Harry Schmidt's War." *Chicago Magazine*, June 28, 2007. https://www.chicagomag.com/Chicago-Magazine/April-2005/Harry-Schmidts-War/.

Smith, Carl D., Adam D. Cooper, Donna J. Merullo, et al. "Sleep Restriction and Cognitive Load Affect Performance on a Simulated Marksmanship Task." *Journal of Sleep Research* 28, no. 3 (2019): e12637. https://doi.org/10.1111/jsr.12637.

Smith, Frank L., III, Nina A. Kollars, and Benjamin H. Schechter, eds. *Cyber Wargaming: Research and Education for Security in a Dangerous Digital World*. Georgetown University Press, 2024.

Smith, Roger. "The Long History of Gaming in Military Training." *Simulation & Gaming* 41, no. 1 (2010): 6–19. https://doi.org/10.1177/1046878109334330.

Smith, Stew. "Avoid These Training Activities When Preparing to Serve." Military.com, September 17, 2021. https://www.military.com/military-fitness/avoid-these-training-activities-when-preparing-serve.

Sobocinski, André. "Teddy Roosevelt, Navy Medicine and the Birth of Physical Readiness." US Navy website, July 18, 2022. https://www.navy.mil/Press-Office/News-Stories/Article/3097256/teddy-roosevelt-navy-medicine-and-the-birth-of-physical-readiness/.

Solomon, Jay. "Some Hamas Killers Were High on Amphetamine, Officials Say." Semafor, November 1, 2023. https://www.semafor.com/article/10/31/2023/hamas-killers-were-high-on-amphetamine-officials-say.

Sonmez, Felicia. "Adm. Linda Fagan Becomes First Woman to Lead U.S. Coast Guard." *Washington Post*, June 1, 2022. https://www.washingtonpost.com/politics/2022/06/01/adm-linda-fagan-becomes-first-woman-lead-us-coast-guard/.

Sotos, John G. "Physical Fitness Programs Don't Fit Today's Fight." *U.S. Naval Institute Proceedings* 145, no. 6 (2019). https://www.usni.org/magazines/proceedings/2019/june/physical-fitness-programs-dont-fit-todays-fight.

South, Todd. "12-Man Rifle Squads, Including a Squad Systems Operator, Commandant Says." *Marine Corps Times*, May 4, 2018. https://www.marinecorpstimes.com/news/your-marine-corps/2018/05/04/12-man-rifle-squads-including-a-squad-systems-operator-commandant-says/.

South, Todd. "31st MEU Tests Artificial Intelligence Sensing Gear to Help Marines, Soldiers See Invisible Threats." *Marine Corps Times*, February 17, 2021. https://www

.marinecorpstimes.com/news/your-marine-corps/2021/02/17/31st-meu-tests-artific
ial-intelligence-sensing-gear-to-help-marines-soldiers-see-invisible-threats/.

South, Todd. "Army Boot Camp Will Soon Include Counter-Drone Training." C4ISR
Net, November 15, 2023. https://www.c4isrnet.com/news/your-army/2023/11/15
/army-boot-camp-will-soon-include-counter-drone-training/.

South, Todd. "Here's the New Gear Being Readied by the Army's One-Stop-Shop for
All Things Soldier." *Army Times*, October 11, 2021. https://www.armytimes.com/news
/2021/10/11/heres-the-new-gear-being-readied-by-the-armys-one-stop-shop-for
-all-things-soldier/.

South, Todd. "Marines Eye Tactical Resupply Drone Prototypes." *Marine Corps Times*,
May 10, 2021. https://www.marinecorpstimes.com/news/your-marine-corps/2021
/05/10/marines-eye-tactical-resupply-drone-prototypes/.

South, Todd. "New in 2019: The Army, Marines Will Put Four Robotic Gear Mules to
the Test in Rugged, Austere Conditions." *Army Times*, January 2, 2019. https://www
.armytimes.com/news/your-army/2019/01/02/new-in-2019-the-army-marines-will
-put-four-robotic-gear-mules-to-the-test-in-rugged-austere-conditions/.

South, Todd. "New Military Simulations for Shooting, Trench War, Drones Unveiled."
Marine Corps Times, November 22, 2023. https://www.marinecorpstimes.com
/flashpoints/2023/11/22/new-military-simulations-for-shooting-trench-war-drones
-unveiled/.

South, Todd. "New War-Gaming Center Will Plug into Marine Units with a Realistic
Experience." *Marine Corps Times*, May 12, 2022. https://www.marinecorpstimes
.com/news/modern-day-marine/2022/05/12/new-war-gaming-center-will-plug
-into-marine-units-with-a-more-realistic-experience/.

Spencer, John. "The Softer American." Modern War Institute (at West Point), July 5,
2016. https://mwi.westpoint.edu/the-softer-american/.

Spillner, Brent. "Artificial Intelligence Is a Human-Centric Endeavor." *U.S. Naval
Institute Proceedings* 149, no. 10 (2023). https://www.usni.org/magazines/
proceedings/2023/october/artificial-intelligence-human-centric-endeavor.

Spirlet, Thibault. "Hundreds of Russian Soldiers Are Going AWOL and Refusing to
Fight as Morale Plummets, UK Intelligence Says." *Business Insider*, August 30, 2023.
https://www.businessinsider.com/hundreds-russian-soldiers-awol-refusing-fight
-uk-intel-2023-8.

Spoehr, Thomas W. *The Administration and Congress Must Act Now to Counter the
Worsening Military Recruiting Crisis*. No. 5283. Issue Brief. Center for National
Defense, The Heritage Foundation, 2022. http://report.heritage.org/ib5283.

Spoehr, Thomas W. "It's Time for a National Security Strategy for Military Recruiting."
Stars and Stripes, August 17, 2023. https://www.stripes.com/opinion/2023-08-17
/national-security-strategy-military-recruiting-11076498.html.

Spoehr, Thomas, and Bridget Handy. *The Looming National Security Crisis: Young
Americans Unable to Serve in the Military*. No. 3282. Backgrounder. The Heritage
Foundation, 2018. https://www.heritage.org/defense/report/the-looming-national
-security-crisis-young-americans-unable-serve-the-military.

Spoehr, Thomas, and Katherine Kuzminski. "Bad Idea: Relying on the Same Old
Solutions to Meet the Military Recruitment Challenge." Center for Strategic
and International Studies, March 10, 2023. https://defense360.csis.org/bad

-idea-relying-on-the-same-old-solutions-to-meet-the-military-recruitment
-challenge/.

Spoehr, Thomas, and Isaac Tang. "Obesity Epidemic Threatens Not Just Public Health, but Also National Security." Restoring America. *Washington Examiner*, August 8, 2023. https://www.washingtonexaminer.com/restoring-america/courage-strength -optimism/obesity-epidemic-threatens-not-just-public-health-but-also-national-se curity.

Stancy, Diana. "Navy to Commission First Sub Designed for Both Men and Women Sailors." *Navy Times*, September 12, 2024. https://www.navytimes.com/news/your -navy/2024/09/12/navy-to-commission-first-sub-designed-for-both-men-and-wom en-sailors/.

Stavridis, James. "US Military's Recruiting Woes Are a National-Security Crisis." Bloomberg.com, July 4, 2023. https://www.bloomberg.com/opinion/articles/2023-07-04/us -military-recruiting-crisis-is-a-national-security-emergency.

Stephens, Jacob. "355th TRS Implements New VR Simulators." Defense Visual Information Distribution Service, October 28, 2020. https://www.dvidshub.net/news /386384/355th-trs-implements-new-vr-simulators.

Stiehm, Judith H. "Women and the Combat Exemption." *The US Army War College Quarterly: Parameters* 10, no. 1 (1980). https://doi.org/10.55540/0031-1723.1229.

Stoet, Gijsbert, Daryl B. O'Connor, Mark Conner, and Keith R. Laws. "Are Women Better Than Men at Multi-Tasking?" *BMC Psychology* 1, no. 1 (2013): 1–18. https://doi.org /10.1186/2050-7283-1-18.

Stokes, Jacob, Colin H. Kahl, Andrea Kendall-Taylor, and Nicholas Lokker. *Averting AI Armageddon*. Center for a New American Security, 2025. https://www.cnas.org /publications/reports/averting-ai-armageddon.

Strebel, Ian, and Matt McKenzie. "Embrace the Nerd: Dungeons & Dragons and Military Intelligence." *War on the Rocks*, August 18, 2023. https://waron therocks.com/2023/08/embrace-the-nerd-dungeons-dragons-and-military -intelligence/.

Subhani, Abdul. "Is the Army Ready to Think Like a Tech Company? Why the Service Needs to Value Coding the Same Way as Shooting." Modern War Institute (at West Point), August 17, 2023. https://mwi.westpoint.edu/is-the-army-ready-to-think -like-a-tech-company-why-the-service-needs-to-value-coding-the-same-way-as -shooting/.

Sussis, Matthew. "Special Operations Command Begins Research on Performance-Enhancing Drugs—National Security Zone." *MEDILL National Security Zone* (blog), February 9, 2018. https://nationalsecurityzone.medill.northwestern.edu /blog/2018/conflicts-terrorism/special-operations-command-begins-research-on -performance-enhancing-drugs/.

Suthoff, Joshua. "Imagining a US Army Drone Corps." Modern War Institute (at West Point), December 19, 2024. https://mwi.westpoint.edu/imagining-a-us -army-drone-corps/.

Suthoff, Joshua. "Reimagining Combat Power for Tomorrow's Battlefield: The Enhanced Brigade Combat Team." Modern War Institute (at West Point), April 18, 2025. https://mwi.westpoint.edu/reimagining-combat-power-for-tomorrows-battlefield-the -enhanced-brigade-combat-team/.

Swain, Rob. "Retention Must Be a Special Evolution." *U.S. Naval Institute Proceedings* 148, no. 12 (2022). https://www.usni.org/magazines/proceedings/2022/december /retention-must-be-special-evolution.

Swick, Andrew, and Emma Moore. "DoD Must Consider Non-Traditional Approaches to Military Service." *The Hill* (blog), February 28, 2018. https://thehill.com/opinion /national-security/376005-dod-must-consider-non-traditional-approaches-to -military-service/.

Tan, Michelle. "Meet the Army's First Female Infantry Officer." *Army Times*, April 27, 2016. https://www.armytimes.com/news/your-army/2016/04/27/meet-the-army-s-first -female-infantry-officer/.

Tangredi, Sam J. "Fighting When the Network Dies." *U.S. Naval Institute Proceedings* 149, no. 3 (2023). https://www.usni.org/magazines/proceedings/2023/march /fighting-when-network-dies.

Tangredi, Sam J., and George Galdorisi, eds. *AI at War: How Big Data, Artificial Intelligence, and Machine Learning Are Changing Naval Warfare.* Naval Institute Press, 2021.

Taylor, Daniel J., Alan L. Peterson, Kristi E. Pruiksma, et al. "Internet and In-Person Cognitive Behavioral Therapy for Insomnia in Military Personnel: A Randomized Clinical Trial." *Sleep* 40, no. 6 (2017): 1–12. https://doi.org/10.1093/sleep/zsx075.

Taylor, William A. "From WACs to Rangers: Women in the U.S. Military Since World War II." *Marine Corps University Journal*, Special Issue: Gender Integration (2018): 78–101.

Tegler, Jan. "Air Hazars: Open Source Flight Tracking Called Threat to Military Aircraft." *National Defense* 107, no. 831 (2023): 22–25.

Temple, Tony. "Bradley Trainer: Atari's Top Secret Military Project." *The Arcade Blogger*, October 28, 2016. https://arcadeblogger.com/2016/10/28/bradley-trainer-ataris -top-secret-military-project/.

Thériault, François, Karyn Gabler, and Kiyuri Naicker. *Health and Lifestyle Information Survey of Canadian Armed Forces Personnel 2013/2014—Regular Force Report.* Edited by Barbara Strauss and Jeff Whitehead. Department of National Defense, 2016.

Thomas, Craig. *Firefox.* Holt, Rinehart and Winston, 1977.

Thompson, Mark. "Army Women: Better Chopper Pilots Than the Guys?" *Time*, February 17, 2014. https://time.com/8404/army-women-helicopter-pilots/.

Tibbetts, Bradley C., Terrell C. Lawson, William A. Hall, Kale D. Sawyer, and Colin J. Gandy. *Heritage Meets Hardware: The Fusion of SOF Tradecraft with Modern Technology.* Futures Seminar Report. Irregular Advantage Initiative. US Army War College, 2025. https://media.defense.gov/2025/Jun/02/2003730500/-1/-1/0/4_HER ITAGEMEETSHARDWARE(USAWCFUTURES_SOCOMJ7FINALPROJECT). pdf.

Tiffany, Kaitlyn. "The Myth of the 'First TikTok War.'" *The Atlantic*, March 10, 2022. https://www.theatlantic.com/technology/archive/2022/03/tiktok-war-ukraine -russia/627017/.

Tiron, Roxana. "US Military Faces Biggest Recruiting Hurdles in 50 Years." Bloomberg Government, September 21, 2022. https://about.bgov.com/news/us-military-service s-face-biggest-recruiting-hurdles-in-50-years/.

Trevithick, Joseph. "A-10 Warthog Pilots Are Using the Digital Combat Simulator Video Game to Train in VR." *The War Zone (at The Drive)*, May 14, 2021. https://www.thedrive.com/the-war-zone/40620/a-10-warthog-pilots-are-using-the-digital-combat-simulator-video-game-to-train-in-vr.

Trevithick, Joseph. "Air Force Colonel Now Says Drone That Turned on Its Operator Was a 'Thought Experiment.'" *The War Zone (at The Drive)*, June 1, 2023. https://www.thedrive.com/the-war-zone/artificial-intelligence-enabled-drone-went-full-terminator-in-air-force-test.

Trevithick, Joseph. "Army Cancels High-Speed Armed Reconnaissance Helicopter Program." *The War Zone (at The Drive)*, February 8, 2024. https://www.twz.com/air/army-cancels-hight-speed-armed-reconnaissance-helicopter-program.

Trevithick, Joseph. "New US Army Manual Shows It's Worried About Russia's Hybrid Warfare Tactics." *The War Zone (at The Drive)*, June 29, 2019. https://www.thedrive.com/the-war-zone/14647/new-us-army-manual-shows-its-worried-about-russias-hybrid-warfare-tactics.

Trevithick, Joseph. "Pentagon Just Made a Massive, Long Overdue Shift to Arm Its Troops with Thousands of Drones." *The War Zone (at The Drive)*, July 10, 2025. https://www.twz.com/air/pentagon-just-made-a-massive-long-overdue-shift-to-arm-its-troops-with-thousands-of-drones.

Trevithick, Joseph. "Tracking U.S. Military Aircraft Online Could Become Much Harder." *The War Zone (at The Drive)*, July 19, 2023. https://www.thedrive.com/the-war-zone/tracking-u-s-military-aircraft-online-could-become-much-harder.

Trevithick, Joseph. "Ukrainian Officer Details Russian Electronic Warfare Tactics Including Radio 'Virus.'" *The War Zone (at The Drive)*, October 30, 2019. https://www.thedrive.com/the-war-zone/30741/ukrainian-officer-details-russian-electronic-warfare-tactics-including-radio-virus.

Tritten, Travis J. "Form-Fitted Body Armor Rolling Out as Combat Roles Expand for Women." *Stars and Stripes*, March 4, 2016. https://www.stripes.com/news/form-fitted-body-armor-rolling-out-as-combat-roles-expand-for-women-1.397554.

Trobaugh, Elizabeth M. "Women, Regardless: Understanding Gender Bias in U.S. Military Integration." *Joint Force Quarterly* 88 (Quarter 2018): 46–53.

Troxel, Wendy M., Regina A. Shih, Eric R. Pedersen, et al. *Sleep in the Military: Promoting Healthy Sleep Among U.S. Servicemembers*. RAND Corporation, 2015. https://www.ncbi.nlm.nih.gov/pmc/articles/PMC5158299/.

Tucker, Patrick. "The Future of Flight Training Is a Gamer Headset That Watches You Play." *Defense One*, December 9, 2020. https://www.defenseone.com/technology/2020/12/future-flight-training-gamer-headset-watches-you-play/170630/.

Tucker, Patrick. "Pentagon Planning Huge Experiment for Its Connect-Everything Concept." *Defense One*, August 7, 2024. https://www.defenseone.com/technology/2024/08/pentagon-planning-huge-experiment-its-connect-everything-concept/398618/.

Tucker, Patrick. "The Pentagon Will Host a 'Top Gun' School for Ukraine-Style Attack Drones." *Defense One*, July 18, 2025. https://www.defenseone.com/technology/2025/07/pentagon-will-host-top-gun-school-ukraine-style-attack-drones/406818/.

Tucker, Patrick. "Soldiers Can Now Steer Robot Dogs with Brain Signals." *Defense One*, March 22, 2023. https://www.defenseone.com/technology/2023/03/soldiers-can-now-steer-robot-dogs-brain-signals/384338/.

Tucker, Patrick. "Special Operations Command Made a Mind-Reading Kit For Elite Troops." *Defense One*, December 11, 2019. https://www.defenseone.com/technology /2019/12/specops-lab-made-mind-reading-kit-elite-troops/161830/.

Tucker, Patrick. "'Swarm Pilots' Will Need New Tactics—and Entirely New Training Methods: Air Force Special-Ops Chief." *Defense One*, May 10, 2024. https://www .defenseone.com/policy/2024/05/swarm-pilots-will-need-new-tacticsand-entirely -new-training-methods-air-force-special-ops-chief/396477/.

Tvaryanas, Anthony P., Lex Brown, and Nita L. Miller. "Managing the Human Weapon System: A Vision for an Air Force Human-Performance Doctrine." *Air & Space Power Journal* 23, no. 2 (2009): 34–41.

Twenge, Jean M. *iGen: Why Today's Super-Connected Kids Are Growing Up Less Rebellious, More Tolerant, Less Happy—and Completely Unprepared for Adulthood—and What That Means for the Rest of Us.* Simon and Schuster, 2017.

UK Ministry of Defence. "Latest Defence Intelligence Update on the Situation in Ukraine." Twitter. July 10, 2023. https://twitter.com/DefenceHQ/status/16782796 48847822848.

UK Ministry of Defence, and Harriett Baldwin. "Xbox Controllers, Hoverbikes and Robotic Trucks Trialled by British and American Armies." Gov.uk, November 14, 2017. https://www.gov.uk/government/news/xbox-controllers-hoverbikes-and-robotic -trucks-trialled-by-british-and-american-armies.

"UkraineWorld on Twitter." Twitter. August 11, 2023. https://twitter.com/ukraine _world/status/1690014709208977408.

Ukrainian Military Training: Forces Partner with Video Game Maker. Al Jazeera English, August 6, 2022. 2:40. https://www.youtube.com/watch?v=Ackgv_gmGnQ.

Ullman, Harlan. *Anatomy of Failure: Why America Loses Every War It Starts.* Naval Institute Press, 2017.

Upadhyay, Akshat. "Virtual and Augmented Reality and Warfare: Fighting War as a Computer Game?" In *Future Warfare and Critical Technologies: Evolving Tactics and Strategies*, edited by Rajeswari Pillai Rajagopalan and Sameer Patil. Observer Research Foundation and Global Policy Journal, 2024.

US Air Force. *United States Air Force Unmanned Aircraft Systems Flight Plan 2009–2047.* Headquarters, United States Air Force, 2009.

US Army. *Army Field Manual FM 21–20: Physical Readiness Training.* Headquarters, United States Department of the Army, 1980. https://www.google.com/books /edition/Physical_Readiness_Training/bU5jQXqmTVIC?hl=en&gbpv=1.

US Army. *Field Manual FM 3–21.8 (FM 7–8) The Infantry Rifle Platoon and Squad.* Headquarters, Department of the Army, 2007.

US Army. *Field Manual FM 7–22: Army Physical Readiness Training October 2012.* Headquarters, Department of the Army, 2012. https://www.google.com/books/edition/FM _7_22_Army_Physical_Readiness_Training/9tZvDwAAQBAJ?hl=en&gbpv=0.

US Army. *Field Manual FM 7–22: Holistic Health and Fitness.* Headquarters, Department of the Army, 2020. https://armypubs.army.mil/epubs/DR_pubs/DR_a/ARN 30964-FM_7-22-001-WEB-4.pdf.

US Army Asymmetric Warfare Group. *Russian New Generation Warfare Handbook.* US Army Asymmetric Warfare Group, 2016. https://info.publicintelligence.net/AWG -RussianNewWarfareHandbook.pdf.

US Army Combat Capabilities Development Command (DEVCOM), Public Affairs Office. "Army, Academia Collaborate on Exoskeleton to Reduce Soldier Injuries." US army website, August 17, 2022. https://www.army.mil/article/259429/army_aca demia_collaborate_on_exoskeleton_to_reduce_soldier_injuries.

US Army Public Affairs. "Army Announces Creation of Future Soldier Preparatory Course." US Army website, July 26, 2022. https://www.army.mil/article/258758/army _announces_creation_of_future_soldier_preparatory_course.

US Army Public Affairs. "Army Establishes New Fitness Test of Record to Strengthen Readiness and Lethality." US Army website, April 20, 2025. https://www.army.mi l/article/284799/army_establishes_new_fitness_test_of_record_to_strengthen _readiness_and_lethality.

US Army Public Affairs. "Transcript: Media Roundtable with Gen. Charles Flynn with USARPAC Update September 13, 2023." US Army website, September 18, 2023. https://www.army.mil/article/270022/transcript_media_roundtable_with_gen_ charles_flynn_with_usarpac_update_september_13_2023.

US Army Special Operations Command. *Breaking Barriers: Women in Army Special Operations*. US Army Special Operations Command, Department of the Army, 2021. https://www.soc.mil/wia/women-in-arsof-report-2023.pdf.

US Department of the Army. "The U.S. Army Holistic Health and Fitness Operating Concept." US Army website, October 1, 2020. https://www.army.mil/e2/downloads /rv7/acft/h2f_operating_concept.pdf.

US Department of Defense. "2020 Qualified Military Available Study." 2020. https:// prod-media.asvabprogram.com/CEP_PDF_Contents/Qualified_Military_Avail- able.pdf.

US Department of Defense. "Defense Department Report Shows Decline in Armed Forces Population While Percentage of Military Women Rises Slightly." November 6, 2023. https://www.defense.gov/News/Releases/Release/Article/3580676/defense- department-report-shows-decline-in-armed-forces-population-while-percen/.

US Department of Defense. *Department of Defense Annual Report on Sexual Assault in the Military: Fiscal Year 2022.* US Department of Defense, 2023. https://www.sapr .mil/sites/default/files/public/docs/reports/AR/FY22/DOD_Annual_Report_on_ Sexual_Assault_in_the_Military_FY2022.pdf.

US Department of Defense. "Deputy Secretary of Defense Kathleen Hicks Keynote Address: 'The Urgency to Innovate.'" August 28, 2023. https://www.defense.gov/ News/Speeches/Speech/Article/3507156/deputy-secretary-of-defense-kathleen- hicks-keynote-address-the-urgency-to-innov/.

US Department of Defense. "Study on Effects of Sleep Deprivation on Readiness of Members of the Armed Forces." *US Naval Institute (USNI) News,* March 2021. https://news .usni.org/2021/03/03/pentagon-report-on-sleep-deprivation-and-military-readiness.

US Department of Defense, Office of the Under Secretary of Defense for Personnel and Readiness. "DOD Instruction 1308.03: DOD Physical Fitness/Body Composition Program." March 10, 2022. https://www.esd.whs.mil/portals/54/documents/dd /issuances/dodi/130803p.pdf.

US Department of the Navy. *Department of the Navy Cyber Strategy.* US Department of the Navy, 2023. https://media.defense.gov/2023/Nov/21/2003345095/-1/-1/0/DEPART MENT%20OF%20THE%20NAVY%20CYBER%20STRATEGY.PDF.

US Department of the Navy. *Unmanned Campaign Framework*. US Department of the Navy, 2021.

US Government Accountability Office. *Coast Guard: Recruitment and Retention Challenges Persist*. GAO-23-106750. US Government Accountability Office, 2023. https://www.gao.gov/products/gao-23-106750.

US Government Publishing Office. "Cyber Strategy, Policy, and Organization: Hearing on S. Hrg. 115–181 at Senate Committee on Armed Services," United States Senate One hundred fifteenth Congress, First Session (2017). https://www.govinfo.gov/content/pkg/CHRG-115shrg28907/html/CHRG-115shrg28907.htm.

US Joint Chiefs of Staff. "Developing Today's Joint Officers for Tomorrow's Ways of War: The Joint Chiefs of Staff Vision and Guidance for Professional Military Education and Talent Management." May 1, 2020. https://www.jcs.mil/Portals/36/Documents/Doctrine/education/jcs_pme_tm_vision.pdf?ver=2020-05-15-102429-817.

US Marine Corps. "Annual Rifle Training Databook (NAVMC 11660 REV 02–12)." 2012. https://www.trngcmd.marines.mil/Portals/207/Docs/wtbn/ART%20NAVMC%2011660%20REV%2002-12.pdf.

US Marine Corps. "Marine Corps Force Integration Plan—Summary." September 2015. https://wiisglobal.org/wp-content/uploads/2013/05/Research-Summary-Final_9Sept15.pdf.

US Marine Corps, Department of the Navy. "Force Design 2030." March 2020. https://www.hqmc.marines.mil/Portals/142/Docs/CMC38%20Force%20Design%202030%20Report%20Phase%20I%20and%20II.pdf?ver=2020-03-26-121328-460.

US Marine Corps Headquarters. "Force Design 2030 Annual Update." June 2023. https://www.marines.mil/Portals/1/Docs/Force_Design_2030_Annual_Update_June_2023.pdf.

US Marine Corps Headquarters. "Training and Education 2030." United States Marine Corps, January 2023. https://www.marines.mil/Portals/1/Docs/Training%20and%20Education%202030.pdf.

US Secretary of Defense. "Combat Arms Standards." US Department of Defense, Office of the Secretary of Defense, March 30, 2025.

US Secretary of Defense. "Unleashing U.S. Military Drone Dominance." July 10, 2025. https://media.defense.gov/2025/Jul/10/2003752117/-1/-1/1/UNLEASHING-U.S.-MILITARY-DRONE-DOMINANCE.pdf.

US Space Force. "U.S. Space Force Vision for a Digital Service." May 2021. https://media.defense.gov/2021/May/06/2002635623/-1/-1/1/USSF%20VISION%20FOR%20A%20DIGITAL%20SERVICE%202021%20(2).pdf.

US War Department. *FM 21–20 Physical Training 1946*. War Department Field Manual FM 21–20. US Government Printing Office, 1946.

Van Horn, Mark. "Big Data War Games: Necessary for Winning Future Wars." *Military Review: The Professional Journal of the U.S. Army*, September 2016, 2–6. https://www.armyupress.army.mil/Journals/Military-Review/Online-Exclusive/2016-Online-Exclusive-Articles/Big-Data-War-Games/.

Vandiver, John. "Army Field-Tests AI System That Shields Wireless Network Use from Foes." *Stars and Stripes*, November 22, 2023. https://www.stripes.com/branches/army/2023-11-22/ai-army-electronic-warfare-12133758.html.

Vardiman, Daniel. "Is Cutting-Edge Military Tech Really Cheaper Than Manpower?" Atlantic Council, September 20, 2022. https://www.atlanticcouncil.org/content-series/automating-the-fight/is-cutting-edge-military-tech-really-cheaper-than-manpower/.

Venable, Heather. "How the Few Become the Valued: Improving Recruitment and Retention of Female Marines." *U.S. Naval Institute Proceedings* 146, no. 10 (2020). https://www.usni.org/magazines/proceedings/2020/october/how-few-become-valued-improving-recruitment-and-retention-female.

Venable, Heather P. *How the Few Became the Proud: Crafting the Marine Corps Mystique, 1874–1918*. Naval Institute Press, 2019.

Vergun, David. "DOD Addresses Recruiting Shortfall Challenges." US Department of Defense, December 13, 2023. https://www.defense.gov/News/News-Stories/Article/Article/3616786/dod-addresses-recruiting-shortfall-challenges/.

Verroken, Michele. "Drug Use and Abuse in Sport." *Best Practice & Research Clinical Endocrinology & Metabolism* 14, no. 1 (2000): 1–23. https://doi.org/10.1053/beem.2000.0050.

Vincent, Brandi. "Hicks Unveils DOD's New 'Replicator' Initiative to Counter China via Autonomous Tech." *DefenseScoop*, August 28, 2023. https://defensescoop.com/2023/08/28/hicks-unveils-dods-new-replicator-initiative-to-counter-china-via-autonomous-tech/.

Vining, Margaret. "Women Join the Armed Forces: The Transformation of Women's Military Work in World War II and After (1939–1947)." In *A Companion to Women's Military History*, edited by Barton Hacker and Margaret Vining. History of Warfare 74. Brill, 2012. https://doi.org/10.1163/9789004206823_008.

Visentin, Lisa and Shane Wright. "ADF Boost Hard to Achieve Without Overhaul of Recruitment and Retention Policies." *The Sydney Morning Herald*, March 10, 2022. https://www.smh.com.au/politics/federal/adf-boost-won-t-be-achieved-without-overhaul-of-recruitment-and-retention-policies-20220310-p5a3bx.html.

Vlamis, Kelsey. "Ukrainian Soldiers Are Using a Handheld Video Game Console to Operate Machine Gun Mounts, per Reports. The US Navy Has Also Utilized Xbox Controllers in the Past." *Business Insider*, May 1, 2023. https://www.businessinsider.com/ukraine-soldiers-video-game-consoles-steam-deck-machine-gun-turrets-2023-5.

Wagenen, Matthew Van, and Arnel P. David. "Lessons from Ukraine Many Don't Want to Hear." RealClearDefense, March 11, 2023. https://www.realcleardefense.com/articles/2023/03/11/lessons_from_ukraine_many_dont_want_to_hear_886750.html.

Wallin, Matthew. "Briefing Note—The Military Recruiting Crisis: Obesity's Impact on the Shortfall." American Security Project, March 2023. https://www.americansecurityproject.org/briefing-note-the-military-recruiting-crisis-obesitys-impact-on-the-shortfall/.

Walsh, Stephen. "Maximize the Two-Seat Super Hornet for the Peer Fight." *U.S. Naval Institute Proceedings* 145, no. 9 (2019). https://www.usni.org/magazines/proceedings/2019/september/maximize-two-seat-super-hornet-peer-fight.

Walton, David. "Are Green Berets Turning Pink? Integrating Women into Special Forces." *War on the Rocks*, September 20, 2022. https://warontherocks.com/2022/09/are-green-berets-turning-pink-integrating-women-into-special-forces/.

Ward, Zachary J., Sara N. Bleich, Angie L. Cradock, et al. "Projected U.S. State-Level Prevalence of Adult Obesity and Severe Obesity." *New England Journal of Medicine* 381, no. 25 (2019): 2440–50. https://doi.org/10.1056/NEJMsa1909301.

Ware, Doug G. "Army's Basic Training Prep Course to Become Permanent Part of Recruiting Strategy." *Stars and Stripes*, September 1, 2023. https://www.stripes.com/branches/army/2023-09-01/army-basic-training-prep-course-recruiting-11237624.html.

Ware, Doug G. "Nearly 70% of US Troops Are Overweight or Obese, Research Report Says." *Stars and Stripes*, October 17, 2023. https://www.stripes.com/theaters/us/2023-10-17/military-troops-obese-overweight-11738212.html.

"Wargaming and Crisis Simulation Initiative Collection." Hoover Institution, Stanford University, n.d. Accessed November 17, 2024. https://wargaming.hoover.org/.

Warr, Bradley J., Patrick Gagnon, Dennis E. Scofield, and Suzanne Jaenen. "Testing and Evaluation of Tactical Populations." In *NSCA'S Essentials of Tactical Strength and Conditioning*, edited by Brent A. Alvar, Katie Sell, and Patricia A. Deuster. Human Kinetics, 2017.

Wasser, Becca, and Philip Sheers. *From Production Lines to Front Lines.* Center for a New American Security, 2025. https://www.cnas.org/publications/reports/from-production-lines-to-front-lines.

Watling, Jack, and Nick Reynolds. *Meatgrinder: Russian Tactics in the Second Year of Its Invasion of Ukraine.* Special Report. Royal United Services Institute for Defence and Security Studies, 2023. https://www.rusi.orghttps://www.rusi.org.

Watling, Jack, and Nick Reynolds. *Tactical Developments During the Third Year of the Russo–Ukrainian War.* Royal United Services Institute for Defence and Security Studies, 2025.

Weatherford, Doris. *American Women and World War II.* Facts on File, 1990.

Weathers, Corie. *Military Culture Shift: The Impact of War, Money, and Generational Perspective on Morale, Retention, and Leadership.* Elva Resa Publishing, 2023.

Weber, Rachel N. "Accommodating Difference: Gender and Cockpit Design in Military and Civilian Aviation." *Transportation Research Record* 1480 (1995): 51–56.

Weber, Rachel N. "Manufacturing Gender in Commercial and Military Cockpit Design." *Science, Technology, & Human Values* 22, no. 2 (1997): 235–53. https://doi.org/10.1177/016224399702200204.

Weir, Kirsten. "The Extra Weight of COVID-19." American Psychological Association, July 1, 2021. https://www.apa.org/monitor/2021/07/extra-weight-covid.

Weiss, Matthew. *We Don't Want YOU, Uncle Sam: Examining the Military Recruiting Crisis with Generation Z.* Night Vision Publishing, 2023.

Weiss, Michael, and James Rushton. "U.S. Could Train Ukrainian Pilots to Fly F-16s in 4 Months." Yahoo News, May 18, 2023. https://news.yahoo.com/exclusive-us-could-train-ukrainian-pilots-to-fly-f-16s-in-4-months-184136820.html.

Werner, Ben. "Military Branches Are Doing More to Recruit Women into Active Duty." *US Naval Institute (USNI) News*, April 13, 2018. https://news.usni.org/2018/04/13/service-branches-want-more-women.

Wess, Mark. "Brain-Machine Interface—the New 'BMI.'" *U.S. Naval Institute Proceedings* 149, no. 5 (2023). https://www.usni.org/magazines/proceedings/2023/may/brain-machine-interface-new-bmi.

Wetzel, Tyson. "'Killer Robots' Are Coming. Is the US Ready for the Consequences?" Atlantic Council, June 17, 2022. https://www.atlanticcouncil.org/content-series /automating-the-fight/killer-robots-are-coming-is-the-us-ready-for-the-conseque nces/.

Wheatcroft, Jacqueline M., Mike Jump, Amy L. Breckell, and Jade Adams-White. "Unmanned Aerial Systems (UAS) Operators' Accuracy and Confidence of Decisions: Professional Pilots or Video Game Players?" *Cogent Psychology* 4, no. 1 (2017). https://doi.org/10.1080/23311908.2017.1327628.

White, Aaron M. "Gender Differences in the Epidemiology of Alcohol Use and Related Harms in the United States." *Alcohol Research: Current Reviews* 40, no. 2 (2020): 01. https://doi.org/10.35946/arcr.v40.2.01.

White, Matt. "No 'Golden Hour'? How Army Medicine Is Changing for the Next War." *Task & Purpose*, June 20, 2023. https://taskandpurpose.com/news/golden-hour -army-medical-training-ukraine/.

Wilkie, Robert L. "DoD Policy for Non-Deployable Service Members." Office of the Under Secretary of Defense for Personnel and Readiness Memorandum. February 14, 2018. https://www.usar.army.mil/Portals/98/Users/151/87/1687/MEMO%20 20180218%20DoD%20Policy%20for%20Non-deployable%20Service%20Members. pdf.

Wilkins, Alex. "US Space Force Scientist Says 'Human Augmentation' Is Necessary in Next Decade." *Metro*, April 29, 2021. https://metro.co.uk/2021/04/29/us-space-force -scientist-says-military-human-augmentation-necessary-14493027/.

Williams, Lauren C. "1/4 of DOD Cyber Jobs Are Vacant. Here's the Plan to Fill Them." *Defense One*, August 4, 2023. https://www.defenseone.com/defense-systems/2023/08 /pentagon-lays-out-plan-boost-its-cyber-workforce/389123/.

Williams, Lauren C. "To Solve National Security Problems, the US May Have to Rethink Higher Education." *Defense One*, November 2, 2023. https://www.defenseone.com /threats/2023/11/solve-national-security-problems-us-may-have-rethink-higher-edu cation/391710/.

Williams, Noel. "Military Force Design in an Age of Accelerating Technologic Change." *Marine Corps Gazette* 108, no. 2 (2024): 10–14.

Williams, R. F. M. "'Our Problem Children': Masculinity and Its Discontents in American Parachute Units in World War II." *Journal of Military History* 87, no. 3 (2023): 675–702.

Wills, John. *Gamer Nation: Video Games and American Culture.* Johns Hopkins University Press, 2019.

Winkie, Davis. "Army Combat Fitness Test Debuts with Major Changes to Scoring April 1." *Army Times*, March 23, 2022. https://www.armytimes.com/news/your-army /2022/03/23/army-combat-fitness-test-debuts-with-major-changes-to-scoring -april-1/.

Winkie, Davis. "Army Expands 'Prep Course' for Low-Scoring Applicants After Pilot." *Army Times*, January 9, 2023. https://www.armytimes.com/news/your -army/2023/01/09/army-expands-prep-course-for-low-scoring-applicants-after -pilot/.

Winnefeld, James. "NMESIS Now." *U.S. Naval Institute Proceedings* 147, no. 11 (2021). https://www.usni.org/magazines/proceedings/2021/november/nmesis-now.

Winnefeld, James A. "AI Surprise in the Black Ditch." *U.S. Naval Institute Proceedings* 147, no. 1 (2021). https://www.usni.org/magazines/proceedings/2021/january/ai-surprise-black-ditch.

Wintour, Patrick. "RAF Urged to Recruit Video Game Players to Operate Reaper Drones." *The Guardian*, December 9, 2016. https://www.theguardian.com/world/2016/dec/09/royal-air-force-recruit-video-game-players-operate-reaper-drones.

Wise, James E., Jr., and Scott Baron. *Women at War: Iraq, Afghanistan, and Other Conflicts.* Naval Institute Press, 2006.

Wolf, Jennifer. "Holistic Fitness During Primary Military Education Attendance." *Strength in Knowledge: The Warrant Officer Journal* 1, no. 3 (2023): 14–17.

Wood, Nathan Gabriel. "The Problem with Killer Robots." *Journal of Military Ethics* 19, no. 3 (2020): 220–40. https://doi.org/10.1080/15027570.2020.1849966.

Woods, Lauren. "Airmen Selected to Participate in New Cyberspace Direct Appointment Program." Air Force Reserve Command, February 14, 2018. https://www.afrc.af.mil/News/Article-Display/Article/1442503/airmen-selected-to-participate-in-new-cyberspace-direct-appointment-program/.

Woodward, David. *The American Army and the First World War.* Armies of the Great War. Cambridge University Press, 2014.

Woody, Christopher. "Russia's War in Ukraine Shows Why Troops Need to Learn to Put Their Phones Away, Top US Marine General Says." *Business Insider*, December 31, 2022. https://www.businessinsider.com/russia-ukraine-war-shows-battlefield-phone-risk-top-marine-says-2022-12.

Woolley, Anita, and Thomas Malone. "What Makes a Team Smarter? More Women." *Harvard Business Review* 89, no. 6 (2011): 32–33.

Woolsey, Christopher. "Mandate Uninterrupted Rest Hours for a Safer Navy." *U.S. Naval Institute Proceedings* 149, no. 11 (2023). https://www.usni.org/magazines/proceedings/2023/november/mandate-uninterrupted-rest-hours-safer-navy.

Work, Robert. "Deputy Secretary of Defense Speech, CNAS Defense Forum." US Department of Defense, December 14, 2015. https://www.defense.gov/News/Speeches/Speech/Article/634214/cnas-defense-forum/.

Work, Robert, and Paul Selva. "Revitalizing Wargaming Is Necessary to Be Prepared for Future Wars." *War on the Rocks*, December 8, 2015. https://warontherocks.com/2015/12/revitalizing-wargaming-is-necessary-to-be-prepared-for-future-wars/.

Wormuth, Christine, Frank Kendall, and Carlos Del Toro. "Uncle Sam Wants You for a Military Job That Matters; We Need Data Scientists, Coders and Engineers as Much as We Need Pilots, Submariners and Infantry." *Wall Street Journal*, October 24, 2022. https://www.proquest.com/docview/2727758691/citation/527B56E49FCD4CA6PQ/1.

Wright, Nicholas, Michael Miklaucic, and Todd Veazie, eds. *Human, Machine, War: How the Mind-Tech Nexus Will Win Future Wars.* Air University Press, 2025.

Yang, Maya. "How Women Played Crucial Roles in Iraq—and Changed US Military Forever." *The Guardian*, March 21, 2023. https://www.theguardian.com/world/2023/mar/21/iraq-war-us-military-women.

Zeiger, Susan. *In Uncle Sam's Service: Women Workers with the American Expeditionary Force, 1917–1919.* University of Pennsylvania Press, 2004.

Zelenskyy, Volodymyr. "I Signed a Decree Initiating the Establishment of a Separate Branch of Forces—the Unmanned Systems Forces—Address by the President of

Ukraine." Official Website of the President of Ukraine, February 6, 2024. https://www.president.gov.ua/en/news/pidpisav-ukaz-yakij-rozpochinaye-stvorennya-okremogo-rodu-si-88817.

Zhao, Lixia, David S. Freedman, Heidi M. Blanck, and Sohyun Park. "Trends in Severe Obesity Among Children Aged 2 to 4 Years in WIC: 2010 to 2020." *Pediatrics*, 2023, e2023062461. https://doi.org/10.1542/peds.2023-062461.

Zihan, Zhang. "Lock and Load." *Global Times*, October 7, 2012. https://www.globaltimes.cn/content/736758.shtml.

Zucchino, David. "Drone Pilots Have a Front-Row Seat on War, from Half a World Away." *Los Angeles Times*, February 21, 2010. https://www.latimes.com/world/la-xpm-2010-feb-21-la-fg-drone-crews21-2010feb21-story.html.

INDEX